READER'S DIGEST

Marvels and Mysteries
of
THE WORLD
AROUND US

READER'S DIGEST

Marvels and Mysteries of
THE WORLD AROUND US

THE READER'S DIGEST ASSOCIATION

PLEASANTVILLE, NEW YORK   MONTREAL, CANADA   SYDNEY, AUSTRALIA

General Consultant: Rhodes W. Fairbridge, Professor of Geology, Columbia University

CONTENTS

Part One

OUR PLANET'S PAST

Chapter One EARTH'S ANCIENT DRAMA

The Earth's Beginning, *Gerald S. Hawkins* 10
The First Stirrings of Life, *George Gamow* 14
Unraveling the Age of the Earth, *E. L. Simons* 17
Long Journey of the Continents, *Ronald Schiller* 19

Chapter Two CREATURES OF THE AGES

Revelations from Fossil Finds, *Ruth Moore* 24
When Dinosaurs Roamed the Arctic, *Edwin H. Colbert* 29
They All Died Out, *A Picture Story* 32
The Riddle of the Quick-Frozen Mammoths, *Ivan T. Sanderson* 34
Death Trap of the Ages, *J. R. Macdonald* 38

Chapter Three MAN'S ANCESTORS

New Findings on the Origin of Man, *Ronald Schiller* 42
Neanderthal Wasn't Always a Brute, *Ralph S. Solecki* 47
Those Mysterious Cave Paintings, *Ronald Schiller* 51

Part Two

AWESOME FORCES OF CHANGE

Chapter One CATACLYSMS FROM BELOW

The Seething Fury of Volcanoes, *Noel F. Busch* 58
The Explosion That Changed the World, *Ronald Schiller* 64
Mount Etna Attacks a Town, *A Picture Story* 69
The Great Killer Quakes, *Ann and Myron Sutton* 72
Perils of the San Andreas Fault, *Sandra Blakeslee* 76

Chapter Two ICE ON THE MARCH

The Power and Glory of Glaciers, *Paul Friggens* 82
Mountain Sculpting with a Chisel of Ice, *John A. Shimer* 87
White Terror of the High Mountains, *James H. Winchester* 91
Antarctic Glaciers—in the Sahara Desert! *Rhodes W. Fairbridge* 95

Chapter Three MONUMENTS TO TIME

Why We Have Mountains, *Lowell Thomas* 100
Wastelands of Wind and Sand, *David I. Blumenstock* 105
The Great Pageant of Erosion, *A Picture Story* 108
Grand Canyon—Rock Calendar of the Ages, *Wolfgang Langewiesche* 112

Part Three

WONDERS OF THE LAND

Chapter One NATURE'S LAST FRONTIERS

Earth's Awesome Belt of Green, *Marston Bates* 120
How Life Defies the Desert, *Edwin Muller* 125
The Rooftop of the World, *Barry C. Bishop,* 130
The Arctic—Unspoiled, But Not for Long, *John P. Milton* 136

Chapter Two MIRACLES OF WATER

There's Magic in a Pond, *Marston Bates* 142
Waterfalls Have Many Moods, *A Picture Story* 146
Great Salt Lake—America's Pocket Ocean, *Wolfgang Langewiesche* 150
We Explored an Underground River, *Lamberto Ferri-Ricchi* 152
The 10,000 Geysers of Yellowstone, *Peter Farb* 157
Iceland's Hot Water Wonderland, *Julian Kane* 160

Chapter Three THE LAND'S SURPRISES

Ayers Rock—Home of the Primeval Spirits, *Victor Carell* 164
The Riddle of Mexico's Strange Stone Spheres, *Robert S. Strother* 168
Another World Beneath Our Feet, *A Picture Story* 172
Down to the Deeps for Gold, *J. D. Ratcliff* 174
The Mystery of the Singing Sands, *Paul Brock* 177
Quicksand—Nature's Terrifying Death Trap, *Max Gunther* 179

Part Four

MARVELS OF THE SEA

Chapter One THE RESTLESS OCEANS

The Eternal Force of the Tides, *Peter Freuchen* 184
Hidden Rivers in the Deep, *J. D. Ratcliff* 188
When Pacific Currents Shift, Life Comes to a Dead Land, *A Picture Story* 193
Face to Face with a Tidal Wave, *Francis P. Shepard* 196
The Wild Whirlpools of Norway, *Olga Osing* 199
The Ways of the Waves, *James Nathan Miller* 202
Strange Landscapes Beneath the Sea, *E. P. Lay* 206

Chapter Two THE SEA AND MODERN MAN

We Made the World's Deepest Dive,
    Commander Don Walsh, U.S. Navy 210
Trapped in an Undersea Avalanche, *Jacques-Yves Cousteau* 214
Safe Havens for Sea Life, *James H. Winchester* 216
We Must Stop Killing Our Oceans, *Gaylord Nelson, U.S. Senate* 220
Once More, Holland Defies the Sea, *A Picture Story* 224
Snug House on the Ocean Floor, *Steven M. Spencer* 230

Part Five

OUR MAJESTIC ROOF OF AIR

Chapter One THE AIRY DOMAIN

The Deep Realm of the Atmosphere, *Theo Loebsack* 236
Storms That Rage Beyond the Sky, *Ralph E. Lapp* 240
What You Don't See on a Clear Day, *Clyde Orr, Jr.* 245
How Polluted Is the Air Around Us? *Wolfgang Langewiesche* 249

Chapter Two ILLUSIONS IN LIGHT

The Sky's Rarest Spectacles, *Richard G. Beidleman* 254
Cold Fire in the Polar Night, *A Picture Story* 259
The World That Isn't There, *Theo Loebsack* 262

Chapter Three WEATHER'S MANY FACES

The Passing Parade Above Us, *Richard M. Romin* 266
The Year Without a Summer, *George S. Fichter* 271
Rainmaking Comes of Age, *Ben Funk* 273
A New Look at the World's Weather, *A Picture Story* 277
World's Most Powerful Storms, *David I. Blumenstock* 280
Inside the Tornado, *Bernard Vonnegut* 285
The Great Waterspout of 1896, *Michael J. Mooney* 289

Part Six

EARTH'S GREAT NATURAL TREASURES

The Mineral Monarchs 294
Artistry by Accident 296
The Rarest of Metals 298
Bedrock for Building 300
Basic Metals for Industry 302
Energy From the Earth 304
A Chemical Cornucopia 306
Soil's Vital Harvest 308
Gifts to Grace Our Table 310

General Index 312

Illustration Credits 318

Acknowledgments 320

Part One

OUR PLANET'S PAST

Is this how all the Earth once looked? Against an overcast sky, a Yellowstone geyser suggests the scene when the crust was forming.

EARTH'S ANCIENT DRAMA

*Though details of the early days of our planet are obscured by time, modern science
has revealed how the Earth was created, how old it is and how life may have started*

The Earth's Beginning

By Gerald S. Hawkins

In the beginning. . . . A scientist cannot continue this sentence with absolute certainty. It would be like asking a child to give an account of his birth or a description of his conception. Religious scriptures explain the creation of the Earth in compelling ways, but no two accounts agree exactly. Some of them, however, do come quite close to the scientist's idea of creation—or, at least, to his reading of the evidence lodged in the Earth's ancient rocks.

In exploring the origin of the Earth we must at the same time try to explain the beginning of the Solar System, for the Earth's past is intimately tied to the history of our nearest neighbors in space. In 1755 the German philosopher Immanuel Kant published his theory of the heavens, postulating that in the beginning there was an immense, cold whirling cloud of dust and gas. This suggestion is accepted readily by astronomers today. Their extremely powerful modern telescopes show remote, dark clouds of dust floating between distant stars —clouds that must even now be similar to the local, swirling cloud that Kant had in mind.

In 1796 Kant's contemporary, the French mathematician Pierre Simon Laplace, took his idea a step further by suggesting how the Solar System might have formed from such a cloud. The immense mass was set spinning by cosmic forces, Laplace hypothesized. At the same time it began to shrink in size under the gravitational pull of its own matter. At intervals, the contracting cloud shed veils of particles into space, which eventually condensed into the planets. Shrinking under the force of its own gravity, meanwhile, the central mass became the Sun. As potent as Laplace's concept was, it fell victim to fundamental physical laws of more recent discovery. Calculations based on these laws show that a

shrinking Sun would spin faster and faster as it grew smaller and smaller, until today it would be rotating at a far greater speed than it actually is.

After Laplace's brilliantly imaginative picture was shown to contain flaws, several other seemingly plausible suggestions were put forward by astronomers. One theory assumed the formation of the Sun first, with no planets. Then, a second star passing close by in space tore out a long stream of material. The planets, it was suggested, might then have condensed around the Sun, with the passing star continuing on its way. Unfortunately, calculations show that such hot material from the Sun would disperse, rather than form planets. Even if by some unknown process planets were to condense, their orbits would be much more irregular than those found in the Solar System today.

Another theory held that in the distant past of the cosmos, or universe, the Sun had a twin companion, and a passing star collided with its twin. Out of the debris resulting from such a collision, planets might possibly form in orbits around the single remaining sun. But the great distances at which the stars are scattered in space make collisions of this type most unlikely. If such a catastrophe did occur, it seems impossible that planets could form directly from the intensely hot and volatile material of the exploding stars. Both the "close encounter" theory and the "collision" theory fail on one further count; neither explains how most of the planets have obtained moons.

Today, cosmologists have gone back to the suggestion of Kant, careful to avoid the pitfall of Laplace. A

*With the Milky Way as a backdrop, the Solar
System is born from a spinning cloud of gas and
dust. While a young Sun glows yellow at the center,
the dust-veiled planets are settling into their
present orbits. The Earth is third from the Sun.*

new theory has taken shape from the combined efforts of astronomers, mathematicians, chemists and geologists. This new hypothesis is called the "nebular" or "proto-planet" theory. It gives unity to so many seemingly disparate details of material reality that a majority of cosmologists have become convinced that it correctly accounts for at least the broad features of cosmic evolution.

Harking back to Kant and Laplace, the proto-planet theory assumes that a large cloud of gas once filled the region of space where the Solar System now exists. This gas consisted of the "cosmic mix"—a mixture of gaseous molecules found everywhere in the universe. In every 1000 atoms, 900 are hydrogen, 97 helium, with the remaining three atoms being heavier elements, such as carbon, oxygen and iron. Slowly the primordial cloud began to turn. Its rotation probably did not develop smoothly; from recent radio-telescopic observation of similar gaseous clouds in distant space, astronomers believe that turbulence must have developed. Indeed, the swirling cloud must have looked something like a whirlpool—with small local eddies forming and re-forming as the entire volume turned in space. A large eddy at the center, contracting more rapidly than the rest of the cloud, formed a dark, more dense object, the "proto-Sun."

In the cold depths of the cloud surrounding the proto-Sun, certain atoms of gas combined to form compounds, such as water and ammonia. Slowly, solid dust crystals began to grow as did metallic crystals, including iron and stony silicates. And, gradually, gravitational and centrifugal forces at work in the spinning cloud flattened it into the shape of an enormous disk. If we could have viewed the events at a great distance, our eyes would have beheld something like a gigantic, revolving phonograph record, with the proto-Sun in the hole at the center.

Within the huge whirling disk, local eddies continued to appear. Some of the swirls were doubtless torn apart in collisions, while others were broken up by the increasingly strong gravitational pull of the proto-Sun. In a sense, each small eddy was carrying on a fight for survival. To hold itself together in the face of such disruptive forces, an eddy had somehow to collect a certain critical amount of substance to provide its own center of gravity. In a kind of cosmic battle within the wheeling system, some local swirls gained material as others lost it. Ultimately a series of large whirling disks developed in the region around the Sun. Each was a proto-planet.

These proto-planets were sufficiently large to hold together under the strength of their own gravitational fields. As each moved through space around the Sun, it acted as a sort of scavenger, sweeping up leftover material from the original cloud.

At this stage thermonuclear fusion began in the core of the proto-Sun releasing large amounts of energy, and the proto-Sun began to shine. It "burned" fitfully at first, a dull red. In time it was to become the golden yellow star that we see today. Remember that the proto-Sun was about one hundred times larger in diameter than the proto-planets. It was this immense difference in size, of course, that caused it to become a star rather than a planet. Its strong gravitational pull was

---

Earth Clock I

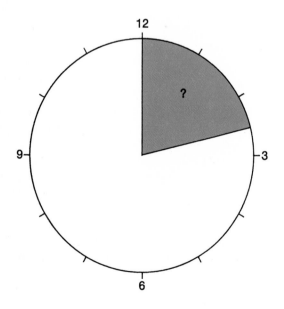

On Earth Clock I, the first in a series of diagrams in Part One, 12 o'clock midnight is considered "zero hour"—the time at which the Earth reached its present size and weight approximately 4.6 billion years ago. As the clock moves forward to today—12 o'clock noon—each hour spans 383.3 million years. A minute equals 6.4 million years.

For almost one quarter of its existence—the interval marked by the "?"—the Earth's crust was cooling and settling down, and the atmosphere and oceans were evolving toward their present states. Indirect evidence provides scientists with a good general idea of what took place in this early period, but as far as is known, no fragments of the Earth's original crust have survived. The oldest known rocks are estimated to be about 3.6 billion years old and show up on the Earth Clock at about 2:36 A.M. At this point the Earth's early geological history begins to come into focus clearly for the first time.

Dark, swirling clouds of dust and gas, such as these silhouetted against a glowing nebula in the Serpens constellation, may be solar systems similar to ours forming in distant parts of the universe.

sufficient to trap light hydrogen atoms in its interior, triggering thermonuclear fusion. Such was not the case with the smaller proto-planets.

Somewhere in the region of the proto-Sun, then, proto-Earth was born as a whirling cloud of icy particles and solid fragments—a cosmic dust storm. Only later did this material collect into a ball, sticking together because of the cohesive attraction of water and ice molecules. As proto-Earth orbited around the Sun, it swept up more material by gravitational attraction. Thus the Earth and the other planets formed by the process of accumulation of cold dusts from the region of space near the Sun.

Gradually radioactive elements within the cold ball of dust that was Earth began to give off heat. After millions of years the Earth's temperature became high enough to melt the material at its center. At that time, the heavy metals—iron and nickel—that were spread throughout the ball began to sink to form the molten core of the planet. Afterward, molten rock frequently broke through fissures to the surface. And slowly, molecules of hydrogen, water vapor and other gases escaped from within to create an atmosphere above the planet's surface. But these light gases did not stay with the Earth for long. A second major source of heat was already in action—the rays of the Sun.

The Sun's radiation was now striking the Earth with full intensity, breaking up the molecular compounds in its primitive atmosphere and scattering them into space. Thus most of the atmospheric hydrogen and other light elements escaped from the Earth. This process eventually left behind a high concentration of the heavier, rarer elements of the universe—elements essential for the formation of rocks, plants and our own bodies. Because of the escape into space over billions of years of such light atoms as hydrogen, the Earth now contains about one thousand times less mass than was present in proto-Earth when it condensed from the dust cloud.

The origin of the Moon remains an enigma to scientists. Did it form at the edge of proto-Earth? Or did it form elsewhere in space as a separate planet that was later captured by the Earth's gravitational field?

Cosmologists favor these two possibilities rather than the older theory that the Moon was ripped out of that part of the Earth that is now the Pacific Ocean basin. And with the advent of manned exploration of the Moon, it seems likely that the scientific enigma of the Moon will one day be solved.

The story of the Earth has almost reached the point where it can be taken up by a geologist. After the Earth stopped collecting debris from its path in space, its surface gradually cooled and became solid. A crust of rock formed; land masses appeared. But the Earth was not yet ready to support life as we know it today; its surface was still too hot for living organisms and the atmosphere was heavy with poisonous methane and ammonia. Molten lava flowed from fissures in the crust, allowing the escape of steam that had been trapped in the Earth's molten interior. In fact, many geologists think that this early volcanic activity brought to the surface most of the water that forms the present-day oceans—water originally trapped in icy dust.

As volcanic activity decreased on the Earth, intense ultraviolet radiation from the Sun broke up a portion of the atmospheric water molecules into separate atoms of hydrogen and oxygen. The Earth's gravitational pull wasn't strong enough to retain the lighter hydrogen atoms, and most of them drifted off into space. The heavier oxygen atoms would have remained. Although some free oxygen was thus liberated in the Earth's evolving atmosphere, the gases methane and ammonia must have remained preponderant for a long time, since most of the free oxygen in today's atmosphere is known to exist as the byproduct of photosynthesis in plants, including the algae of lakes and oceans.

Year by year the Earth became cooler as it radiated

heat and proto-Sun faded to the intensity of brightness we know now. Soon the Earth's atmosphere had cooled enough to cause water vapor in the air to condense and fall back to the surface as rain. At first, the raindrops spattering on the hot surface boiled back in a hiss of steam. Eventually, though, the Earth cooled sufficiently to permit pools of water to collect over the surface. Soon the cooling atmosphere must have begun to yield tremendous amounts of rain. All the water in the seven seas may have descended in one long continuous deluge. Gradually the shallow areas in the wrinkled crust filled, and oceans appeared on the face of the Earth.

Although scientists are generally convinced that the Earth on which we live has passed through the stages of development outlined in the previous paragraphs, no one, of course, can vouch for the exact chronology. Probably, proto-Earth reached its present size and shape some four and a half billion years ago. After this, one and a half billion years may have passed before conditions on the Earth became hospitable to early forms of life. The evolution of life, of course, is a separate drama. Here we have attempted to suggest how Nature set the stage.

The First Stirrings of Life

By George Gamow

Nature worked hard for billions of years to build complex organisms like the one who wrote these lines and those who will read them, beginning with simple chemicals formed on the Earth when it was quite young. What started the chain of events that led to the immense variety of living beings and to the almost unbelievable complexity of their individual structure?

The problem of the origin of life on our planet begins with the question of how proteins and nucleic acids, the two chemicals basic to all living organisms, could have originated naturally on the surface of the Earth. Without much difficulty a good organic chemist today can synthesize all the substances vital to life found in a living cell. But how could these have arisen spontaneously?

A brilliant idea about how this could have happened was expressed several years ago by the distinguished American scientist Harold Urey. Urey's idea is based

After the Earth's crust hardened, heavy clouds of volcanic gas and water vapor formed, as in this artist's conception. Cooling brought on great rains which pooled in low spots to form the oceans.

on the modern nebular theory of the origin of the Solar System, according to which proto-planets originally possessed extensive atmospheres of hydrogen and hydrogen compounds, such as methane, ammonia and water vapor. Chemical elements in these compounds (hydrogen, carbon, nitrogen and oxygen) are exactly those that form amino acids—the basic "building blocks" of long protein molecules. Urey theorized that, when subjected to ultraviolet radiation from the Sun and electric discharges from thunderstorms in the atmosphere, the molecules of these simple compounds could unite to form more complex amino-acid molecules.

To confirm his ideas, Urey asked one of his students, Stanley L. Miller, to carry out an experiment by putting a mixture of hydrogen, methane, ammonia and water vapor into a test tube and subjecting it for several days to an electric discharge. When the contents of the test tube were analyzed, they revealed the presence of several amino acids normally found in proteins, thus constituting a brilliant confirmation of Urey's hypothesis. Presumably, during the early existence of our planet, when its atmosphere still consisted of hydrogen

and hydrogen compounds, amino acids were continuously produced in that atmosphere. These substances were slowly precipitated to the surface, forming concentrated solutions on the fringes of the ocean waters. This process, then, provided one chemical component essential for life.

Much less is known about the origin of nucleic acid, the other component fundamental to life as we know it. The molecular chains of nucleic acid contain atoms of phosphorus, which are not likely to be found in the atmosphere. Also, the synthesis of nucleic acids requires high temperatures rather than ultraviolet radiation or electric discharges. One bold hypothesis suggests that nucleic acids were produced as the result of the activity of rain-washed volcanoes, but experimental evidence for this hypothesis is not yet conclusive.

The next problem, of course, is how such solutions of proteins and nucleic acids in the ocean waters could have evolved to form the first living organisms capable of reproduction. There is hardly any doubt that a Darwinian "struggle for existence" principle operated very early in the evolution of life on the Earth. In fact, we can trace Darwin's evolutionary principle back past the vague borderlines of life to simple inorganic reactions. If a mixture of powdered iron and silver is exposed to oxygen, more iron oxides than silver oxides will be produced because the oxidation of iron proceeds at a faster rate than that of silver. Similar, if more complicated, evolutionary chemical processes must have been occurring among protein molecules dissolved in the waters of primordial oceans. Those molecules whose reactions were intrinsically faster had an obvious edge on the slower ones.

This early development of organic matter remains hidden from us behind a heavy curtain of mystery because early biochemical reactions could not leave fossil traces in the rocks of those periods. Neither do we have a record of when and how organic molecules acquired the ability to produce other molecules with the same chemical properties. All we can surmise is that such developments must have occurred during the geological eras that preceded the formation of our oldest known fossil—a primitive bacterium some 3.2 billion years old found recently near Barberton, South Africa.

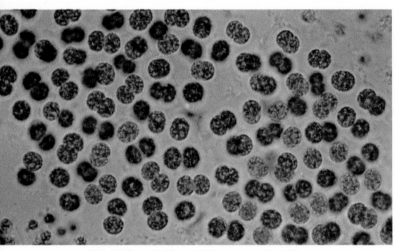

On the borderline of life, one-celled blue-green algae (above) bear a close resemblance to non-living spheres (below) which form when amino acids strike hot volcanic lava, then mix with rainwater.

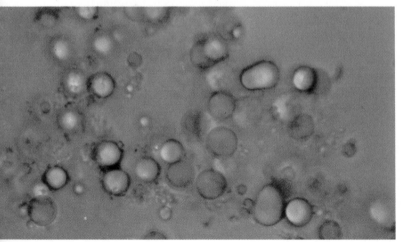

There is not much chance, either, of finding extensive evidence of the organisms that preceded the first fossils, since earlier forms of life must have been miniature, soft-bodied organisms not much different from organic molecules. In fact, if through some miraculous device we were able to go back in time to a point about three billion years ago, the primordial pools of water and rocky slopes of the land masses would appear to be lifeless. Only through minute examination would we discover that life was already present on the surface of the planet, and that numerous micro-organisms of many different kinds were even then hard at work in their fight for existence. At this early stage in the evolution of our planet its surface was still warm, and a large part of the water that now fills the ocean basins existed in the atmosphere, forming a thick layer of heavy clouds. No direct sunlight could penetrate this heavy atmosphere to reach the surface of the Earth, and any life able to arise in such damp darkness must necessarily have been limited to micro-organisms that could survive entirely without sunlight.

Some of these primitive organisms must have fed on organic substances dissolved in the water around them. But others grew accustomed to purely inorganic food. This second class of "mineral-eating" organisms can still be found in the "sulfur and iron bacteria," which obtain their vital energy through the oxidation of inorganic compounds of sulfur and iron. The activity of such bacteria has played quite an important role in the development of the Earth's surface. For instance, the iron bacteria may possibly be responsible for many thick deposits of bog iron ore, the main commercial source of iron in the world.

As time went on, the surface of the Earth grew cooler and cooler. More and more water accumulated on the surface, while the heavy clouds blocking the Sun gradually thinned out. Under the action of the Sun's rays, now readily piercing the atmosphere to reach our planet, some primitive micro-organisms were slowly developing the substance chlorophyll. This let them use the energy of solar radiation to convert carbon dioxide in the air into simple compounds necessary for their growth. Thus the possibility of "feeding on the air" opened up new horizons for the development of organic life—eventually culminating in the present highly developed and complex forms of the modern plant kingdom.

But some of the primitive organisms chose another way of development. Instead of synthesizing their food from carbon dioxide in the air, they obtained carbon compounds in "ready-to-use" form by feeding parasitically on plants. Some of these parasitic organisms soon developed the ability to move—a great advantage in the competition for food. Not being satisfied with a purely vegetable diet, some members of the parasitic branch of living things then began to eat one another, and the need to catch prey—or to flee when pursued—led to even better means of moving around, culminating, after hundreds of millions of years, in the advanced locomotive adaptations of the animal world.

Rapid motion through the water could not be achieved by soft-bodied, easily deformed animals. Such swiftness requires rigid streamlined shapes, and muscles working in tandem with rigid "moving parts." Rigid body parts also afford protection from attack, while at the same time providing better weaponry for attacking others. In the Earth's primeval oceans—where life was a struggle in which the fittest survived—these advantages resulted in the evolution of the soft, jellylike forms of the animal world into heavily armored creatures with powerful claws.

The development of rigid parts proved of great value to animals, and incidentally to modern paleontologists too. Whereas information about the soft-bodied organisms of the past can be gleaned only from their occasional imprints in soft sand—imprints through mere chance preserved to the present time—animals that possessed rigid shells or skeletons can be studied from their fossils almost as well as if they were living today.

The historical record of life on the Earth begins, properly speaking, at the time when animals began to develop rigid parts and bodies. While modern museums are full of shells and skeletons that permit us to visualize the life forms of the more recent past, the very first living organisms are a secret lost in time.

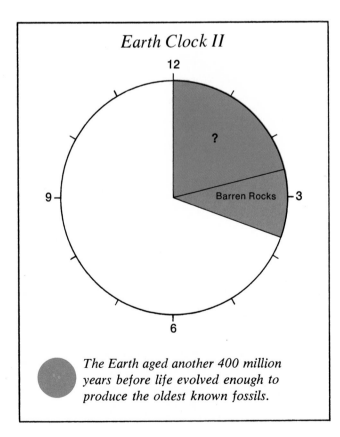

Earth Clock II

The Earth aged another 400 million years before life evolved enough to produce the oldest known fossils.

Unraveling the Age of the Earth

By E. L. Simons

For thousands of years man has asked the question: How old is the Earth? The far-ranging investigations by which he has attempted to arrive at an answer have often been as interesting as the riddle itself. Not until the past three decades, however, have researchers possessed the precise dating techniques and instruments needed to track down the Earth's true antiquity. And it is only since the late 1950s, really, that accurate dates have been calculated for events in the prehistoric past.

Previously, there had been no lack of estimates of the Earth's age—with chronologies appearing in the sacred books of many religions. In India, for instance, the Brahmins considered the Earth, and time, as eternal. A list of dynasties of gods and demigods spanning 36,000 years of the Earth's existence was drawn up by the Egyptian scholar Manetho in the third century B.C., while Hebrew scribes dated creation only at about 4000 B.C. The Hebrews may have based their story of Genesis partly on earlier Babylonian myths, although some of these implied an older time for the Creation.

In 1654 Archbishop Ussher of Armagh, Ireland, calculated from Old Testament genealogies that the Earth had been created in 4004 B.C. One of Ussher's contemporaries claimed to have determined the precise day and hour of the Earth's appearance—October 23 at nine o'clock in the morning. However, Hebrew scholars, studying the same Old Testament sources, gave as the year of the world's creation 3761 B.C. And with this date the traditional Hebrew calendar begins.

Many modern theologians challenge the very notion of attempting literal interpretations of scriptural sources. Indeed, as far back at the fifth century A.D., St. Augustine proposed that the six days of creation might signify the logical stages rather than the literal time sequence of creation. Even the more conservative Christian scholars today usually agree that we can think of the six days of creation as symbolic of six general periods, or eras—rather than specific 24-hour periods.

In any case, as scientific knowledge grew in the 18th and 19th centuries, scholars recognized that the Earth must be much more than a few thousand years old. Hence, scientists began a series of ingenious inquiries to unravel the actual age of the Earth. In 1715 the British astronomer Edmund Halley proposed that the evidence scientists sought lay in the oceans. Let's assume, Halley said, that the oceans originally consisted of fresh water precipitated from the atmosphere. The salt content of the present-day oceans would thus be owed to the salts that the rivers had washed into the seas during the intervening eons. By dividing the amount in tons of sodium salts now present in the oceans by the average amount in tons that is added each year from the world's great rivers, one could calculate the age of the oceans. Although scientists tried Halley's approach, their estimates of the age of the seas proved too low. We now know that the rate at which salts are carried into the oceans by the rivers has fluctuated greatly over past ages. Moreover, salts are believed to collect in the oceans from other causes as well—including volcanic activity and lava upwellings through fissures in the ocean floor.

Another early procedure suggested for estimating the Earth's age involved measuring the annual rate at which sediment is deposited on the ocean floors. This rate divided into the total estimated thickness of ocean sediments would give an age in years. Today, we know that several problems stand in the way of such an approach. Among other things, recent discoveries relating to "plate tectonics" suggest that ocean-bottom sediments are continually being shifted about and destroyed as the sea floors move over the globe.

In 1799 William Smith, an English canal surveyor, formulated one of the basic concepts upon which the science of geology rests—a way of telling the comparative ages of the strata of rock. He pointed out that the fossils found in layers of sedimentary rock "always succeed one another in the same order." Fossils thus became the key that allowed geologists to identify the relative ages and sequences of rock layers regardless of their location. But such dating processes were only of a comparative nature, pinpointing one rock layer as older or younger than others. No absolute date for the age of any rock could be gained thereby.

By the 19th century a number of geologists had gone off on another tack. They had begun trying to estimate how much time had passed since life first developed on the Earth. In 1867, one of the fathers of the science of geology, Sir Charles Lyell, conjectured that 240 million years might be enough time to account for the great successional changes that had occurred among the Earth's animal and plant populations over the ages. But a contemporary of Lyell estimated that only 60 million years was necessary. Naturalist Charles Darwin leaned toward Lyell's estimate, arguing that 60 million years could hardly account for the whole history of life and the evolution of modern organisms.

At about this time an even younger age for the Earth was proposed by the British physicist Lord Kelvin. Assuming that the Earth had originated as a molten-hot body, Kelvin calculated the time necessary for it to cool to its present state with only a molten core. Kelvin finally put the planet's age at only 20 million years. So small an age seemed impossible to most of the geologists and paleontologists of Kelvin's day.

Kelvin, it turned out, had not taken into account an extra source of heat not fully understood in his day; namely, the heat that warms the Earth's interior rock,

This fist-sized chunk of lunar rock was called the "Genesis rock" because it may be a fragment of the Moon's original crust. Since Earth and Moon probably formed in the same way, ancient lunar rocks yield evidence of the Earth's geological history as well as the Moon's. Below: Apollo 15 *astronaut Jim Irwin—who helped find the rock—works with the lunar "Rover" on the Moon's surface.*

which is produced by radioactive elements trapped under the surface. Moreover, the Earth almost certainly did not originate as a molten body. Modern theory postulates that proto-Earth gradually accumulated from a cloud of cool dusts and gases, only warming up as the result of the heat released by the radioactive elements trapped in its interior. Had Kelvin been aware of these factors, he would have known that the Earth must be much older than 20 million years.

Today scientists know that the Earth is far older even than Charles Lyell and Darwin had imagined. Modern estimates of the Earth's age are based upon a kind of "geological clock" that exists in the Earth's radioactive rocks. In 1905 the American chemist Bertram Boltwood pointed out the universal presence of the element lead in uranium-bearing rocks. Boltwood observed that in any one region of the Earth the ratio between the lead and the uranium found in local radioactive rocks is usually highly uniform. Boltwood hypothesized—correctly—that lead in various forms is the end product of the process of radioactive decay, with uranium isotopes gradually being transformed into lead isotopes. Boltwood proposed that if lead were the final stage of this slow process of decay, the ratio of lead to uranium should be the same for radioactive minerals of equal age. He suggested that the age of such rock could be precisely determined from the relative proportions of its lead isotopes and its still-unchanged uranium isotopes, if only the rate at which uranium disintegrates were known. Subsequent research revealed that the rate of radioactive decay for uranium is extremely slow.

More than 4.5 billion years are known to be required for half the original uranium atoms in any given sample to decay. This figure is called the "half-life" of uranium.

So far, a number of ancient rocks have been dated by the uranium-lead and other, newer radioactive dating methods. Among the oldest are rocks from Antarctica, South Africa, Australia, the Soviet Union and North America's Canadian Shield. Radio-chemical analysis suggests that the oldest of these rocks were formed more than three billion years ago. Rock layers of this great age are known to lie over beds of rock that must have solidified even earlier in the Earth's past. As yet, geologists have not been able to date these older rocks. Possibly they were derived from the primordial crust of the Earth. Up till now, however, no rock has been identified as part of the Earth's original crust.

How much older might the original crust of the Earth be? Many scientists believe that meteorites are the remains of a planet or planetoids that formed at the same time as the Earth and the rest of the solar system, and later broke apart. If so, then the age of the Earth, as indicated by uranium-lead ages obtained for fragments of meteorites that have fallen from the skies, is near 4.6 billion years.

Long Journey of the Continents

By Ronald Schiller

Crisscrossing the high seas in recent years, the ship *Glomar Challenger* has been logging voyages of exploration as important as any undertaken since Columbus sailed in 1492. Her scientist crew is drilling into the ocean bottoms and examining sedimentary mud and rocks millions of years old. Each time the crew pulls up her three-and-a-half-mile-long drill pipe it confirms one of the most astonishing scientific discoveries of our time: The world's continents are adrift on the Earth's pliable inner mantle of rock.

This revolutionary discovery has led to a concept known as "global-plate tectonics," challenging all traditional views that rest on the old notion of a stable planet. The new concept pictures a turbulent, dynamic world where, during 4.6 billion years of geologic history, oceans have opened and closed like accordions and continents have been buffeted about like hulks on a stormy sea. This revolutionary insight also answers riddles that have baffled man since the dawn of natural science: How the continents, oceans, mountains and islands were formed; the reasons for volcanoes and earthquakes; and why marine fossils are embedded on Himalayan peaks, to name but a few such enigmas.

Admittedly some of the pieces of the geological puzzle are still missing, and scientists do not agree on all details. But the general outlines of the picture have been verified—and accepted by most scientists as geologic truth. "Global-plate tectonics can no longer be referred to as a mere theory," says Maurice Ewing, dean of American oceanographers. "Scientifically it is as significant as Darwinian evolution or Einstein's laws of energy and motion—and as important to mankind."

The concept of which Dr. Ewing speaks advances these views of the geologic process:

1. Far from being the solid, indestructible shell once imagined by geologists, the Earth's crust consists of a number of separate plates (ten major ones, subdivided into varying sizes), made of rock 40 to 60 miles thick, which float on the hot, viscous mantle beneath them.

2. The Earth's land surfaces, which rest on these plates (as do the ocean floors), were once welded together in a single super-continent. Some 200 million years ago this super-continent began to split up. Eventually its pieces formed the present seven continents and the major islands we now know. These land masses have slowly drifted apart like packages on a moving conveyor belt.

3. The crustal plates are being built up at their edges by molten rock welling up from deep fissures in mid-ocean; at the same time these plates are being propelled

across the globe by forces arising from deep within the Earth, in various directions, at the geologically wild speeds of from one-half inch to six inches a year.

4. Phenomenal things occur as the plates jostle one another for room: When a moving continental plate (mainly granite) meets an ocean-floor plate (consisting of dense, less-buoyant basalt), the continent rides over the sea-floor plate like a titanic bulldozer; it scrapes off the layers of sediment deposited on the sea floor over many millions of years, as well as slices of crustal rock. This debris piles up along the edge of the continent like a rumpled blanket, forming mountain ranges. The ocean-floor plate is forced down at a steep angle into trenches at the edge of the continent. It melts from the heat of the friction, forming deep underground pockets of white-hot lava. The trapped lava is forced up through crevices, erupting on the surface to create volcanoes. The collision, separation and shearing of the crustal plates also cause seismic disturbances.

How did geologists arrive at this seemingly fantastic conclusion? For generations, schoolchildren studying their atlases have noticed that if South America and Africa were brought together and twisted slightly they would fit like the pieces of a jigsaw puzzle. In 1912 the German geologist Alfred Wegener theorized that the continents actually were once joined. He pointed out that the rock formations along the bulge of Brazil and Africa's Gulf of Guinea are enough alike in age and structure to have been torn from the same geologic fabric. He noted that identical fossil plants and fresh-water animals, which could not have survived a trip across thousands of miles of salt water, have been found in South America, Africa, Australia and even distant India. But his idea that these regions must once have been joined was discredited because no one could conceive of any mechanism that could propel vast continents through the Earth's solid crust. Moreover, it was thought, such moving land masses would have left behind gigantic wakes of displaced rock on the sea floors. Despite intensive searches, no ripple of such disturbances was discovered.

A second mystery was the strange sparsity of sediment on the ocean floors. Geologists had calculated that the sediment formed by microscopic marine organisms—together with dust blown or washed into the sea —should have blanketed the ocean beds over the ages to a uniform depth of at least 12 miles. Yet they found practically no sediment in the center of the world's oceans and only a half-mile thick veneer near the borders adjacent to the continents. Then, in the late 1850s, telegraph engineers, laying the transatlantic cable, found submerged mountains in mid-ocean. Similar ridges were later found in the Pacific and elsewhere.

Nearly a century afterward, in the 1950s, oceanographers discovered that these ocean ridges form a continuous 40,000-mile chain that winds around the globe like the seam on a tennis ball. And down the center of this gigantic serpentine ridge run deep, hot rifts, oozing lava. Where the ocean floors are splitting apart, lava is welling up, forming new ocean-floor material as it hardens. Scientists speculated that the ocean floor might be moving out from the ridges, eventually plunging into the deep troughs bordering the land.

This startling idea received striking support in 1963 from a dazzling piece of deduction by F. J. Vine and D. H. Matthews of Britain's Cambridge University. Geologists had first discovered that many times in the Earth's history the magnetic poles had reversed polarity —that is, during certain ages the iron particles in the Earth's rock pointed south instead of always pointing north. What caused this phenomenon is not yet known; but, by measuring the extent to which the radioactive elements in the rock have decayed and by determining the age of fossils embedded in such rock, geologists have learned to date and read these magnetic reversals like rings in a tree. (The last magnetic flipover occurred 700,000 years ago.) If the ocean floors actually were born from lava at the ridges and were spreading apart, reasoned the Cambridge scientists, then there should be an identical series of magnetic reversal bands on either side of the mid-oceanic ridges.

When oceanographers took to the sea in ships, towing magnetometers, they found magnetic reversal bands exactly like those predicted. Since scientists can tell when each reversal occurred, they are able to determine not only the age of any particular segment of ocean floor but the direction and speed at which it is moving in the crust. They learned, for instance, that the Atlantic Ocean floor is widening, pushing Europe and North America apart, at the rate of one inch a year.

With the announcement of these findings, many mysteries were solved. Now scientists know why there are mountain ranges under the sea, why undersea cables snap, and why there is so little sediment on the ocean bottoms. In the last instance, the sea floor moves too fast to allow sediment layers to accumulate to any great depth; sediment is either plastered up against the continents or carried down into the trenches. And by tracing the movement of the sea floors back in time, oceanographers were able to figure out where the present continents were once located.

Still needed to reconstruct the history of our planet was the precise sequence of these dramatic events. Their chronology could be obtained only by digging into the ocean floors and recovering sediment and crustal rock, then determining the composition and ages of this material. Five American institutes, headed by California's Scripps Institution of Oceanography, banded together in this brilliant scientific effort aboard the *Glomar Challenger*, one of the strangest craft ever to put to sea. A 142-foot-high oil derrick sits amidships, and a 20-foot-wide hole extends through the bottom— through which a miles-long string of steel pipes is low-

As the continents moved apart, the Arabian Peninsula (right) split away from Africa, creating the Red Sea and the Gulf of Aden. The closely matched coasts were photographed on a 1966 Gemini space flight.

ered. Since anchors are useless at the enormous depths where the *Challenger* drills, a steady position over the selected site is maintained by means of twin screws aft and a total of four lateral thrusters—one on either side of the bow and stern—which allow the vessel to move sideways when necessary. The ship's thrusters are commanded by computers which, by reference to sonar beacons on the ocean floor, can hold the ship within 40 feet of the drill site even in the roughest seas.

It used to be that once the *Challenger*'s drilling bit wore out, the hole in the ocean floor would have to be abandoned. But now, thanks to newly developed underwater sensors, motors and a wide-mouthed metal funnel that is placed on the sea bed, the drillers are able to pull up the string of pipes, screw on a fresh bit and find their way back into the same 10-inch-wide hole miles below on the bottom.

Since August 1968 the *Challenger* has been covering

the world's oceans, drilling hundreds of holes, the deepest of which reaches almost 4000 feet. From these holes it pulls up long cores of bottom mud and bedrock, some of which is as much as 160 million years old. As each core reaches the deck, it is sliced into five-foot lengths and whisked to well-equipped laboratories on board ship to be photographed, X-rayed, metered for radioactivity and analyzed for age. Next the cores are packed in plastic and sent to refrigeration vaults in the United States for further study. Among the crew, excitement runs high. Says the project's scientific manager, Melvin N. A. Peterson, "You feel like an explorer rummaging through the ruins of a lost civilization.

Although some details of the Earth's past will be revised each time the *Challenger* returns to port, most geologists can now agree that the Earth's land surfaces were indeed joined 200 million years ago in a single super-continent. They call this primeval land mass "Pangaea" (Greek for "all lands"). The heart of this mother land appears to have been located along the Equator. Japan lay near the North Pole, India near the Antarctic. According to a reconstruction made by American geologists Robert S. Dietz and John C. Holden, the first separations were a gigantic east-west crack in the Earth's crust and a rift that appeared between the South American-African mass and Antarctica-Australia. India was liberated and started bolting north. The 200 million years required for the plates to split apart and reach their present locations represent a phenomenally short time from the geologic point of view —equal to only a little over 30 minutes, if all the Earth's history were compressed into the 12 hours on the face of a clock.

To arrive where it is today, North America drifted northwest. The Eurasian plate, twisting 20 degrees clockwise, moved off to the north. Eurasia was pursued by Africa, which rotated counterclockwise. In the process, South America broke loose to the west, while Greenland and Northern Europe were parting company. The last two continents to separate from one another were Antarctica and Australia, the latter migrating northward to warmer climates. The most spec-

How the Continents Moved

1. The world 200 million years ago probably contained this single supercontinent, Pangaea. Computers were fed profiles of today's submerged continental slopes—the true boundaries of continents—to reconstruct Pangaea. Note that Australia is joined to Antarctica; India fits between Africa and Antarctica.

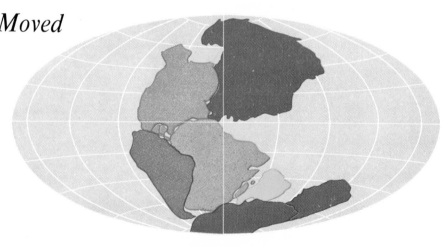

2. By 135 million years ago, Pangaea had split along an east-west fault just above the Equator. North America was separating from Europe, creating a small ocean basin, and South America had begun cracking free of Africa. India was racing northward while Australia remained attached to Antarctica.

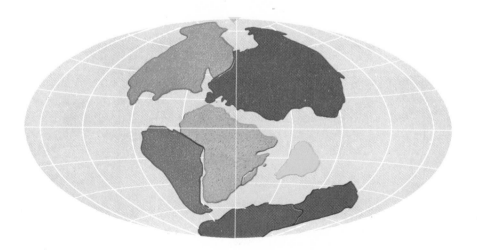

tacular journey of all has been made by India, which, once it tore loose from Africa and Antarctica, traveled 5500 miles north in 180 million years, ramming into the belly of Asia. The marine sediments between these two continents were deformed like molding clay in a vise to create the world's highest mountains, the Himalayas. (Thus, climbers who reach the top of Mount Everest stand on the former ocean floor!)

Almost as violent were the happenings in the Mediterranean; here, the African plate crashed into Europe. The collision erected in Europe the mountain ranges of the Pyrenees, Apennines and Alps. On the other side of the globe, the Americas, Asia and Australia had started plowing toward one another across the vast Pacific. The resulting disturbances in the trenches under the leading edges of these plates erected the Andes Mountains of South America, the Sierras along the western coast of North America, and the island arcs of the Aleutians and Japan.

Just how often in the past the oceans have opened and closed, and the continents collided, split asunder and welded together again, geologists are now trying to determine. About the future they can already speculate with considerable confidence. Assuming the plates keep moving in their present directions, scientists predict that the Atlantic Ocean will continue to widen, while the Pacific Ocean shrinks. With Africa once again creeping toward Europe, the Mediterranean seems doomed eventually to become a shallow pond. The Himalayas will grow, and India will cease burrowing under Asia and will slide eastward. Australia will proceed northward at two inches a year.

North and South America will pursue their present westward voyage. The peninsula of Baja California and a sliver of the California coast west of the San Andreas Fault, which lie on a separate plate from the rest of North America, will tear away from the mainland, heading northwest. But Los Angeles residents should be in no hurry to pack their bags. Scientists calculate that it will take ten million years more for the city to sail past San Francisco, and another 50 million years before it slides into the Aleutian trench.

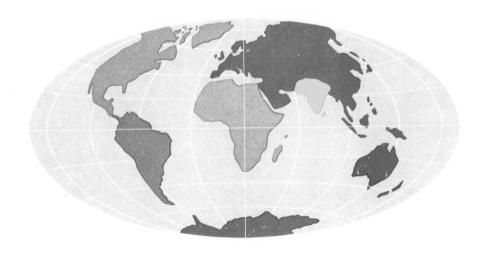

3. Today the globe looks like this. The Americas, after drifting west, are separated from the Old World by the newly formed Atlantic Basin. While Africa crept north, India collided with the underside of Asia. Australia broke away from Antarctica and drifted to its present position, pushing New Guinea up ahead of it.

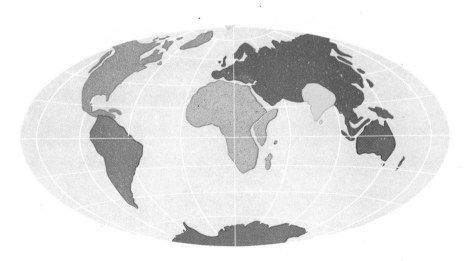

4. Present trends suggest this map in 50 million years: The Atlantic will be larger, especially the South Atlantic, and the Pacific smaller. India will move farther east; Australia will almost touch Southeast Asia. East Africa and part of California will set out to sea, and new land will rise in the Caribbean Sea.

Chapter Two

CREATURES OF THE AGES

Long before man, an astonishing parade of living things spread over the Earth—leaving a trail of fossils and buried bones from some of Nature's most spectacular achievements

Revelations from Fossil Finds

By Ruth Moore

The time was March 1835. The road climbing the precipitous western slope of the Andes Mountains had turned into a narrow zigzagging track, and the mules transporting the great naturalist Charles Darwin's first expedition over the rugged Peruvian peaks had been halted for a needed rest. As Darwin surveyed the impressive vista—the jagged pinnacles, the deep valleys, the giant condors wheeling in the brilliant sky—his attention was suddenly drawn to a small, shiny object in a nearby rock wall. Up close, it proved to be a seashell. Then he saw numberless shells protruding from the same pale band of limestone.

Darwin quickly dismounted and started picking loose the shells. He knew that the elevation was around 13,000 feet. His mule-drivers had been complaining about the cold, and in that rarefied air Darwin was suffering shortness of breath. The naturalist could have reaped a large harvest had he remained for several days. But the discovery had come late in the southern summer; if snow began falling, the expedition, already suffering under the severe conditions in the high Andes, might find itself trapped.

As Darwin gathered his fossil shells, he saw that some were similar to ones he had previously collected on the beaches of the Pacific Ocean far below. Presumably they had at some time in the past rested on the ocean bottom. Through some unknown process of upheaval, the once low-lying beds had been elevated to a height of 13,000 feet. Darwin concluded that the Andes had not been wholly created by the molten outpourings of volcanoes, as geologists had hitherto believed. Today, we know that Darwin was right. The friction of the Earth's slowly drifting crustal plates has in many regions wrinkled the ocean floors and pushed up great

mountain ranges along the edges of several continents.

The fossils Darwin collected on this high mountain helped to change his ideas and ultimately those of the world about the age of the Earth. For such shell-bearing sediment to have been uplifted from sea floor to mountain crest would have taken millions and millions of years, Darwin recognized. And Darwin knew that the Andes shells were not an isolated case. Similar fossils had been discovered previously in the Alps and other mountain ranges.

The fossils buried in the rocks could be read, in a sense, like a geological calendar of the past, recording the succession of eras in the Earth's history, each with its characteristic forms of life. As long as there were no violent disturbances, the oldest stratum of sedimentary rock, containing the oldest shells, would lie at the bottom of any series of sediment layers. The younger ocean sediments, with younger fossils, would necessarily be deposited on top of the older layers on the ocean bottoms. If the strata were undisturbed, the order of the ages would be as clear as numbered pages in a book.

In addition to the evidence fossils give of past geologic processes, they offer concrete records of organisms that no longer exist on the Earth. Down through the ages, the remains of countless creatures have drifted to the ocean bottoms. Although most were perishable, many possessed hard parts; their impressions were cast and preserved in the sands, sediments and silts of the ocean floors—and were buried under the continuous rain of other organic debris settling to the sea floors. On land, fossil creatures have been preserved in tar pits, bogs, swamps, caves, stream beds, ice sheets and other sites.

For centuries before Darwin's Andean discovery, of course, many other people had found shells and bones at various places on the Earth. The ancient Greeks had picked up seashells far inland from beaches, reasoning that the sea must once have rolled in over the land where they lay. Occasionally a huge fragment resem-

A burial vault of slate preserved the fossil remains of this ancient shark for 350 million years. As its body decayed on the ocean bottom, organic matter was replaced by chemicals that hardened into rock.

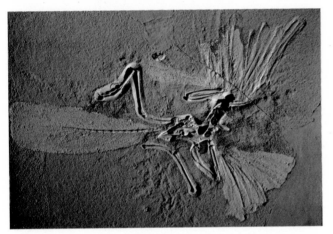

The delicate wing and tail feathers of Archaeopteryx make it a true bird despite its reptilian skeleton.

Soft sediments pressed this fossil sumac so gently that networks of veins show up clearly in its leaves.

A moth's fragile structure is preserved in a fossil treasure discovered in an ancient lake bed.

This skull of a distant relative of the horse was found in a Paris quarry more than a century ago.

bling the bone of a giant animal might be discovered —only to be attributed by the Greeks to a mythical monster.

In following centuries men continued to find the fossil remains of sea creatures on dry land. Few, though, knew of the Greek explanation of changing sea levels in the past, and even fewer accepted the Greek theory. Some learned men believed the fossils to be foreign bodies grown on the Earth from seeds that had fallen from the stars; other scholars argued they must have formed in the ground in chance imitation of life. Still other thinkers deduced that they were the works of Satan—buried in the ground to deceive curious men.

In 17th-century Europe the puzzle posed by fossil remains took on a new significance. Church scholars were shocked as they saw the Bible being subjected to various readings and translations. To uphold the long-accepted Biblical authority upon which society and religion rested, religious leaders undertook to prove by contemporary scientific methods that the miracles—and particularly the account of creation—were true. Scientific certainty was to support scriptural revelation.

Certainly the disconcerting evidence of strange fossil creatures on the Earth had to be accounted for. John Ray (1627-1705), a Cambridge University scholar, was too good a naturalist to attribute fossils to seeds from remote stars or the work of the devil. He saw that the fossil shells he collected inland were, in fact, not at all unnatural; they were exactly like those washing up on the beaches even then. Other inland finds were the skeletons of fish that he knew inhabited the ocean depths.

Ray accepted church doctrine that the marine fossils found on land were evidence of the Biblical Flood. But why were they not spread more evenly across the Earth, he wondered, instead of being amassed or lumped within layers of rock? And why did the fossil troves sometimes contain the remains of species that were not known to exist on the Earth? Hadn't all present-day terrestrial animals descended from those Noah rescued in the ark?

During the 40 days and nights of the scriptural deluge, Ray concluded, the so-called waters of the abyss—reservoirs thought to exist in the bowels of the Earth and to connect with the waters of the oceans—must have filled to overflowing. Under these tremendous pressures, the floodwaters burst forth, he wrote, "at those wide Mouths and Apertures made by the Divine Power breaking up the Fountains of the Deep." The concentrations of fossils at certain sites on land were thus satisfactorily explained. All appeared to be once more in order, even though Ray did not explain the remains of creatures nonexistent on the Earth.

Less than a century later, new fossil finds made it impossible to ignore these strange anomalies any longer. The French naturalist Baron Georges Léopold Cuvier

Nineteenth-century naturalist Georges Cuvier tells a meeting of the French Academy of Sciences about his startling discoveries of extinct creatures.

(1769-1832) began to find the fossilized bones of elephants, flying lizards and other fantastic animals in the soil around Paris. People in Paris rushed out to the gypsum quarries where he was unearthing the bones to witness the spectacle. Cuvier caused an even greater sensation when he reconstructed many of the skeletons to suggest the appearance the animals had in life.

"Is Cuvier not the greatest poet of our century?" exclaimed the popular French novelist Honoré de Balzac. "Our immortal naturalist has reconstructed worlds from blanched bones. He picks up a piece of gypsum and says to us, 'See!' Suddenly stone turns into animals, the dead come to life and another world unrolls before our eyes."

The lost menageries that Cuvier was suddenly producing, as though possessing powers of a sorcerer, were not a random assortment, by any means. The animals of the past—like those of the present—can be classified as birds, mammals, reptiles and so forth.

As he unearthed these ancient remains from their beds of rock, Cuvier saw that the sea fossils always lay within their own stratum, while the fossils of land creatures occupied a separate layer. Between the marine and terrestrial groupings might fall a stratum that held no fossils whatsoever. Cuvier theorized that this part of France must have lain under the sea; during this time the stratum containing marine fossils was deposited as sediment. Afterwards, the waters had receded; hence, the fossils of land animals appeared in a separate sediment layer laid down by lakes and rivers. Many such successions had occurred in the region.

At almost the same time as Cuvier was causing a stir in France, similar discoveries were being made in England. In 1811 Mary Anning, a carpenter's daughter, found the skeleton of a 21-foot marine reptile in a crumbling cliff on the south coast. William Smith, an astute geological observer then employed as a surveyor and an engineer in digging Britain's new canals, saw that in the newly excavated banks of canals lay strata of rock, each characterized by a certain type of fossil remains. He realized that rock strata, judging by their fossil content, seemed to be laid down everywhere on the Earth in an orderly sequence. Smith prepared a map of Britain, showing, in his words, that "the same species of fossils are found in the same strata, even at a wide distance."

Through their observations Cuvier and Smith helped to develop geology as a real science. Although most other scientists of the time were willing by then to grant that fossils were the remains of actual animals, few suspected that those remains and their associated rock layers gave a clear record of the past—or that they could be used to determine the order of events in the Earth's history.

Shortly thereafter, in 1860, while digging in a limestone quarry in Bavaria, Germany, workmen came upon a remarkable fossil imprint. Nothing like it had ever been seen before by anyone. It was a cast of a creature the size of a chicken, with sharp teeth, an elongated head, an extended neck and strong hind legs —all reptilian characteristics. But the cast also showed, in exquisite detail, the unmistakable impression of feathers on a wing that had claws. If the impression of the feathers had not been preserved, who could have guessed at their presence? Was this toothed animal with claws bird or reptile?

Later two more casts of the creature were found.

Presumably these birdlike reptiles had fallen into the waters of the coral lagoon that had covered this region in the remote past and been buried in sediment of tiny marine shells. As the skeletons of the small marine organisms settled over the bodies of the reptile-birds, the latter were preserved with their features cast in sharp detail. Later, this soft ooze of bottom sediments hardened into limestone so that the cast had lasted some 135 million years or more—long after the lagoon had dried up.

This prehistoric creature of the air was named *Archaeopteryx* (ancient bird). It is now known to be one of the earliest and most primitive of a group that evolved into our modern birds.

Archaeopteryx was soon followed by numerous fossil discoveries, each enlarging man's perspective of the Earth's past. Smooth-grained, limy lagoon floors and the dried-up bottoms of ancient shallows disclosed the crawl-trails—sometimes even footprints—of early crawling and wading forms of life. In beds of rock now raised high and dry, fossil hunters came unexpectedly upon the meandering, scratchy paths traced by insects millions of years in the past. Here, perhaps, the trail of a tiny crab was afforded a permanence far exceeding the lifetime of humanity; while there, the footprints of extinct birds were found perfectly cast in solid, enduring stone. Other rocks held evidence of the tracks of long-departed dinosaurs.

At sites such as the chalk cliffs of Dover and elsewhere along the English Channel, beds of limestone thousands of feet thick were found actually to be fossil graveyards. Such deposits had formed from the calcium-rich skeletons and shells of small marine organisms collecting in ocean sediment over tens of millions of years. Some of the shells—like those Darwin had gathered high in the Andes—were discovered intact

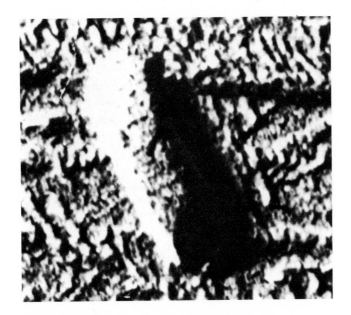

Oldest Fossils Ever Found

One of the Earth's two oldest, accurately dated life forms is this ancient, microscopic bacterium from South Africa which looks here like a raised capsule under the powerful magnification of an electron microscope. (The other: a spherical alga from the same fossil beds.) This fossil was found in a type of sedimentary rock known as chert. In South Africa, layers of chert thousands of feet thick were laid down by primeval seas and now stand raised up in mountains. Radioactive dating of the rocks established the bacterium's age at about 3.2 billion years. Similar in appearance to certain rod-shaped modern bacteria, the microorganism lacks a cell nucleus. It has been named *Eobacterium isolatum: eo* from the Greek word for "dawn," denoting the dawn of life.

amid the chalk. In the vast majority, physical change had occurred, transforming them into the chalk itself. Occasionally the original structure of shell, bone or plant tissue had been replaced by minerals precipitating from the waters—forming petrified remains. Thus, the sea floors and the sedimentary rocks of the continents were recognized as vast burial grounds of the past.

Only recently has the extent of the record on the ocean bottoms been comprehended by scientists. In the past decade or so, oceanographers have been drilling from shipboard into ocean sediments and their underlying rocks. The drilling rigs of the special exploratory ship *Glomar Challenger* have now been lowered through as much as three miles of water to reach the top of sediment deposits. Then, drilling into the sediment, researchers have brought back drill cores from almost 4000 feet—four-fifths of a mile farther down. These tubular cores consist of many layers of sedimentary material, whose cross sections depict the history of the sea floors in periods of tropical warmth and glacial cold and in their gradual movements outward from the mid-ocean ridges. Furthermore, for rather long periods there is a continuous record on the sea floors of the evolution of life. On land, this is seldom so, because of the disruptive forces of wind and water erosion which strip away accumulating soils and strata.

For many years the records discovered in the rocks had stopped short of disclosing evidence of early man. Cuvier had argued that there would prove to be no such thing as fossil man, and Darwin himself could cite no fossils that seemed to lie in their development between the apes and man. But in 1868 a fossil skull of a man with a pronounced low brow, and with other traits seemingly more primitive in development than those of modern man, was found in the Neander Valley in Germany. Some authorities insisted that the Neanderthal skull was only that of a deformed individual of the species *Homo sapiens*. Darwin's friend and supporter, the British naturalist Thomas Huxley, stated that Neanderthal man could not be regarded as an intermediate evolutionary form; rather, Huxley believed, he had been an advanced human being, even though an unusual one. But Huxley did pose the question whether in some older strata some future paleontologist might not find the fossilized bones of a more manlike ape, or a more apelike man, than any remains then known.

In 1887 Eugène Dubois, a young Dutch physician, uncovered on the island of Java in what was then the Dutch East Indies the fossil bones of a specimen he called *Pithecanthropus erectus* (the upright ape-man). Java man, as he has since become known, possessed a flattish skull somewhat apelike in appearance. The shape of his thigh bone, though, indicated that he had walked upright, like a human.

Public outrage ran high at Dubois' claim to have found an early ancestor of man. Many people could not accept the possibility that modern man had descended from such primitive stock. Shortly after the discovery of Java man, however, a similar fossil skull was found in a cave at Chou-kou-tien, near Peking. It was called the Peking man. The two pieces of evidence at such widely separated points in Asia suggested that a species of man more primitive than *Homo sapiens* had once ranged over much of the Orient.

In 1924 Dr. Raymond Dart, a professor of anatomy at South Africa's University of Witwatersrand, identified a fossil from a limestone quarry at Taung as the skull of a six-year-old child with a brain case no larger than that of a young ape, but with other clearly human traits. Not until 20 years later were enough adult skulls of this same type recovered by another South African paleontologist, Dr. Robert Broom, to confirm that these early creatures were not slightly unusual chimpanzees but higher forms of primates more closely allied to man, despite their small brain case and small stature. These creatures are known as Australopithecines (southern apes). The potassium-argon method of radioactive dating has found some of the Australopithecine remains to be as old as 2.5 million years.

It is in this same region of Africa where the Australopithecines once flourished that more advanced and recognizably manlike beings eventually are believed by anthropologists to have arisen, probably as descendants of the Australopithecines. If the Australopithecines cannot quite qualify as the first men on the Earth, their descendants do.

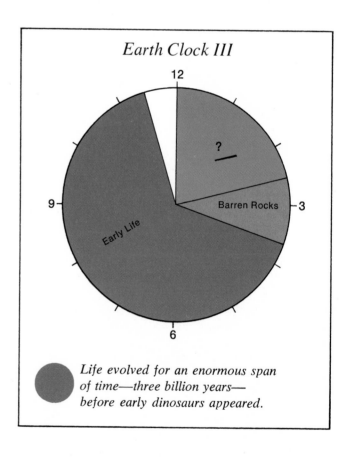

Life evolved for an enormous span of time—three billion years— before early dinosaurs appeared.

When Dinosaurs Roamed the Arctic

By Edwin H. Colbert

In the late afternoon of August 3, 1960, a party of geologists from several countries was exploring the top of a vertical sandstone cliff in the remote islands of Spitsbergen halfway between the northern tip of the Scandinavian Peninsula and the North Pole. (The sandstone here was formed in the Cretaceous Age, the last period of the Mesozoic Era that constitutes the Age of the Reptiles.) The scientists were participating in a field excursion led by Professor Anatol Heintz of the University of Oslo. Two members of the group, Professor Albert F. de Lapparent of Paris and Robert Laffitte, climbed down the cliff to the shore. As they looked back up at the wall towering above them, they saw on its surface a number of huge footprints. Within a few moments the rest of the group scrambled down the cliff to look at the tracks.

They counted 13 footprints, each distinctly three-toed, and each between 25 and 30 inches long! Seven of the prints formed a 45-foot trackway obviously made by an animal walking on its hind legs. The other footprints were scattered in various directions. There was no doubt in the minds of the viewers that they were looking at the footprints of a large dinosaur.

The geologists thought at first that the impressions were those of a gigantic, meat-eating dinosaur similar to *Tyrannosaurus*. But careful examination of the prints showed less prominent claws. It was therefore concluded that the tracks were probably made by large, blunt-toed, plant-eating dinosaurs. Subsequent study of the prints by Professor de Lapparent convinced him that these were made by the Lower Cretaceous dinosaur *Iguanodon*, that lived about 100 million years ago.

The discovery of these dinosaurian footprints was both exciting and frustrating. In the words of de Lapparent, "As this discovery was entirely unexpected, we were unable to make castings. . . . We did not even have a piece of chalk to show up the outlines of the prints. . . . After having measured the footmarks and made sketches, we were obliged to leave, as the *Valkyrien* [the excursion ship] was waiting to depart. . . ." The tracks were so important that plans were made to return to make casts of them. Accordingly, a second Norwegian-Swedish expedition set out the following year, under the leadership of Dr. Natascha Heintz of the Norwegian Polar Institute.

If the members of the 1960 field conference were surprised by their discovery of these large dinosaurian tracks in Spitsbergen, it is fair to say that since then many paleontologists throughout the world have been equally astonished. *Iguanodon* is a dinosaur hitherto

Iguanodon, *whose footprints surprisingly turned up in the Arctic islands of Spitsbergen, stood 15 feet tall and walked erect on two ponderous hind legs to forage on the lush tree foliage of a tropical world 100 million years ago. Additional fossil evidence indicates that* Iguanodon *was widespread in Britain and northern Europe during a time of mild climate ideally suited to the cold-blooded reptiles.*

Casts of Iguanodon *dinosaur footprints, found on cliffsides in Spitsbergen's Festningen Fiord, dry in the Arctic summer sun. The cliffs' once-level strata were pushed upright by geological forces.*

known from England and northern Europe, where numerous skeletons and some footprints have been unearthed and described during the past century and a half. To find indications of this large dinosaur in Spitsbergen means that in early Cretaceous time there must have been some sort of land connection between what is now an Arctic island and the European continent. What is of particular importance is that this discovery extends the range of any dinosaurs far north of previous limits. The spot where the tracks were found is only about 800 miles from the North Pole. Hitherto, the most northerly records for dinosaurs had been about 2500 miles from the Pole. We have long known that dinosaurs were spread across the globe during Cretaceous time, but the discovery in Spitsbergen has extended their northern range to much greater limits than previously suspected.

The northward extension of Cretaceous dinosaurs, interesting though it is in widening the recorded range of these reptiles during the height of their evolutionary development, raises certain questions about the environment and climate in which they lived. We may assume,

if our knowledge of modern reptiles has any bearing on reptiles long extinct, that the giant dinosaurs of Mesozoic times were tropical and subtropical animals. We may also assume that dinosaurs, like modern reptiles, were animals having no internal temperature controls; their body temperatures supposedly were closely correlated with the temperatures of their environments. If these assumptions are correct, the dinosaurs must have lived in tropical and subtropical climates, as do modern crocodiles. Certainly they were far too large to burrow underground to escape cold winters, as lizards and snakes do. So it seems probable that the footprints of large dinosaurs were made, and their bones were buried, in lands of perpetual summer.

Consequently, the discovery of *Iguanodon* tracks in Spitsbergen reinforces and even extends the idea, long held by many geologists and paleontologists, that the Cretaceous world was largely tropical. Mild climates allowed large dinosaurs to exist from the tips of the southern continents and from Australia through the middle latitudes and north into what is now Canada and northern Eurasia. It would seem to have been a world in which there were no polar ice caps and in which there were probably poorly defined temperature zones. If there were any cool regions as we know them, they must have been toward the Poles.

But there is a problem. *Iguanodon* evidently was a plant-eating dinosaur that consumed large amounts of vegetation to maintain its great bulk. Plants, however, require daily amounts of sunlight in order to flourish. And in Spitsbergen, located halfway between the Arctic Circle and the North Pole, the Sun does not rise at all for four months during the winter. Thus with Spitsbergen at its present location *Iguanodon* could not have found food there during a third of each year, even if the climate was considerably milder than it is today.

How is such a paradox to be explained? Recent discoveries appear to have confirmed the theory of continental drift, which supposes that the present land masses were once combined into a single ancestral continent, Pangaea, which split into fragments that drifted through time to their present positions. According to this theory, the Spitsbergen area was once a part of the Eurasian land mass that drifted north toward the end of Cretaceous time. When *Iguanodon* lived, the Spitsbergen area may have been located far enough south to have sufficient daylight to provide the plant life it needed for food.

In recent years the complementary theory of "polar wandering" has attracted much attention and gained many adherents. The study of magnetism in rocks seems to indicate that in former geologic ages the poles were not situated where they are now. According to some authorities the North Pole at the time of the dinosaurs was at a point in northern Siberia. In this case, Spitsbergen would have been located roughly in the same latitude as present-day Oslo and Stockholm—again, far enough south for plant life to provide food. Thus the theories of continental drift and polar wandering coincide with the evidence we have of a warmer Earth during Cretaceous time to explain how dinosaurs could have lived on Spitsbergen.

Iguanodon was first described in 1825 by Gideon Mantell, a rather eccentric physician-scientist, who spent much of his life collecting and studying fossil bones from the Wealden beds in southern England. To Mantell, the bones of the Weald revealed an England of ancient ages quite unlike the England of 19th-century days—one of tropical aspect, inhabited by gigantic reptiles. Through the years this concept has been extended by the successors of Mantell until today it encompasses the world. And it has grown by the accumulation of separate discoveries, one by one, and year after year. The Spitsbergen footprints constitute one of the latest and most interesting studies that began almost a century and a half ago. They extend the tropical environments of Cretaceous time, and take us back to a vanished world.

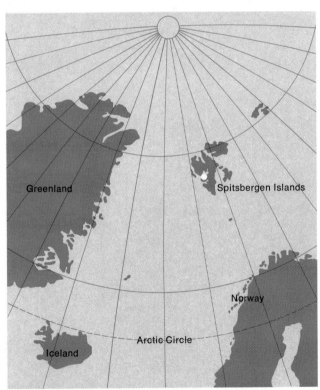

The location of the Iguanodon *footprint discovery lies above the Arctic Circle in Spitsbergen, now situated only 800 miles away from the North Pole.*

The three-toed rear foot of Iguanodon, *as outlined in this cast of its footprint, measured more than two feet from the heel to the end of the middle toe.*

They All Died Out

Flying, crawling, running, swimming—the great reptiles adapted perfectly to their environment and reigned supreme for 135 million years before they became extinct

STEGOSAURUS
Plated for Defense

ALLOSAURUS
Early Flesh-eater

CORYTHOSAURUS
Swamp Dweller

PARASAUROLOPHUS
Aquatic Plant-eater

ARCHELON
Ancestor of Turtle

ICTHYOSAURUS
Reptilian "Porpoise"

PLESIOSAURUS
Predator of the Sea

MESOSAURUS
Ancient Alligator

The dinosaurs, the rulers of the reptiles, made their entrance on this globe when the environment was a paradise for cold-blooded creatures. Continental climates were generally more tropical than now. Reptile species multiplied widely, and around 200 million years ago the stage was set for the dinosaurs.

The first primitive dinosaurs tended to stand up and run on their hind legs and were flesh-eaters that preyed on other reptiles. Many later evolved into plant-eaters and some of these reverted to four-footed locomotion. Others developed in an astonishing direction—toward flight. From the early flesh-eaters sprang such giant predators as *Allosaurus*, predominant around 150 million years ago. *Allosaurus*

walked upright and measured nearly 40 feet from the tip of his snout to the tip of his tail. His head was about three feet long, his massive jaws armed with saberlike teeth. *Allosaurus* was succeeded by *Tyrannosaurus*, a bigger killer that remains the largest flesh-eater ever to stalk the Earth. *Tyrannosaurus'* skull and jaws were twice as massive as those of *Allosaurus*, and its overall length was about 50 feet. It carried its head some 20 feet above the ground.

Certainly, among such formidable carnivores, daily survival required extraordinary defensive measures on the part of plant-eaters. Thus *Stegosaurus'* back bristled with a double file of bony plates; its lashing tail was armed with spikes. *Ankylosaurus'* armored shell

PTERANODON
Largest Flying Reptile

TRICERATOPS
Horned Dinosaur

TYRANNOSAURUS
Largest Flesh-eater

DIPLODOCUS
Nearly 90 Feet Long

TRACHODON
Late Plant-eater

ANKYLOSAURUS
Armored Dinosaur

BRONTOSAURUS
35-Ton Heavyweight

was edged with sharp spines that must have discouraged any prying. And *Triceratops* looked like a knight ready to joust with two long, lancelike horns projecting from a bony shield that protected the back of its neck.

Diplodocus and *Brontosaurus* belonged to a group whose members attained lengths of nearly 90 feet and weights exceeding 50 tons. They waded in swamps, often with only their tiny heads exposed, the buoyant effect of the water helping to support their bulk. *Pteranodon*—its wingspan of 25 feet making it the biggest flying creature in the annals of this planet—is thought to have used its foot and wing claws to climb trees. Then, spreading its wing membranes, it swooped down in a silent glide to take unsuspecting prey.

No one is sure why the dinosaurs became extinct. Some 70 million years ago they began dying off; after a few million more years, dinosaur evidence vanishes totally from the fossil record. Some scientists speculate that certain species of newly evolved mammals caused their demise by eating their eggs; others suggest they perished in an epidemic. Most authorities, though, point to the onset of colder climates, which limited the dinosaurs' food supplies. In winter today, reptiles survive by burrowing into the soil to hibernate, which the giant reptiles could hardly do. Why colder temperatures? Continental drifting was carrying major land masses generally poleward, and new mountain ranges cut off warm air from the sea.

33

The Riddle of the Quick-Frozen Mammoths

By Ivan T. Sanderson

Nobody, as far as I have been able to ascertain, seriously wants to quick-freeze an elephant. But the idea seems to have piqued the curiosity of some people in the frozen-foods industry since I started asking if they could tell me how to do such a thing. The reason for my question is simply that we already have lots of quick-frozen elephants; the flesh of some has retained its original meatiness, and I want to know how the job was done.

About one seventh of the entire land surface of the Earth, stretching in a great swath around the Arctic Ocean, including the northern regions of Europe, Asia, Alaska, Canada and Greenland, is permanently frozen. The "permafrost"—or frozen ground—throughout most of this territory is covered with a layer of soil called muck. This blanket of muck thaws in summer. It is usually composed of mud or silt, but also includes a high proportion of black organic matter, all bound together with ice.

The list of animals that have been taken out of this muck, whole or in fragments, is long. It includes extinct woolly mammoths and woolly rhinoceroses, wild horses like those still existing in Asia, extinct giant oxen and an extinct species of giant saber-toothed cat resembling a huge modern tiger. In Alaska, giant bison, wolves and mountain lions have been exhumed.

The riddle is: When, why and how were these creatures quick-frozen in the Arctic soil?

When western scientists first became aware of the existence of these amazing specimens, many merely assumed that over the centuries "the animals fell into the ice." Those who suggested that it is impossible to fall into solid ice were silenced by dismal accounts of Swiss mountaineers falling into crevasses in glaciers.

It came to light, however, that there are not—and never were—any glaciers in Siberia where many of the frozen animals were discovered, except on the upper slopes of a few mountains. Moreover, the animals were never found in mountains, but always on the plains a little above sea level. Further, it was pointed out that none of the animals has ever been found in ice. All have been discovered in the muck.

It was then suggested that the animals had fallen into rivers, and were washed downstream and deposited in deltas and estuaries—later being covered over with layers of silt. That explanation seemed all right, until further investigations showed that the animal remains are not found in deltas or estuaries, but in the plateaus that occur all over the tundra between the river valleys.

Next, the so-called "mud theory" became popular. There are certain kinds of clays found on the tundra, in the mud of which men may sink and find themselves unable to use their legs. Russian scientists suggested that a few feet of this substance could hold a mammoth until a gigantic blizzard blew up and froze him in the goo forever. But there are always some spoilsports, even in the business of explaining frozen beasts, and they pointed out that no such special clay has ever been found holding any of the animals in question.

Is there no evidence, then, of what may have caused those animals unearthed from the Arctic soil to freeze as they did?

About 60 years ago, a mammoth was found sticking headfirst out of a bank of the Berezovka River in northern Siberia. This Berezovka carcass was discovered in a sort of squatting position, raised up on one foreleg. The head had been largely eaten down to the bone by wolves, but much of the rest of the animal was perfect.

Most important, the lips, the lining of the mouth and the tongue were preserved. And on the tongue, as well as between the teeth, were portions of the animal's last meal, which apparently it had not had time to swallow. This meal was composed of delicate sedges and grasses and—most interesting—buttercups.

The Berezovka carcass could not conceivably have lain around throughout the arctic summer months (when buttercups bloom even today), later to freeze with the onset of winter snows. If it had, much of its body would have decayed—or have been devoured by scavengers—before it eventually froze.

Freezing meat is not quite so simple a process as one might think. To preserve it properly, it must be frozen *very* rapidly. If it is frozen slowly, large crystals form in the liquids in its cells. These crystals burst the cells, and the meat begins to deteriorate.

At −40° Fahrenheit it takes 20 minutes to quick-freeze a dead turkey, 30 minutes to do a side of beef. But these are mere bits of meat, not huge, live mammoths clothed in fur with an internal body temperature comparable to that of other large mammals.

Unless there were *tremendous* cold, moreover, the center of the mammoth would remain comparatively warm for some time, probably long enough for decomposition to start. Meanwhile, the actual chilling of the flesh would be slow enough for large crystals of ice to form within cells, rupturing them and causing the flesh to deteriorate. Neither event occurred with most of the mammoths, one of which has been found by the radiocarbon dating method to be around 10,000 years old. The flesh of many of the animals found in the muck is remarkably well preserved. Still palatable to dogs and wolves, it was sampled with no ill effects by a group of Russian paleontologists and mammoth steaks brought to London on ice were consumed by members of the Royal Society.

34

Further, several studies indicate that mammoths did not originate as Arctic animals, nor did they thrive under Arctic winter conditions. The Asian elephant, which is a close relative of the mammoth and just about the same size, needs several hundred pounds of food daily just to survive. For more than six months of the Arctic year, there is nothing for a creature to eat.

Yet there were tens of thousands of mammoths. And little flowering buttercups, sedges and grasses were found in the mouth and stomach of the Berezovka mammoth. Buttercups will not grow even at 40° Fahrenheit, and they cannot flower without fairly long daily periods of sunlight. Therefore, either the mammoths lived below the Arctic in winter, making annual migrations north for the short summer, or the part of the Earth where their remains are found today was warmer at the time of their death. Or, both these possibilities may have been the case.

Here, then, is an amazing picture: Vast herds of enormous, well-fed beasts congregating in sunny pastures, browsing delicately on flowering buttercups. Somehow, a number of these beast die. One—the Bere-

zovka mammoth—perishes without any visible sign of violence. With bits of food still in his mouth, he freezes so that the cells of his body are preserved for thousands of years.

Is there any scientific evidence of sudden climatic changes in the ancient Arctic—evidence that can explain the fate of the Berezovka mammoth?

Perhaps. Fossils of plants that require sunlight every day of the year have been found in Greenland and Antarctica. In polar regions today there are six months in the winter when the Sun doesn't appear above the horizon. This alone proves that at some time in the past either the poles were not where they are now, or those portions of the Earth's surface that lie about the poles today were elsewhere at one time.

Astronomers and geophysicists concur in the opinion

Uncovered during a landslide of thawing mud in northern Siberia, the carcass of a woolly mammoth attracts wolves to a feast. Its meat is still edible after being frozen for about 10,000 years.

When the Berezovka mammoth was discovered sticking out of a river bank in Siberia, the flesh of its skull was partly eaten away by wolves, but its body and huge front legs were perfectly preserved.

that the axis of the Earth itself cannot have suddenly shifted, because the Earth is like a huge, spinning gyroscope. Even if we assume a force great enough to cause a sudden shift in the Earth's axis of rotation, the resulting stresses would tear the planet to pieces. It is possible, though, that the Earth's crust alone might have shifted.

Could the shifting crust possibly account for a sudden plunge of temperature in mammoth country sufficient to freeze these great beasts? Modern oceanographic exploration gives the answer: Sea-floor spreading and continental drifting is going on today at rates of up to two inches a year, but the intense friction of the massive crust would prevent a lightning-swift shift in land masses large enough to produce sudden climatic change. So another hypothesis is laid to rest. The great mammoths could not have been caught off-guard by a shift of continents.

Undoubtedly something special is needed to explain this enigma. What does the record of geological history tell us? The Great Ice Age was not just a simple world freeze-up, followed by a thaw. Records in buried sediment show that there were 16 or 17 cold cycles during

the last two million years. Astronomers have correlated these climatic changes with slight variations in the Earth's orbit and its position in space with respect to the Sun. In other words, there appear to be regularly recurring cycles every 50,000 to 100,000 years when the Earth's surface receives either more or less heat from the Sun. The peak of the last cooling cycle was 25,000 years ago. Thus the ice sheets of the most recent glaciation, developing slowly, reached their maximum growth over the next 10,000 years. Then the warming trend set in. As the glaciers melted, the world sea level started rising, and the icy tundras of Siberia and the other Arctic lands began to give way to wet, swampy conditions, richly clothed in summer vegetation.

The warming solar radiation reached its peak 10,000 years ago, scientists believe, and for the next 5000 years the Earth experienced what is often called the "climatic optimum." Lush vegetation developed in the Arctic. Forests reached nearly a thousand miles north of their present limits and the migratory animals followed in their quest for foodstuffs. The Asian ancestors of today's Eskimos migrated to the Arctic, as hunters following the herds.

The presence of the warm-blooded mammoth in this region at this time, 10,000 to 5000 years ago, is thus readily explained. But how did some of the beasts get trapped there? Scientists working on Arctic oil-drilling operations and on the proposed Alaska pipeline, have recently been discovering more and more about permafrost, the permanently frozen ground of the northlands. In some places, they find the deep-freeze effects reach 3000 feet below the surface. Vast volumes of frozen earth lie beneath deceptive blankets of soil.

What happens to this permafrost after an ice age? Although it melts, it does so very slowly. A patch of permafrost, left over from the last Ice Age, was recently discovered in an extinct volcanic crater in Hawaii. Beneath flat plains of the Arctic coast, where the surface soils warmed rapidly and were soon richly mantled with vegetation, the melting ice patches remained frozen for thousands of years. As ice caverns gradually gave way to subterranean streams, the ground was undermined by honeycombs of caves and hollows. Slowly it melts beneath the cover of soil. Hollows develop. The technical term "thermokarst" describes these conditions—the word "karst" implying solution caves, and "thermo" the heating that causes the ice to melt.

What happens then when a mammoth, following his search for rich sedges and delicious buttercups across the land, walks unknowingly over one of these ice caves? Catastrophe! The cave roof collapses and the beast is imprisoned. In the case of the poor Berezovka mammoth, his hind legs plunged into a hole in a cave roof, his front legs pawing the air. Of course, he was hopelessly trapped. His carcass is now carefully preserved in the Zoology Museum in Leningrad, U.S.S.R., where a visitor can see for himself.

But what caused the beast to freeze? North of the Arctic Circle the days are long and bright in summer, and the nights long and dark in winter. Travelers and explorers tell of the shocking drop in temperature that accompanies the first blizzard of the approaching winter. They speak of beautiful, balmy days in late summer, days when the buttercups are blooming on the tundra. Abruptly over the northern horizon appears a dense black cloud. More than a cloud, it is a cold front of terrifying potential. It sweeps down across the tundra, and in hours the Arctic paradise becomes a subfreezing deathtrap. The other animals have already started on their southerly migrations, but not so our poor, mired Berezovka mammoth. Gradually weakening, and unable to free his hind legs, he cannot escape. Within hours, the mercury plummets almost 100 degrees, and the great mammal is engulfed in swirling snow. Under a snowdrift, then, his life is snuffed out.

All winter the carcass sits there, a solid mass in a natural deep-freeze locker. Come the next summer, it might be expected to rot away. Most mammoth bones and teeth, like those found recently in the United States near New York's Hudson Valley, are, in fact, found loose or scattered by river floors, their carcasses long since decomposed. But in the Arctic the melting of the soil over the permafrost creates a special type of hazard known as "solifluction"—the flowing soil. Once this soil is saturated by meltwater, the water cannot drain off because of the underlying permafrost. On a rise, the whole gooey mass will slide downhill. Even on a quite gentle slope, the liquid soil flows slowly downward, forming marshes at the bottom. And in such marshes lay our trapped mammoths, still frozen in early summer, eventually being covered over by more soil accumulations from the Arctic slopes. Before the rising summer temperature can thaw their flesh, the beasts are deeply buried. Many of these buried animals have now been discovered.

Since about 5000 years ago the world has been gradually cooling. In Canada, for example, the mean summer temperature in the vicinity of Ottawa is now a few degrees cooler than it was in the climatic optimum, with the winters being somewhat longer. Of course there are ups and down in the long-term climatic trends. The first 40 years of the present century saw a warming trend, but for the past three decades we have been in a cooling cycle. And within a few thousand years, geologists predict a return of the great ice sheets to the latitudes of New York, London and Berlin. Who knows what other creatures may meet a fate like the mammoth—one misstep, and frozen for centuries.

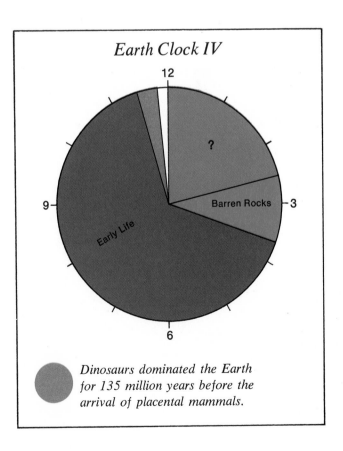

Earth Clock IV

12
?
Barren Rocks — 3
9 —
Early Life
6

Dinosaurs dominated the Earth for 135 million years before the arrival of placental mammals.

Death Trap of the Ages

By J. R. Macdonald

The quiet of the basin is shattered by trumpets of fear and rage as a great imperial mammoth struggles to free himself from the deadly pool of asphalt. Pacing the solid bank and snarling at each other, small packs of sabertooths and dire wolves wait for the end—and an evening meal. Overhead, great teratorns circle on their 12-foot wingspread in the patient vultures' deathwatch. One sabertooth, bolder—or perhaps hungrier—than the rest, leaps toward the struggling mammoth. He lands on its side, slips, and catches himself with his front claws, but his hindquarters sink into the pool. The teratorns continue to circle; struggling animals are not for them.

Rancho La Brea claims two more victims. Later, when their struggles cease, one of the teratorns, settling on a carcass, dips a wing into the pool and becomes additional bait for this self-perpetuating trap.

Between 10,000 and 40,000 years ago, near the end of the last Ice Age, this was an oft-repeated scene on what is now the "Miracle Mile" section of Wilshire Boule-

vard, just east of Beverly Hills, California. Today the sticky asphalt still oozes to the surface through natural fissures in the rocks. No longer do sabertooths, dire wolves or mammoths come to drink the water covering the sticky pools. Today there are only sparrows, pigeons and rabbits.

Europeans first took note of this site on August 3, 1769, when Gaspar de Portolá wrote, "For three hours we proceeded on a good road; to the right of it were extensive swamps of bitumen which is called *chapapote*. We debated whether this substance, which flows melted from underneath the earth, could occasion earthquakes."

For thousands of years these swamps of asphalt have been fed by springs of oil, and in the slightly moister climate of the last Ice Age they were covered with water which lured thirsty animals to their doom. Coming to drink in the area, they went to the pools, where

a misstep entangled them in sticky asphalt. Their struggles and cries attracted predators and scavengers which, in turn, became entangled in the asphalt, adding more bait to the trap.

Near the end of the last Ice Age, the Los Angeles basin looked very much as it did when the first Europeans came to California. The rainfall averaged only a few inches more a year, the native trees crept a little further down the hillsides and out onto the plain to form a savanna; and the sky was hazy, not with smog but with fog and haze trapped by the temperature inversion that often holds eye-burning smog in the basin today.

The wildlife, however, was completely different. The only way to picture North America in the geologic past is to think in terms of African game. The species were different, but the herds of herbivores and packs of stalking carnivores looked much the same.

In any animal community the herbivores, eaters of grass and leaves, make up the bulk of population. Since a carnivore such as a mountain lion requires one deer every week or ten days for food, obviously there must be many more deer than mountain lions. This economy normally prevailed in the geological past, but among

the millions of bones recovered from the La Brea tar pits the ratio is reversed.

There, carnivorous mammals outnumber the herbivores ten to one; and among birds, 75 percent are hawks, eagles, falcons, condors or vultures. Vast numbers of carnivorous animals were attracted to the asphalt pools by the sight of an easy meal and, in turn, became trapped themselves. So sabertooths, dire wolves and coyotes are the most common fossils in the pits.

The scene is the La Brea tar pools in California at the end of the last Ice Age. Floundering helplessly in the sticky ooze, an imperial mammoth is attacked by a pair of sabertooths, one of which has already become trapped itself. Beside a fallen camel a teratorn vulture fights to free a wing from the tar while a pack of dire wolves closes in.

There are no dinosaur remains. Dinosaurs became extinct nearly 70 million years before the first drop of sticky oil bubbled to the surface at Rancho La Brea.

Certainly the most spectacular mammals at Rancho La Brea were mammoths and slightly smaller mastodons. Arriving across the Bering Strait land bridge to Asia some two million years ago, mammoths quickly spread throughout the continent and diversified into many species. The largest that roamed southern California was the imperial mammoth, which sometimes stood 13 feet high at the shoulder. The scarcity of mammoths in the asphalt pits may be due to their greater strength, higher intelligence or smaller numbers. Mastodons and their near relatives were earlier immigrants to the New World, entering about 15 to 20 million years earlier. They were smaller than mammoths, with elongated heads and, occasionally, small tusks on lower jaws.

To those familiar with American and African wild animals, ground sloths seem the most bizarre creatures of the time. Standing slightly over four feet high, when walking on all fours, these massive, bulky animals made their way into North America from South America when the two continents were rejoined after 60 million years of isolation. The strange construction of the sloths' feet and limbs gave them their unusual gait.

Their ancestors were South American tree sloths which had become adapted for hanging upside down from branches. When some were forced out of forests and onto the plains, they re-adapted for ground living, but their twisted feet did not become straightened again. Instead, they retained the "curl" developed by their ancestors. Thus the ground sloth walked on the knuckles of his front feet and the outer sides of his back feet. This stance caused modification of the limb bones, which made them unique among mammals. There were three kinds of sloths in the Rancho La Brea area, but some paleontologists believe this was not their normal habitat. Perhaps the ones trapped in the asphalt were passers-by who just stopped for a drink.

The only other unusual herbivore in the area was the camel, which had a body about like that of the modern camel, but with somewhat longer legs. It is not generally known, but the camel is a native of North America and a fairly recent emigrant into Eurasia. The earliest member of the family dates back 40 million years.

The other large herbivores are familiar to all of us—horses, bison and deer. The bison was somewhat larger than the modern species, and there were a few giants whose horns had a span of nearly twelve feet.

Many of the carnivores are old friends of the modern forest and plain: foxes, badgers, skunks, weasels, mountain lions, lynx, black bears and grizzly bears. These have changed little, and if we had their entire carcasses to study we might find they differed from their modern counterparts only at the subspecies level.

Subterranean oil seeps up through a fissure and mixes with sand to form the animal traps of Rancho La Brea. Water deceptively masks the pool's surface.

The carnivore that seems strangest to us is the sabertooth. The largest known assemblage of sabertooth fossils comes from Rancho La Brea. Through the years, they must have flocked in vast numbers to feed on animals trapped in the asphalt. Paleontologists don't agree whether sabertooths made their living by hunting active prey. In the Rancho La Brea area they hunted the weak, the young, the dying, the sick and the old. Their relatively weak hind legs, their massive forelegs, and their great sabers were well suited for dispatching the weak and cutting up carrion.

All the sabertooths were well adapted for their way of life, and their design was frozen almost as soon as they appeared. It is said imitation is the sincerest form of flattery, and in the case of sabertooths, it must be so.

Several times during the age of mammals, lineages of true cats adapted to this mode of life and became "false sabertooths." In South America, where all mammalian carnivores were once marsupials (animals which carry their young in pouches), there was a marsupial sabertooth that nearly outsabered Northern Hemisphere types. The sabertooth, perhaps, symbolizes Rancho La Brea to most people, not only because of its numbers, but because of its exotic form and its (perhaps unwarranted) reputation as a mighty hunter.

Outnumbering sabertooths were dire wolves. Packs of them roamed southern California by the thousands. Slightly smaller than the northern timber wolf, the dire wolf ranged through North America for thousands of years. Fossils have been found as far east as Kentucky and as far south as Mexico City. Leg proportions indicate it was not as swift a runner as the modern timber wolf, and thousands of individuals in the asphalt pits suggest the great packs were scavengers when given the chance. The common coyote also ranged the Los Angeles basin and frequented Rancho La Brea. Only minor differences separate it at the subspecific level from the living species, the familiar singer of the American West.

Two carnivores stand out among the fossils recovered from the tar pits because of their size, although few individuals are known. The great cat structurally resembled the jaguar, although the males were nearly a quarter larger than the biggest living cats. Ranging North America during the closing days of the last Ice Age, this magnificent species was certainly the mightiest hunter of its day.

Exceeding the great cat in size was the short-faced bear. On all four feet it stood a foot higher than the modern grizzly bear, and was more massively built than the Kodiak bear. Its teeth suggest a more carnivorous way of life than any living bears.

With two exceptions, the birds of Rancho La Brea would seem familiar to all but the most highly trained observer. Many of the species or their close relatives are still living nearby today. The two exceptions are the La Brea stork, unknown in California today, and the huge condorlike vulture, teratorn.

Teratorns are the largest known ancient flying birds, having wingspans of 12 feet or more. The number of these scavengers found in the pits makes one think at times the sky was clouded with circling teratorns.

"The Death Trap of the Ages" was a place that would have offended the eyes, the ears and the nose. The vast numbers and kinds of animals roaming the area would have made it a nature lover's paradise, although the pace of death in the pools of asphalt would sicken a humanitarian. However, the existence of this trap has given us an unparalleled view into a small area in the geologic past which will probably never be excelled.

Today, the center of Rancho La Brea is a county park, named for the donor of 23 acres, Captain G. Allan Hancock, a descendant of the original California family that uncovered the first La Brea fossils during asphalt mining operations in 1875. Many of the pits have been filled, but others are fenced to protect pets and children from the still active asphalt seeps. The Los Angeles County Museum of Natural History administers the park. It maintains a display in an observation pit with attendant tour guides and it is developing a display of fiberglass reconstructions of the extinct animals of the area.

There are now reconstructions of several mammoths, sabertooths and other animals. Eventually it is hoped to have about 60 reconstructions in a re-created Pleistocene environment that will give the public a vivid and instructional glimpse of the wildlife of the past.

Lakes That Bubble and Ooze

Natural asphalt lakes are found in many parts of the world. One of the largest sites is on the island of Trinidad, where more than 10 million tons have been mined for use in building materials and pavement for highways and airport runways. Asphalt starts out deep in the ground as crude oil. Heat, pressure and oxygen drive off the volatile constituents of the oil and transform it into asphalt. Then, if fissures exist in overlying rock, the asphalt escapes to the surface, often accompanied by frothy but foul-smelling bubbles of gas as in the Trinidad puddle at right. Under the tropical sun, the asphalt forms a crust hard enough to stand on. This is then cut into blocks and loaded aboard ship for worldwide export.

Chapter Three

MAN'S ANCESTORS

When an apelike creature in Africa first picked up a stone and chipped it into a tool, he raised the curtain on the most exciting evolutionary drama the world has ever seen

New Findings on the Origin of Man

By Ronald Schiller

Two phenomenal discoveries made recently in East Africa have sent violent shock waves through the anthropological world, for they challenge the validity of long-cherished theories concerning the origin and evolution of the human race.

One was the finding in Kenya of a human skull and other bones in a layer dated about 2.6 million years ago. According to current evolutionary doctrine, these findings are extraordinary. Most textbooks state that the first primate that can be called man was *Homo erectus*, who did not evolve until around a million years ago. Yet the bones dug out of deposits in the East Rudolph Basin in Kenya are more than twice as old, and they are far more modern in shape and appearance than those of our million-year-old presumed ancestor.

The second discovery was in a cave in southern Africa on the border between Swaziland and Natal. Men of modern type may have inhabited the cave as long as 70,000 years ago. The books had told us that probably the only men in existence at that time were beetle-browed, bandy-legged Neanderthalers. But the remains unearthed in Border Cave, including the skeleton of an infant, are unmistakably those of our own species, *Homo sapiens*, who was not supposed to have appeared until around 35,000 years ago.

Equally disconcerting are the artifacts found with the fossils. They indicate that men were using tools aeons longer than had been imagined, while the evidence of the cave proves that men had developed intellects and had embarked on the road to civilization many millennia earlier than had been believed possible. The Border Cave dwellers had already learned the art of mining. They manufactured a wide variety of so-

phisticated tools of bone and stone, including agate knives with edges still sharp enough to slice paper. They could count and kept primitive records on fragments of bone. They also held religious convictions, believed in an afterlife and treated their dead with reverence, for the body of the infant had been carefully and ceremoniously buried, with a seashell pendant it may have worn in life. They must have spoken a well-developed language, for such abstract ideas as immortality cannot be conveyed by grunts and gestures.

Neither of these epic African discoveries was accidental. Anthropologist Richard Leakey, convinced that the dates ascribed to man's appearance on Earth were far too recent, was actually looking for the creature he found in Kenya. And it was inspired detective work by prehistorians Adrian Boshier and Peter Beaumont that led to the discovery at Border Cave. Back in 1964 engineers opening an iron mine on Bomvu Ridge in Swaziland came across stone implements of curious design, and Boshier and Beaumont were engaged to explore the site. "We hoped they might find something interesting," the mine manager recalls, "but we did not expect they would rewrite history."

In 18 months the young researchers located 10 ancient filled-in pits, some as deep as 45 feet, from which a bright-red ore called hematite had been dug, as well as two underground caverns, hacked into the steep slope of the ridge to a depth of 44 feet, from which glittering specularite ore (another form of hematite) had been extracted. These are some of the richest deposits of Stone Age relics ever uncovered, yielding more than a half-million mining tools, including thousands of cleavers, picks, hammers, wedges, and chisels heavily bruised from use. From the archaeological and geologic evidence the earliest strata of the diggings have been estimated to be 70,000 to 80,000 years old.

But a mystery still remained. Why would ancient men go to such enormous effort to extract ore for which they had no practical use? Certainly not for the iron in it—not until around 1500 B.C. was the secret of smelting it for the metal discovered.

Boshier and Beaumont found the answer in anthropological literature. The red hematite, literally "resembling blood" in Greek, was prized by early man entirely for its color. He used it, as some cultures still do, as a cosmetic to adorn the body in ceremonies, as a medicine and, most significantly, in fertility and burial rites. To our remote ancestors, as to present-day users, blood and life were synonymous. They observed that death often followed loss of blood and that when women stopped menstruating they could no longer create new life.

Thus, to ensure rebirth in the afterworld, it was essential to provide the corpse with a substitute, and the red hematite—the "blood of Mother Earth"—has been employed for this purpose throughout human history, everywhere in the world. Thus, it was not physical need but concern for the spirit that led to mining, man's first great industrial enterprise, 70,000 or more years ago.

In 1970, having discovered the reason for the mines, Boshier and Beaumont began to look for the miners themselves. Their quest led them to start digging in Border Cave. The cave had been investigated in 1934. In 1940 a local prospector in search of bat guano dug a pit in the floor of the cave, from which he recovered various pieces of fossilized human bones. During the following two years scientists revisited the site and found the infant skeleton lying in a shallow grave in a Middle Stone Age stratum. But since radiocarbon dating had not yet been developed and the bones were of modern type, they were not regarded as being of any great antiquity and evoked little interest. The grotto's earth had remained undisturbed for nearly 30 years when Boshier and Beaumont plunged their trowels into it.

The cave proved to be "the most perfect preservation ever recorded at an archaeological site extending to such antiquity." In 50 active days, before supplies and money ran out, they unearthed some 300,000 artifacts and charred animal bones, many of creatures long extinct. Charcoal from an overlying ash level, more recent than the stratum in which the child's skeleton was discovered, proved to exceed the limit of radiocarbon dating, which is around 50,000 years. Thus the burial had occurred more than 50,000 years ago. But exactly how much earlier is difficult to say. Stone implements and ground ocher appear right down to bedrock, nine feet below the surface, suggesting that the cavern had been occupied for 100,000 years.

"Practically everything we found was three times as old as the books said it should have been," Boshier observes. The sensational discovery of stone arrowheads places the invention of the bow prior to 50,000 years ago, whereas most archaeologists had previously dated its appearance in Europe at only 30,000 B.C. The

Milestones Toward Man

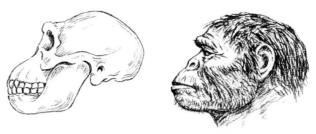

Australopithecus, 5,000,000 years ago

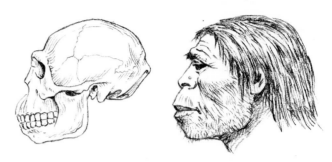

Homo erectus, 500,000 years ago

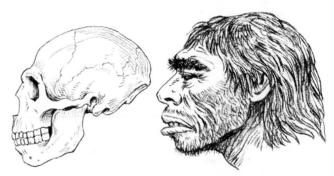

Neanderthal, 200,000 years ago

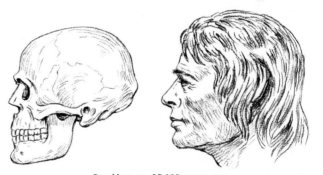

Cro-Magnon, 35,000 years ago

Fourteen million years ago Ramapithecus, *a creature with a larger brain than any of the apes, made his four-legged way across the African plains. Some 3 to 5 million years ago, scientists believe, a line of his descendants divided into four branches of creatures more or less comfortable with two-legged locomotion. Three of the branches, including* Australopithecus, *became extinct. The fourth,* Homo erectus, *evolved into* Homo sapiens. *The extinction of the subspecies* neanderthalensis *left only our own subspecies,* sapiens. *One race, Cro-Magnon, settled in Europe.*

43

Archaeologists search for human bones and artifacts in Border Cave on the Swaziland-Natal border in southern Africa. The cave was apparently occupied for about 100,000 years.

One of the human skeletons found in Border Cave awaits treatment for preserving it and further study to determine its age.

recovery of carefully notched bones, which may have been used to record the phases of the Moon, from a 35,000-year-old level, indicate that man had learned to count at that remote period. The atmosphere of the cave is such that it has perfectly preserved layers of twigs, leaves, grass and feathers, brought in as bedding, which have been found in levels dating to beyond 50,000 years.

What the evidence of Border Cave proves, to cite Boshier and Beaumont, is that "As early as 100,000 years ago man had developed an interest in happenings beyond the hard everyday needs of survival. No longer content to be nature's guest, he sought to master it. He had begun to question the purpose of existence and the nature of human destiny, to seek causes and fabri-

said, had played no part in this evolutionary scheme. The area was regarded as a geographic cul-de-sac into which men had drifted at some comparatively recent date in prehistory. Most scientists believed that even the skull of a creature named *Australopithecus africanus* belonged to an extinct ape with no right to a place on man's ancestral tree.

This hypothesis held sway until it finally crumbled under a wealth of discoveries made by anthropologists Louis and Mary Leakey in Tanzania's Olduvai Gorge. In 1959 they astonished the world by finding a nearly 2-million-year-old skull of a cousin of *Australopithecus*. In 1960 the Leakeys found part of the braincase and lower jaw of a similar prehuman, together with chipped stone tools he had undoubtedly used. They called this one *Homo habilis* ("handyman"). Later on the same level they unearthed fragments of a still more advanced being, *Homo erectus*, the first man definitely known to have used fire. He was of the same species as Java and Peking man but more than a half-million years older. Since then new discoveries have come so thick and fast from eastern and southern Africa that it is difficult to keep track of them all.

Today the theory that man originated in Asia has collapsed. Africa is now favored as the most likely birthplace of the human race, where each of the successive species evolved and from which they radiated out to the rest of the world. But there is great disagreement as to exactly who begat whom, for it is apparent that several species, both subhuman and human, coexisted simultaneously or overlapped each other in time. Anthropologists no longer depict the ascent of man as a chain, with some links found and others still missing. They view it rather as a tangled vine whose tendrils loop back and forth as species interbred to create new varieties, some of which progressed while others died out. The drama of human evolution is being rewritten with every major new discovery. But though there are still great gaps in the script, and pieces that don't fit, the most widely accepted scenario now runs as follows. The first primate to embark on the road to humanity was *Ramapithecus* (also a Leakey find, also in Kenya), who turned up 14 million years ago. He was succeeded 5 million years ago by *Australopithecus*, still subhuman but possibly a tool user. "Handyman" appeared 2 to 3 million years later. Then, 1 million years ago, came *Homo erectus*. He was a confirmed traveler, if a slow one. He could well have taken half a million years to reach Java and Peking and many thousands of years to reach Europe, where he left his tools in ancient riverbeds from Spain to Hungary. The next-to-last actor on the scene was Neanderthal man, who, though credited with being the first *Homo sapiens* ("wise man"), may actually have been an evolutionary side branch, since his facial and skeletal characteristics were more primitive in many respects than those of species that preceded him.

cate explanations. This was the birth of intellect and the ascendancy of reason."

Although no additional human remains of any great antiquity have been found in the great cavern, only 5 percent of its deposits have been excavated, and nearby lie similar caves that have never been touched. Archaeologists are confident that when funds are available to continue the digging many more skeletons will be uncovered, providing incomparable evidence of man's early history.

The landmark revelations from southern Africa and Kenya, added to many others emanating from Africa in recent years, undermine the last vestiges of a hypothesis that was accepted as scientific fact only a generation ago. On the evidence of 500,000-year-old skulls that had been found in Java and near Peking, many scientists were convinced that man had originated in Asia and had subsequently migrated westward to the Middle East. Africa south of the Sahara, they

Some anthropologists surmise that he evolved about 200,000 years ago and was firmly settled in Europe about 100,000 years ago.

The only character missing from this cast until recently was its star, *Homo sapiens sapiens*, the species that survived all others to dominate the Earth. If the Border Cave findings are borne out by further study and more discoveries, anthropologists will have to reconsider the scenario and recognize these people as our ancestors. It may be possible to trace our origins even further back to the 2.8-million-year-old bones discovered by Richard Leakey, son of the famed husband-wife team. Although the braincase of Leakey's Kenya creature is small, "Its whole shape is remarkably reminiscent of modern man's," says Leakey. "The leg bones are practically indistinguishable from those of modern man." It may be that we did not evolve from any of the previously known human types but descended in a direct line of our own.

It may be years before prehistorians can fully evaluate the significance of these latest discoveries, but from the evidence it seems clear that modern man evolved on Earth far earlier than has been realized and that it may well have been in the darkness of an African cave that the miracle of civilization had its genesis.

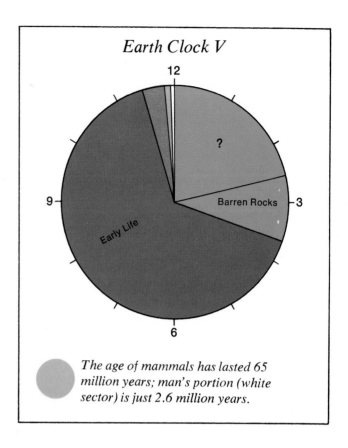

Earth Clock V

The age of mammals has lasted 65 million years; man's portion (white sector) is just 2.6 million years.

These Rocks Are Really Tools

At first glance, the tapered hand axes from the bottom of Olduvai Gorge, Tanzania, look more like ordinary rocks than the contents of someone's tool kit. But these hand-held chopping tools were painstakingly shaped by early East African *Homo habilis* (handyman) about 1,750,000 years ago, according to archaeologists Louis and Mary Leakey, who discovered them buried with ancient human remains. Thus these elongated stones—rounded or blunted to fit comfortably into the palm of a hand—are the oldest known examples of man's toolmaking in existence.

Close examination of the primitive artifacts revealed the outlines of fracture scars along the contours of points and cutting edges, showing that someone repeatedly chipped away flakes of rock to get the proper shapes. Probably the primitive toolmaker began by picking up hand-sized pieces of easy-to-fracture rock—such as flint, quartz or chert—from a stream bed. Choosing stones already worn by the water to resemble ax heads, the toolmaker put on the finishing touches by using another rock as a flaking tool. Hand axes—or choppers—were probably used in heavy tasks, such as butchering meat, while the sharper flakes likely served early man as knives for cutting hides.

Neanderthal Wasn't Always a Brute

By Ralph S. Solecki

The top of a skull was perched on the edge of the yawning excavation in the huge cavern. Except for its heavy brow ridge, the skullcap looked like a gigantic egg, soiled and broken. When fully exposed on the narrow excavation shelf, it was an awesome sight—obviously the head of a person who had suffered a sudden, violent end. The bashed-in skull, the displaced lower jaw and the unnatural twist of the head were mute evidence of a horrible death.

As we unearthed the rest of the skeleton, we found that this individual had been killed on the spot by a rockfall. His bones were broken, sheared and crushed on the underlying stones. A large number of rocks must have fallen on him within a split second, throwing his body backward down the slight slope; at the same instant a block of stone severed his head and neck from his trunk.

This was "Nandy," as we affectionately called him, a member of *Homo sapiens neanderthalensis* who was killed about 48,000 years ago. In our scientific reports we labeled him Shanidar I, because his were the first adult human remains that we identified as Neanderthal from this cave near the village of Shanidar, high in the mountains of Kurdistan in northern Iraq.

Large, airy and convenient to a water supply, Shanidar Cave is even today sometimes used as a seasonal home by Iraq's local Kurdish tribesmen. I had led our first expedition to Shanidar Cave in 1951, to look for Old Stone Age cultural artifacts—ancient tools, weapons, and so forth, produced by *Homo sapiens*. Human remains, much less Neanderthal remains, were not the original goal. Yet, in the course of four expeditions from 1951 to 1960, we were to uncover nine Neanderthal skeletons.

Ever since the first Neanderthal remains were found more than 100 years ago in the Neander Valley in Germany, these extinct people have represented a special problem to anthropologists. Primarily through the writings of one man, Marcellin Boule, who was a greatly respected French authority in the field of human paleontology, the man reconstructed on the basis of the Neander skull was soon cast in the role of a brutish figure, slow, dull and bereft of sentiment.

Although we now know much more about Neanderthal man—there have been at least 155 individuals uncovered in 68 sites in Europe, the Near East and elsewhere—he is still difficult to place with certainty on the tree of human evolution. Some anthropologists feel that he represents a "dead-end" branch on this tree. In any case, his 165,000 years on Earth more than double

the longevity as yet achieved by modern man who replaced him. But he enjoyed only one-fifth the time span of *Homo erectus* who preceded him.

The classical hypothesis, now abandoned, was that Neanderthal man was an ancestral stage through which *Homo sapiens* passed. A second theory is that Neanderthal man was a species apart from *Homo sapiens*, contemporary but reproductively isolated, as wolves are from foxes. The third is that Neanderthal man was a subspecies of early *Homo sapiens*, forming a distinct geographic race, just as there are distinct races of man living side by side on the Earth today. On the whole, the evidence appears to indicate that the Neanderthal did not gradually change into *Homo sapiens*, but was replaced by invaders among the latter. The greatest difficulty for a specialist in human paleontology is that there is a real scarcity of skeletal finds to which he can point with confidence as *Homo sapiens* of an age comparable to that of the Neanderthals.

There was, however, no scarcity of Neanderthals at Shanidar Cave. Prior to the discovery of Nandy, or Shanidar I, we had recovered the remains of an infant. Only later was it identified as Neanderthal by our Turkish colleague, Dr. Muzaffer Senyürek of the University of Ankara. When we found the remains, we had little reason to suspect that we had uncovered a Neanderthal child.

But not so with Nandy. "A Neanderthal if I ever saw one," is the comment in my field notes for April 27, 1957, the day we found him. Although he was born into a savage and brutal environment, Nandy provides proof that his people were not lacking in compassion.

According to the findings of T. Dale Stewart, the Smithsonian Institution physical anthropologist who studied the remains of the Shanidar Neanderthals, Shanidar I lived for 40 years, a very old age for a Neanderthal—equivalent to an age of about 80 today. Moreover, he seems to be a prime example of the arts of rehabilitation as practiced in his day. His right shoulder blade, collar bone and upper arm bone were undeveloped from birth. Stewart believes that this useless right arm was amputated early in life just above the elbow. Moreover, he must have been blind in his left eye, since extensive bone scar tissue mars the left side of his face. And as if all this were not enough, the top right side of his head had received some damage that had healed before the time of his death.

In short, Shanidar I must have been at a distinct disadvantage in an environment where even men in the fittest condition had a hard time surviving. That Nandy made himself useful around the hearth (his remains were found near two hearths) is evidenced by his unusually worn front teeth. Presumably, in lieu of a right arm, he used his jaws to assist in grasping. But Nandy could barely have foraged or fended for himself, so we must assume that he was accepted and supported by his people. A stone heap we found over his skeleton and nearby food remains show that even in death he was an object of some esteem, possibly respect born of close association with his fellows in a hostile environment.

The discovery of Shanidar I was for us a major, and unexpected, event. The discovery, about a month later on May 23, 1957, of Shanidar II was, by comparison, an overwhelming experience.

The initial find was made by Phil Smith, then a Harvard University graduate student, who laid bare the great eye sockets and broken face of another Neanderthal. My first impression of Shanidar II was of the horror a rockfall could do to a man's face. The lower jaw was broken, the mouth agape. The eye sockets,

crushed by the stones, stared misshapenly from under a warped heavy brow ridge, behind which was the characteristic slanting brow of the Neanderthal.

Apparently, his demise had not gone unnoticed by his companions. Sometime after the thunderous tumult of the crashing rocks had subsided, they must have returned to see what had happened to their cave mate. Judging by the evidence, a small collection of stones was placed over the body and a large fire lit above it. In the hearth, we found several stone points, which together with several split and broken mammal bones nearby, suggest the remains of a funeral feast. It appears that, when the ceremony was at an end, the hearth was covered over with soil while the fire was still burning.

Later that year, 1957, we unearthed the bones of a

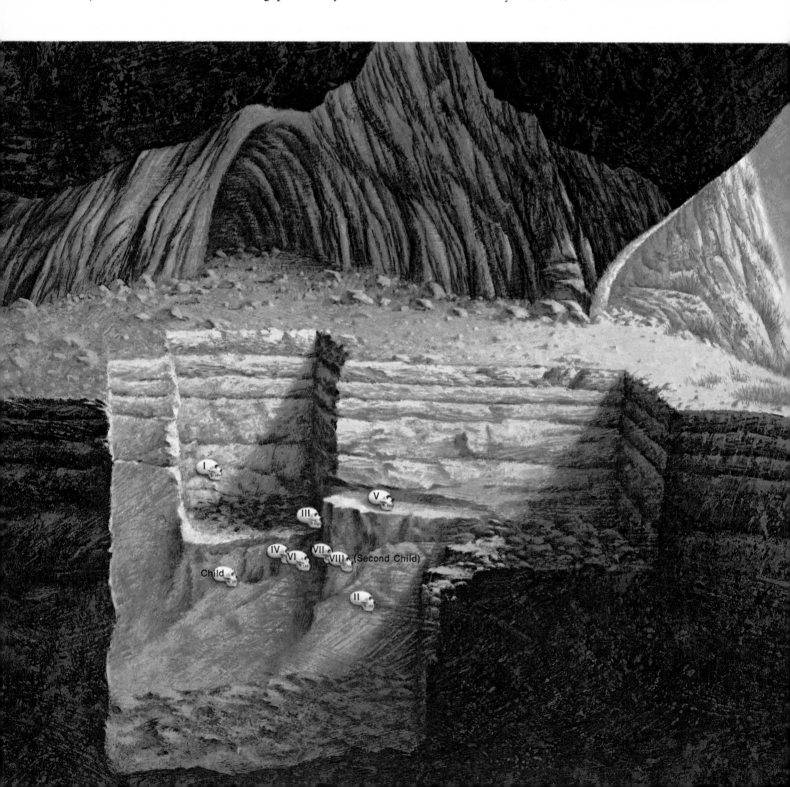

third Neanderthal in the cave. Like the other two, Shanidar III had been accidentally caught under a rockfall and killed instantly. Yet, one of his ribs showed a strange cut. X-rays taken later revealed that his rib had been wounded by a rectangular-edged implement and that his wound had been in the process of healing for about a week when he died. Most likely, Shanidar III had been disabled in a conflict with unfriendly neighbors, and was recuperating when he was killed by the rock slides. Clearly, all the dangers of the caveman's life were by no means left outside when he crossed the threshold of his home.

On August 3, 1960, in our fourth and last season at Shanidar, we uncovered the fragile and rotted bones of Shanidar IV. While Stewart worked to uncover these remains, I started to explore the stones and soil closer to the spot where three years before we had found Shanidar III. In my first trowelings, I turned up several bones that appeared to be animal remains. At second glance, one did not look like an animal bone; it looked human. Later I dug up a rib bone that Stewart quickly authenticated as human. But it was not until I uncovered a human molar tooth that we confirmed the presence of Shanidar V. This was becoming too much.

In the next four days, we found several more bones of this fifth Neanderthal, including the scattered fragments of a skull. It appeared that he, too, had been killed by a rockfall, perhaps the same one that had killed Nandy.

There was yet another discovery to be made. Stewart was clearing the ground around Shanidar IV when he encountered some crushed pieces of an upper arm bone near the skull. "It doesn't make sense," said Stewart. "It's not in anatomical position." His immediate reaction was that he hated to think that there was yet another Neanderthal in the cave. Furthermore, we had already obtained two similar bones for Shanidar IV, the correct number. And now a third? Here, certainly, was Shanidar VI.

Thus in the space of only five days, we had discovered three different Neanderthal skeletons. Before us lay the problems of preserving, recording and transporting these remains safely to the Iraq Museum in Baghdad. As we feverishly carried on these activities, we discovered—in some loose material associated with Shanidar VI—more bones. They later proved to be those of yet another Neanderthal, Shanidar VII. These two, VI and VII, were females. We also retrieved from the ground some bones of a second baby.

Neanderthal remains from Shanidar Cave, Iraq, are shown here in the positions in which they were found. Shanidar I, top, and Shanidar II showed evidence of burial with some kind of funeral ritual. Shanidar IV, one of the four closely grouped skulls, was apparently interred with wildflowers.

A ray of sun spotlights Shanidar I's pronounced brow ridges and broken, battered skull, freshly exposed from the hard-packed debris of the cave.

The skeleton remains of Shanidar IV (a male), VI, VII and VIII (the baby) all appeared to lie in a niche bounded on two sides by large stone blocks. The nature of the soft soil in combination with the position of the stone blocks leads me to believe that a crypt had been scooped out among the rocks, and that the four individuals had then been interred there and covered over with earth.

As part of our archeological routine, I had taken soil samples from the area where we found Shanidar IV and Shanidar VI, as well as samples of other soils at the cave site. These were sent to Paris for pollen analysis by Mme. Arlette Leroi-Gourhan, an expert on fossil plants.

Under her microscope, several of the prepared slides showed not only the usual kinds of pollen from trees and grasses, but also pollen from flowers. Mme. Leroi-Gourhan found clusters of flower pollen from at least eight species of flowers—mainly, small, brightly colored varieties. They were probably woven into the piny branches of a shrub, evidence of which was also present in the soil. Surely, no accident of nature could have caused the presence of these remains so deep in Shanidar Cave. The burial of Shanidar IV had been graced with bouquets of flowers!

Someone in the last Ice Age, it seems, had ranged over the mountainside with the mournful task of collecting flowers for the dead. Here was a discovery unprecedented in archeology. Today, it is customary that flowers be placed with the cherished dead. To find flowers in a Neanderthal burial that took place some 60,000 years in the past is another matter, of course, making

Twenty-five feet down, anthropologist T. Dale Stewart works to free the bones of a Neanderthal skeleton from the walls of the Shanidar excavation pit. Debris was removed from the pit by buckets.

all the more pointed our curiosity about these people.

Regarding their livelihood, we can certainly say the Neanderthals of Shanidar were hunters, foragers and gatherers. They most likely made seasonal rounds in their wilderness domain, returning between times to the shelter of Shanidar Cave.

In the cave, the animals they hunted are well documented by the presence of the bones of wild goat, sheep, cattle, pig and land tortoise. More rare are bear, deer, fox, marten and gerbil. It should be noted that the most common animals represented are the more docile types, the gregarious, grazing mammals, such as sheep. It is likely that the Neanderthals caught them by running the herds over cliffs or into blind canyons where they could be slaughtered. In fact, there are several such canyons within easy distance of Shanidar Cave.

Although we cannot rule out the possibility that Neanderthal may have done his hunting as a "lone stalker," the evidence is strong that these people lived in a communal setting. It would be more natural for them to have engaged in communal hunting. And the discovery that their lame and disabled were the recipients of care is excellent supporting testimony for a life style of communal living and cooperation.

By means of radioactive-carbon dating techniques, I estimate that the cave was first occupied about 100,000 years ago. For perhaps the next 2000 generations, over a period of some 60,000 years, groups of Neanderthals —probably numbering 25 members at a time—must have taken up seasonal residence in Shanidar Cave.

Throughout this period of known climatic changes, the Neanderthals altered little their means of adapting to the varying conditions around them. For instance,

their tools remained much the same throughout—mostly flaked stone tools, including points, knives, scoopers and perforators, all struck from local flint. We found only a few fragments of bone tools. Yet with this meager assortment of implements, Neanderthal man was able to survive and to prosper in his own way.

Then, around 40,000–35,000 B.C., the Neanderthals were gone from Shanidar Cave, replaced by a wave of *Homo sapiens*. As yet, we have no skeletal remains of these first *H. sapiens* invaders. But we already have ample evidence that they possessed a brand new stone tool kit. Using the same raw materials available to their predecessors, they fashioned a wider variety of stone tool types, as well as a variety of bone tools. They also possessed a woodworking technology such as the Neanderthals never developed.

By 35,000 B.C., the Neanderthals seem to have disappeared entirely from the world. We may well ask, what did the *Homo sapiens* successors have that their Neanderthal predecessors lacked? To my way of thinking, there were probably two things that weighed heavily in the balance. One was language. Jacquetta Hawkes, a British student of language and prehistory, feels that although the Neanderthal was a comparatively skillful toolmaker, his tool kit is conspicuous for its lack of invention and adaptability in design. Probably, Jacquetta Hawkes suggests, he was handicapped in not possessing a fully articulate language. This was the new weapon that we think his successor possessed, which is reflected in his diversified tool kit. With his greater articulateness, *H. sapiens* was able to describe and demonstrate the details of useful toolmaking innovations to his peo-

Those Mysterious
Cave Paintings

By Ronald Schiller

The skeleton of Shanidar I, crushed by a rockfall 48,000 years ago, lies revealed in the excavation.

ple, including the children who would carry on the group's activities in the next generation.

The second critical cultural achievement of Neanderthal's successor was, in my opinion, his ability to keep track of events with respect to their recurrence in the future. Thousands of notational sequences have been found on engraved bones and stones dating as far back as 30,000 years. These markings have puzzled archeologists since the time of their discovery more than 100 years ago. Recently, though, Alexander Marshack, a research fellow at Harvard University, has shown that these curious notations must have served as a kind of farmer's almanac tied in with a primitive lunar calendar. Some of the notations are illustrated with pictures representing natural events, giving their possessor a device to remind him when to expect the changes of the seasons and the movement and dispersal of game.

So, men armed with these remarkable abilities and artifacts overtook and presumably eliminated the Neanderthals. But in the millions of years of evolution that began with the apelike hominids of Africa, not until the Neanderthals do we find evidence of the first stirrings of man's social and religious sensibilities. His capacity for tenderness is seen in the obvious care with which the lame and crippled were treated—in the burials, and the flowers. Although the Neanderthal has long been ridiculed and rejected as a symbol of subhuman brutality, in the light of recent findings at Shanidar we can no longer feel so smug in our assumptions about him.

For which of us would refuse to acknowledge a cousin of proven sympathetic character, one who laid his dead to rest with flowers?

One afternoon in the summer of 1879, amateur archeologist Marcelino S. de Sautuola was on his knees digging in the rubble at the mouth of the recently discovered cave of Altamira, which lies near Santillana del Mar on the north coast of Spain. Suddenly, from the depths of a side chamber, came the muffled scream of his nine-year-old daughter, Marie: *"Toros! Toros! Papa, come quickly!"*

Alarmed, he dropped his pick and darted into the cavern where the frantic little girl stood in the half light, pointing at the ceiling. Raising his lantern, Sautuola felt his hair stand on end as he saw not toros (bulls), but prehistoric bison, magnificently painted in shades of brown, red, yellow and black. In fact, the 60-by-30-foot ceiling was crowded with the shaggy bison—17 of them, in incredibly lifelike poses, standing, pawing the ground, lying down, curled in sleep, bellowing, rolling in the dust or pierced with spears—amid ferociously charging wild boars, a horse, a trembling doe and a wolf. Exploring farther into the labyrinth, Sautuola found dozens of other painted and engraved animals, including great-antlered stags, giant cattle, a cave lion and a woolly mammoth.

This discovery at Altamira was to fling open the door on an unsuspected era of human history. Don Marcelino concluded that the cave animals were ancient. Most of the animals so vividly depicted were either long since extinct or had disappeared as natives of western Europe centuries before. Artifacts that he had dug out at the cave's entrance on prior visits dated from the Paleolithic Era—the Old Stone Age—which ended with the melting of the last great glaciers around 10,000 B.C.

He described his discovery, and showed reproductions of the cave paintings, at a meeting called by archeologists in Lisbon in 1880. The Lisbon congress immediately branded the cave paintings forgeries. Never, asserted the learned men, could such sensitive art have been created by savages scarcely above the level of apes. Sautuola was accused of having painted the pictures himself, perhaps as a plot to discredit the "new science" of prehistory. Sautuola died an object of ridicule in 1888.

Sautuola's vindication was not long in coming, however. In the years following his death, similar discoveries of prehistoric art and artifacts were made at a number of other sites. In 1902, young Abbé Henri Breuil, who was to win fame as the "priest of the painted caves," visited Altamira. There, petrified ani-

mal bones bearing engraved pictures of animals on them, undisturbed for millennia and almost identical to the figures on the ceiling, had been uncovered on the floor. The authenticity of the murals could no longer be doubted. The Abbé hailed Altamira as the "Sistine Chapel of prehistoric art."

In the decades since then more than 100 other grottoes decorated with Old Stone Age paintings, engravings and sculptures have been discovered in northern Spain, the French Pyrenees and the Dordogne region of southern France, along with a few in southern Italy and the Ural Mountains, which lie between Europe and Asia. Nowhere, however, did the paintings look as fresh or compare in grandeur with the masterpieces of Altamira. In 1940 a group of schoolboys near Lascaux, in France, discovered a cavern that proved to hold a veritable menagerie of prehistoric beasts. Among the lively figures in this Louvre of Paleolithic art are prodigious bulls 13 to 17 feet long, galloping ponies, a magnificent frieze of antlered stags which seem to be swimming across a river, exquisite yellow-and-black "Chinese horses" (so called because they resemble those of the Tang dynasty), giant elk with surrealistic antlers and a mythical monster with what looks like the body of a pregnant hippopotamus but with long, straight horns jutting from its square head (it had been called "The Unicorn").

In striking contrast to the superbly rendered animals are the crude portrayals of the human form. With only a few known exceptions—important because they demonstrate that the cave artists were not simply incapable of drawing the human form accurately—the men are depicted in the most rudimentary caricatures, their figures nothing more than scrawls comparable to the work of small children. One interpretation is that the realistic depiction of the human form was banned by a powerful religious taboo.

Despite 70 years of intense study, and the use of every technique known to science, archeologists can answer with confidence only a few questions about cave art and the people who created it. The answers they have arrived at, however, are startling.

How old is Europe's cave art? Although the various cave murals resemble each other so closely in style and feeling as to represent a recognizable "school," radiocarbon tests show that they were put on the walls at different times between 30,000 and 10,000 B.C.—a span about four times longer than all of written history! During the earliest eras most of the murals were crudely done. But later on enough attention was paid to detail so that modern zoologists can refer to some of the figures as anatomical charts when they reconstruct

As vibrant and colorful as a modern work of art, this mural of a bull and two red cows was painted 15,000 years ago in a cave at Lascaux, France.

extinct beasts from fossil bones. Cave painters underwent one period reminiscent of Picasso, during which an animal's horns were shown front view, although its head was in profile, and the ears might be stuck on the cheek, neck or wherever else the artist fancied.

Who were the cave artists? The predominant inhabitants of western Europe at the time were Cro-Magnon men. Cro-Magnon man belongs to the same species as modern man, *Homo sapiens*. His name derives from the Cro-Magnon rock shelter in the Dordogne region of France, where his remains were first uncovered in 1868. The Cro-Magnon people were generally smaller than we are, with straight limbs and high foreheads. Wearing animal skins as clothing, they were hunters and fishermen and gatherers of fruits, berries and shellfish; they had not yet learned to plant crops or domesticate cattle. They probably did not know how either to ride horses or use them as beasts of burden. Although the climate was cold and life in this environment was often brutal and short (remains indicate that few men reached 50, or women 35), the Cro-Magnons are believed to have been people of superior sensitivity and intelligence, if only on the basis of their cave art. Their songs, dances and culture may have been as remarkable as their visual art. Archeological evidence suggests that they believed in life after death since they placed food and tools in graves to accompany the deceased on their journeys. They may have believed that animals had souls too; for example, in one cave, an engraving shows a small horse leaping out of a larger one which is dying of a spear thrust.

Why were these works of art made? Archeologists tend to agree that cave art had serious religious purposes. Since hunting large animals was a formidable task for men armed only with primitive weapons, experts surmise that prehistoric man developed a highly complicated ritual of "sympathetic magic" intended to cast a spell over the beasts that were his quarry. The most potent magic may have entailed painting or engraving a picture of the beast—in the belief that once its likeness was captured on a cave wall, it could no longer resist man's power. By thus capturing the creature's spirit in a painting, the artist might also acquire an animal's outstanding physical characteristics: speed of foot, if the image were of a horse or deer; strength, if it were a bison or mammoth; courage, if a wild boar or lion. What's more, the artist might impale the beast with a painted spear to assure success in the hunt, and also employ the figure as an instructional chart to show young hunters the animal's vulnerable points. Since shortages of game animals must have frequently threatened the food supply, many archeologists theorize that a principal function of the paintings was to increase the fecundity of the herds; hence, the curious designs that surround many of them may be fertility symbols.

How did the artists work? Frequently, the first step was to engrave the outline of the figure on the cave wall with a pointed piece of flint before coloring it in. Paleolithic painters had no green or blue pigments in their palettes. Black and violet pigments, though, were readily available from earth containing manganese oxides. Black pigments from charcoal and the soot of burnt fat may have been used. The brown, red, orange and yellow hues came from iron ore, which paleolithic man ground to powder between stones, then mixed with animal blood, plant juices or animal fat, to make paints. The paints were laid on thickly in a variety of ways: By finger; with brushes made of fur, feathers, or the chewed ends of twigs; with pads of lichen and moss; or by blowing the pigments onto the walls through hollow reeds or animal bones. Sometimes the ocher was mixed with tallow and rolled into slender crayons, the remnants of which have been found at Altamira.

The Cro-Magnon artist often achieved three-dimensional effects by taking advantage of the contours of the cave surface. A small natural hole in the rock might be converted into a glaring eye; or a larger nick in the rock might be limned in red to represent a wound. Odd-shaped bulges were made into animal heads, humps or haunches, and stalagmites into legs.

But whether rendered flat or in relief, the figures are usually painted in bold, sure strokes, with few signs of modification or correction. Indeed, some prehistoric artists may have gone to school to learn their craft. In Limeuil, in southwestern France, 137 stone sketch sheets have been found, many of them poorly executed, with the details corrected as if by a teacher's hand.

How did these ancient works of art manage to survive to this day? The answer is that the majority didn't. Those that have endured have all been found in dark caves in which the temperature and humidity remain constant, the ventilation is good but not excessive, the moisture content of the air just sufficient to keep the colors from drying out and scaling off. (The grease paint applied thousands of years ago to some figures can still be smudged by a rub of one's finger!) Most important is that the entrances of these caves were sealed off by rockfalls in ages past, thus protecting them from any damage by intruding people. (The perspiration, body heat and micro-organisms recently brought into the Lascaux cavern by thousands of visitors, plus the use of electric lights, did more damage to its paintings in 15 years than they had suffered in the previous 15,000. As a result, Lascaux was closed to the public in 1963.)

Actually, much of the credit for the preservation of cave art must go to the artists themselves. For their works, even in prehistoric times, were probably *not* intended to be seen by the general public. The caves, or at least those locations in which the art appears, are believed to have been religious sanctuaries. As such, they are generally difficult and often dangerous to reach. Furthermore, the artists frequently went to inordinate lengths to place the murals in inaccessible places. For instance, the splendid rhinoceros and mammoth paintings and engravings of Rouffignac, in France, lie just over a mile from the mouth of the cave. And there are murals and engravings in Font-de-Gaume (also in France) at the end of a twisting tunnel so narrow that they can be seen only by a lean man pulling himself forward on his stomach.

With the withdrawal of the last great glaciers, around 10,000 B.C., Paleolithic cave art seems to have come to an abrupt end. Presumably, as the temperatures rose, the people moved out of the warm caverns into wooden shelters along rivers and lakes. Thereafter they learned to cultivate crops and to domesticate animals, gradually losing their dependence on wild game for food. Thus it was no longer necessary to appease the spirits of the caves, where man had once sought shelter from the harsh elements and many of the animals still hibernated. By the time the first traders from the eastern Mediterranean reached western Europe, around 1500-1800 B.C., the most recent cave paintings

were some 9000 years old. The skills, the styles and the inspiration that had created them were already forgotten. The first and one of the greatest eras of artistic expression had passed forever.

It is unfortunate that reproductions do so little justice to prehistoric cave paintings. Their fascinating texture, sense of mystery and enormous vitality do not fully emerge unless seen where the artists placed them, in the dim light of the caves themselves. On my first visit to Altamira, while lying on my back in a rocky niche, I had been gazing up at the three-dimensional bison for perhaps five minutes when my guide murmured in Spanish, "Do you see it yet?"

"See what?" I asked.

"Keep looking," he said.

I stared for another few moments until, suddenly, it hit me. "They're breathing!" I gasped. "I can see their muscles quivering!"

"Yes," the old man nodded, with a smile. "Fifteen thousand years old and still alive. That is the miracle of Altamira."

Why Did the Cave Man Hide His Art?

The prehistoric artists of Europe went to extreme lengths to keep their work from being seen. In the Font-de-Gaume cave (below) in France's Dordogne region, Cro-Magnon artists of 10,000 years ago painted their pictures on the walls of a low, narrow tunnel. A few figures adorn the entrance (arrow), but the best paintings lie along a 165-yard gallery so narrow that only one person at a time can squeeze through. Paintings at the end can be reached only by crawling. Cro-Magnon was a hunter; thus he may have painted game animals to invoke good luck in secret without

competing tribes finding out about it. Also, the art may have been tucked away in dark recesses for use in initiation mysteries for adolescent boys about to join the hunt. But no one has proved these theories.

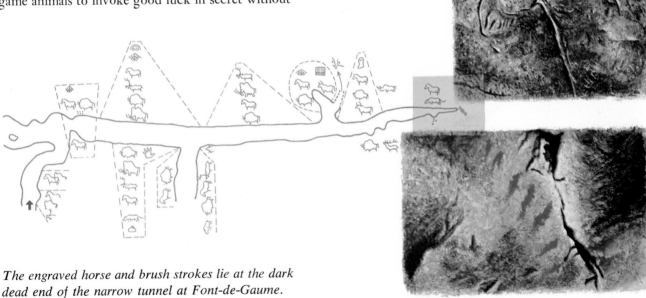

The engraved horse and brush strokes lie at the dark dead end of the narrow tunnel at Font-de-Gaume.

Part Two

AWESOME FORCES OF CHANGE

*The spectacular eruption of
Italy's Mt. Etna in 1971 opened
a window on the fiery furnace
that rages beneath us everywhere.*

Chapter One

CATACLYSMS FROM BELOW

The ground tilts wildly, fissures open, mountains explode—and man trembles before upheavals in the Earth's crust that are well understood but completely uncontrollable

The Seething Fury of Volcanoes

By Noel F. Busch

By far the most destructive volcanic eruption of the 20th century, as measured by its toll of human lives, occurred on the West Indian island of Martinique in May 1902. By way of warning, there had been a few minor earthquake tremors in April. The days passed, and then, at ten minutes before eight o'clock on the morning of May 8, four loud explosions rocked Martinique. At the same time the crater of Mount Pelée discharged an enormous glowing cloud of burning gas and ash. Moments later more fiery gas, dust and ash shot from vents in the side of the cone, and moved down the mountain at incredible speed. This discharge swept over the city of St. Pierre in a little more than a minute, killing all but one of its 30,000 inhabitants.

The eruption of Mount Pelée will remain on record for its tragic impact on mankind. In terms of the physical forces unleashed, however, the Earth has known even more violent volcanic upheavals in recent decades. One of the more savage of these occurred on March 30, 1956, when Mount Bezymyannaya (Mount Nameless) in Kamchatka, Siberia, erupted and expelled 2.4 billion tons of rock—enough to bury a city the size of Paris under a 32-foot layer of debris. As reported by witnesses in a town 125 miles east of Mount Bezymyannaya, a huge black cloud blanketed the horizon and reached a height of 22 miles. It showered down ashes as far as 250 miles away. Yet this titanic explosion caused no loss of life. Why? Primarily, because it happened in a sparsely inhabited area. But partly, too, because Russian volcanologists had foreseen the explosion for some months.

Beginning in September 1955, a series of violent earth tremors had been recorded by Russian equipment in the majestic Klyuchevskaya mountain range, where the U.S.S.R. maintains a volcanological observatory. Scientists found that, when they graphed these tremors, their lines intersected at the supposedly extinct 10,121-foot Mount Bezymyannaya. Preliminary eruptions of volcanic ash began on October 22, forming huge cauliflower-shaped clouds. By November 17 of that year the ashfall in the village of Klyuchy, 28 miles from the mountain, was so dense that cars had to use their headlights in daytime. Smoke continued to pour out of the crater until the climactic explosion the following March. Afterward the disturbance gradually subsided, finally coming to a stop eight months later.

Mount Bezymyannaya is part of a volcanic "Ring of Fire" that lies along the boundaries of the vast Pacific Ocean basin. This line of volcanoes runs along the Pacific coast of South America from Chile north through Peru. Its peaks dot Mexico and the west coast of North America, climaxing in the huge, inactive cones of Mount Shasta, California, Mount Hood, Oregon and Mount Rainier, Washington. In Alaska the line curves westward through the Aleutian Islands, then south through the peninsula of Kamchatka in Soviet Asia. From the Soviet Kurile Islands next to the Sea of Okhotsk, the volcanic chain passes through the Japanese islands, the Philippines, the East Indies, New Guinea and New Zealand. Of the world's 529 active volcanoes, 421 lie along this huge rim or inside it. The arc of volcanoes that runs through the Caribbean islands is generally considered a spur.

Another but much less dangerous chain of volcanoes follows the Mid-Atlantic Ridge, from Tristan da Cunha in the South Atlantic Ocean through the Azores to Iceland, which alone contains 28 recently active volcanoes. The Mediterranean basin includes Vesuvius, Etna, Stromboli and Vulcano. The Pacific ring, however, contains the most powerful volcanoes. On this

Sangay volcano in central Ecuador has erupted almost continuously since its discovery by Spanish explorers in the early 16th century. Sangay's smoke-plumed cone is now more than 17,000 feet high.

perimeter of smoke and flame are densely populated Indonesia with 78 active volcanoes, Japan with 49 (many close to great cities) and the United States with 37 (mostly in Alaska and the Aleutians).

In Japan the scientific study of volcanic phenomena—volcanology—is of great practical importance. Japan has the largest number of volcanologists—some 60 of the world's leading specialists. One of the best study facilities is the U.S. Federal Volcanological Observatory on Kilauea, Hawaii. Kilauea, 4090 feet above sea level, is a gigantic dome formation rising 20,000 feet from the ocean floor. A neighboring peak, Mauna Loa, at 13,680 feet, is presently the world's biggest active volcano. In fact, Mauna Loa and another neighbor, Mauna Kea, can be considered the most formidable mountain peaks in the world, since they rise some 30,000 feet directly above the ocean floor. (By way of comparison, the summit of Mount Everest stands roughly 15,000 feet above the terrain of a surrounding plateau.)

For study purposes, Mauna Loa and Kilauea are ideally situated on the island of Hawaii. They exist in a climate that permits year-round observation and are readily accessible to scientists—being only half an hour's flight from Honolulu. Both peaks are constantly active; their numerous eruptions are often predictable and usually gentle. Scientists working in the observatory on the lip of Kilauea's crater feel as safe as the tourist guests in Volcano House farther away, where sulphur fumes blend with the scent of flowers.

Hawaii's immense volcanic formations are of a type that geologists call shield domes. They differ markedly from the pointed cones typified by Mount Pelée, Mount Vesuvius and Fujiyama. Their rounded flanks are composed of the solidified lavas from slow upwellings of magma (molten rock and gas) rather than from heaps of debris ejected from the Earth by violent eruptions. At Kilauea, for example, the flanks of the mountain contain numerous small craters that eject comparatively gentle magma founts, with the lava outpourings spreading down the mountainsides. Such lava flows piling up over centuries result in the distinctive domelike shape.

The Hawaiian archipelago, stretching 1600 miles across the mid-Pacific, consists of the peaks of numerous undersea mountains which apparently emerged from the ocean in a chronological sequence from northwest to southeast. The island of Hawaii, the most recently formed member of the chain, is bigger than all the other islands combined. It is thought to have been built up from the ocean floor by volcanic processes taking no more than one million years; thus Hawaii's story is a brief one in terms of the Earth's total age of 4.6 billion years. And its volcanic building process is still going on, in full view of Kilauea's volcanologists and some 1000 sightseers a day.

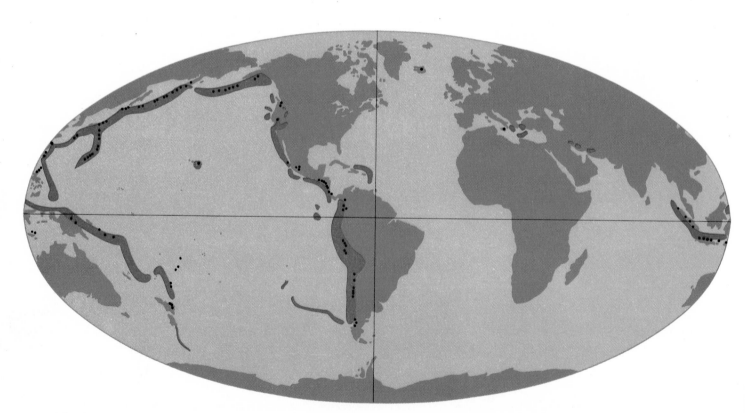

The Earth's most active volcanoes (black dots) lie in well-defined belts, such as the "Ring of Fire" around the Pacific Ocean, where faults in the crust also produce most of the world's earthquakes.

Born of fire and lava in the North Atlantic near Iceland in November 1963, the infant island of Surtsey demonstrated the ageless power of volcanoes to add new land to the surface of the Earth.

Perhaps the most remarkable volcanic formation on Hawaii is Mount Kilauea's giant caldera, or collapsed crater floor. A bowl two miles in diameter, it contains a hot lava lake that at times bubbles and steams like soup in a saucepan. Here mild eruptions occur quite frequently, the lava harmlessly overflowing from the main vent into the caldera. But on occasions when surface water seeps into the mountain's crevices, forming steam and sudden pressure, spectacular upheavals ensue. On November 14, 1959, a lava fountain spouting from a rift in one of Kilauea's subsidiary craters soared to a height of 1900 feet—probably the highest such jet ever witnessed anywhere.

As for Mauna Loa, its eruptions are numerous. Some have been confined to its summit crater, but others have split the mountain's side to flash fire and gush lava from openings far below. On June 1, 1950, a river of lava liberated from a long vent in the mountain's flank flowed for 23 days at a velocity of almost six miles an hour, swamping one village. Had the disaster struck New York or some other major city of the world, the outpouring of lava—600 million cubic yards, or about an eighth of a cubic mile—would have been enough to submerge totally a four-block area of skyscrapers.

The lava flows of Mauna Loa and Kilauea originate in great subterranean reservoirs of magma. These are forced up into the base of the mountains from the hot layers of rock of the Earth's mantle, which lies under the Earth's crust. Any increase in the magma reservoirs puts pressure on the mountain walls, perceptibly altering the angle at which the sides slant toward the summit. This phenomenon, detectable only with sensitive instruments, is the subject of continuing study by the Kilauea observatory.

Until the end of the 19th century, volcanology had hardly achieved the status of a science. Then came the violent explosion of Krakatoa.

A small island between Java and Sumatra, Krakatoa had long been recognized as the stump of an old volcano. Its three craters, however, had been inactive for 200 years and were assumed to be extinct. When explosive sounds were heard from one of Krakatoa's craters on May 20, 1883, the noises aroused more curiosity than alarm. In the next few days an excursion steamer visited the island, bringing a party of sightseers whose more mettlesome members climbed the peak and there observed a column of steam roaring out of a vent 30 yards wide.

Early in August several explosions occurred that were heard all over western Java. Apparently the volcano walls collapsed and sea water poured into the openings, creating steam. Then at 10:02 A.M. on August 27, 1883, the steam caused the whole mountain to erupt. It proved to be the most violent explosion recorded in modern times. Four hours afterward the detonation was heard 2968 miles away. Tidal waves caused by the blast reached a height of 120 feet, flooding and destroying some 300 towns and killing 36,000 people. Ashes—including great quantities of powdered pumice stone—darkened the skies for 275 miles around. Three days after the explosion, ashes rained down on the decks of ships 1600 miles away. Fine dust remained in the earth's upper atmosphere for over a year afterward, causing brilliant red sunsets and sky glows as far away as the East Coast of the United States, where fire engines were called out in some cities to extinguish what looked like conflagrations in nearby areas.

The disastrous eruption at Krakatoa, followed within two decades by that at Mount Pelée, gave volcanology its major impetus. Much of what the geologists learned

In a cluster of volcanoes along the western rim of the "Ring of Fire" stands Java's steaming, 7841-foot Bromo (second in foreground). It is one of 35 active cones among Java's 112 volcanic peaks.

was summed up by one investigator: "Volcanism everywhere has unity. Gas is the prime mover." In volcanic disasters, gas is always the activating component. When gas, including superheated steam, predominates in a mixture of molten magma, it can cause an explosion of fiery vapor, technically known as a *nuée ardente*, as happened at Mount Pelée. Or, as at Krakatoa, such a mixture can blow out the insides of the volcano, so that the whole mountain collapses. Where the gas is a less volatile or less plentiful element, volcanic eruptions will take the form of more docile lava flows. In Iceland, for example, the lava generally oozes intermittently from a series of vents, so gently as to build up in almost level layers, or beds.

On February 20, 1943, near the village of Paricutín, 200 miles west of Mexico City, a Mexican farmer, Dionisio Pulido, his wife, their son and a neighbor were working in their cornfield. A small aperture in the ground that had been there as long as anyone could remember had expanded into a crack several feet long. Now, as they watched, the ground began to shake

with thunderous noises while a cloud of steam, smoke and sparks poured out of the rift. By five P.M. the villagers of Paricutín could also see the thin column of smoke rising from the direction of the cornfield.

By the next morning a cone of ashes 30 feet high lay surrounding the aperture. Word went out, and over the following few days scores of geologists began to arrive. By the end of a week the cone had reached a height of 500 feet, and from its peak, amid an awesome bombardment of rocks, smoke ascended 3000 feet. At night the red-hot stones flying from the vent looked like giant skyrockets. The fireworks were visible for 50 miles around, and the roar could be heard still farther away. By late March a 20,000-foot column of smoke at the site was scattering ashes as far away as Mexico City.

The volcano's development reached a climax in July 1943, when the lava pool in the crater rose to within 50 feet of the top. Lava fountains rose frequently from the peak. In the fall new vents appeared at the base of the cone. During the following year this lava overran the nearby town of San Juan Parangaricutiro, leaving only its church steeple visible. By this time the newly born volcano, Paricutín, had reached a height of 1200 feet above the surrounding terrain. Over the next several years, the lava flow gradually diminished; only in 1952 did the volcanic activity cease entirely.

There have been other cases where men have witnessed the birth of a volcano. But, since 70 percent of the Earth's surface is covered by the oceans, eruptions are more frequent underwater than on land. Most sea-floor upheavals go undetected unless they produce new islands—as happened in 1963 when a new volcanic island, since named Surtsey, suddenly surfaced in the North Atlantic Ocean near Iceland.

What are the underlying causes of volcanism on the Earth? Today scientists are approaching consensus on the view that the volcanoes along the borders of the dynamic Pacific Ocean basin are largely the product of plate movements in the Earth's crust—movements that create friction and collisions between the continental plates at the edges of the Pacific basin and the plates of the spreading sea floor. Other volcanoes, such as those of Iceland, are thought to result from up-wellings of the Earth's molten subcrustal rock at rifts in the mid-ocean ridges; here new crustal rock is continuously being added to the sea floors. Presently geologists are confident that the more they can discover about plate movements in the Earth's crust, the more light will be shed on the causes and origins of volcanoes.

A slope of volcanic ash and cinders leads Japanese pilgrims to the quiet, sacred crest of Fujiyama.

Wave upon wave of hardened lava spreads out in tortured patterns from Hawaii's Kilauea volcano.

An Old Volcano's Icy Heart

One of Africa's most surprising natural features is a small glacier near the Equator. It lies at the top of legendary Mount Kilimanjaro, whose 19,340-foot central peak, Kibo, is the continent's highest point. To those crossing the sweltering Equatorial plains at its feet, Kilimanjaro's crest appears as a glittering, snowy dome. Actually it is the dormant cone of an old volcano. The volcano's caldera—or sunken crater floor—is a mile and a half wide and 600 feet deep. Inside this deep bowl's steep walls stand towering, solitary blocks of ice several stories tall, like the one at left. Daily surface melting in the Equatorial sun and nightly refreezing in the mountain air etch the fluted surface of the ice. Kilimanjaro is one of many East African volcanoes, now dormant, that erupted millions of years ago, fed by lavas rising along the Great Rift Valley— the deep break in the crust that extends from Africa into Asia Minor. Kilimanjaro's ice is dwindling—possibly signaling renewed volcanism within the peak.

The Explosion That Changed the World

By Ronald Schiller

It happened long ago on a peaceful summer's day in the Aegean Sea between the rugged coasts of present-day Greece and Turkey. The wind was blowing from the northwest. The beautiful island of Stronghyli (Santorini), some 70 miles north of Crete, lay basking in the sun. Its harbor was crowded with ships. Its terraced vineyards were heavy with fruit. In the warm springs that gushed from the sacred mountain in the center of the island, people bathed, and in the steam fissures on its slopes they consulted the oracles.

Suddenly the 4900-foot mountain heaved, roared, then blew up in a volcanic eruption of unimaginable violence. When the fiery rain finally stopped, the central portion of the island dropped into a deep hole in the sea. The pieces that remained—called the islands of Santorini today—were buried under volcanic ash. The explosion and its aftereffects were enough to change the course of history.

Archeological evidence has long indicated that a series of catastrophic events—in fact, the cataclysm out of which Western civilization emerged—took place around the 15th century B.C. But did the Santorini eruption occur at the precise time, and was it of sufficient magnitude, to have had such enormous consequences?

When Santorini blew up in the 15th century B.C., *gigantic tidal waves radiated from the Aegean Sea, lashing shores throughout the eastern Mediterranean.*

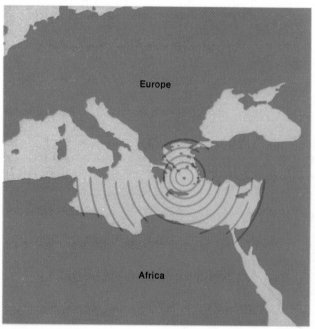

Europe

Africa

In 1956 an accidental discovery was made by Prof. Angelos Galanopoulos, of the Athens Seismological Institute. On the island of Thera, one of the remnants of Santorini that had not sunk under the sea, he visited a mine from which volcanic ash is removed for use as cement. At the bottom of the mine shaft he discovered the fire-blackened ruins of a stone house. Inside were two pieces of charred wood and the teeth of a man and a woman. Radiocarbon analysis disclosed that they had died in approximately 1400 B.C.—the 15th century B.C. And the volcanic ash that covered them was 100 feet thick; the eruption that laid it down may indeed have been the greatest in history.

Just how violent was the Santorini explosion? For comparison, scientists turn to the Krakatoa eruption in the East Indies in 1883. That volcanic island cracked at its base, allowing an inrush of cold sea water, which mingled with hot lava. The irresistible pressure of expanding steam and gas blew the top off 1460-foot Krakatoa, sent a fiery column of dust 33 miles into the air and hurled rocks 50 miles. The dust circled the earth for many months. When the eruption had spent its force, the empty shell of the volcano collapsed into a 600-foot-deep crater in the sea, creating enormously destructive tidal waves. The roar shook houses to a distance of 480 miles and was heard almost 3000 miles away.

The explosion of Santorini followed the same pattern, geologists say—except that it must have been many times more violent. The aerial energy released was equivalent to the simultaneous explosion of several hundred hydrogen bombs, according to Galanopoulos. It buried what remained of the island under 100 feet of burning ash. The wind spread the Santorini ash over an 80,000-square-mile area, largely to the southeast, where it still lies as a layer of the seabed, from several inches to many feet thick.

When the volcano had emptied itself, the hollowed-out mountain dropped into its magma chamber, 1200 feet below sea level, creating a vast crater or "caldera," into which the ocean poured. Tidal waves were set up, estimated to have been one mile high at the vortex. Roaring outward at 200 miles per hour, the waves smashed the coast of Crete with successive walls of water 100 feet high, engulfed the Egyptian delta less than three hours later, and had enough force left to drown the ancient port of Ugarit in Syria.

These are the calculations of the Santorini explosion's physical effects. Its historical effects may have been even more profound.

Western civilization traces its aesthetic, intellectual and democratic traditions back to classical Greece. At the time of the Santorini explosion, however, Greece was inhabited by relatively unsophisticated Helladic tribes. In contrast, the Minoan civilization, centered in a dozen cities on Crete with outposts on Santorini, was

The explosive force of the Santorini eruption produced precipitous slopes of volcanic ash on Thera, a remnant of the original island, which must still be checked by retaining walls (lower left).

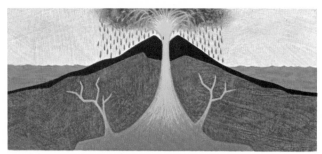

The Santorini explosion began as a simple volcanic eruption when magma welled up and broke the surface.

Then the magma subsided, creating a vast, gas-filled chamber whose rock roof was riddled with cracks.

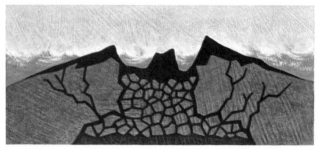

Finally the island collapsed in on itself. The sea rushed in, piled up, then subsided in giant waves.

already highly advanced. The Minoans employed a sophisticated form of writing. They enjoyed a variety of sports, including boxing, wrestling and bull games in which contestants vaulted over the horns of the charging animals. They used flush toilets, air-conditioned their houses by channeling cool breezes into them and created superb vases, ornaments and wall paintings. Their ambassadors and merchant fleets ranged the oceans of the ancient world.

Late in the 15 century B.C., at the height of its strength, this brilliant civilization abruptly vanished. Excavations indicate that all of the Minoan cities were wiped out at the same time, all the great palaces destroyed, their huge building stones tossed around like matchsticks.

Until the recent geologic discoveries, the obliteration of Minoan civilization was an intriguing mystery, attributed to revolution or invasion. However, scholars, led by Professors Dragoslav Ninkovich and Bruce C. Heezen of Columbia University's Lamont-Doherty Geological Observatory, are now convinced that the destruction was caused by the eruption of Santorini—by the holocaust itself, by its aerial shock waves and by the ensuing tidal waves. The heavy fallout of volcanic ash filled Crete's fertile valleys, destroyed the crops and rendered agriculture on the island impossible for decades. Almost the entire Minoan race perished.

There were scattered survivors—those who managed to reach the high mountains, those who were on distant voyages at the time. Archeological evidence indicates that most of these people fled to western Crete, and from there northward to Mycenae on the nearby shores of Greece. Although battered by tidal waves, Greece had not suffered from the volcanic fallout, thanks to the northwest wind.

Fragments of Santorini's old crater give a broken, circular shape to the present-day island of Thera. The collapsed crater now has a maximum depth of 1300 feet and embraces an area of 32 square miles.

The results of the Minoan migration were quickly apparent in the flowering of Mycenaean civilization, about 1400 B.C., when the written history of Greece begins. The refugees introduced the Greeks to their alphabet, art, archery and games. They taught them to work in gold and probably helped them build the great tombs and palaces that are the glory of Mycenaean culture.

Neither the vanished civilization nor the catastrophe was forgotten. These lived on for hundreds of years in various legends, including the story of Atlantis.

According to Plato, who recorded the incident later, Solon, the Athenian lawmaker, on a visit to Egypt in 590 B.C., was told by Egyptian priests that in the ancient past "there dwelt in your land the fairest and noblest race of men which ever lived; of whom you and your whole city are but a seed or remnant. But there occurred violent earthquakes and floods, and in a single day and night of rain all your warlike men in a body sank into the earth, and the island of Atlantis disappeared beneath the sea."

Atlantis, by this account, was an island kingdom. It reputedly had an area of 800,000 square miles—too big to fit into the Mediterranean—and Plato placed it in the ocean beyond the Pillars of Hercules (the Strait of Gibraltar), thereby giving the Atlantic its name. It was destroyed, according to Plato, 9000 years before Solon's time.

Archeologists point out many factual impossibilities in Plato's account of the lost Atlantis. Galanopoulos believes that Solon simply misread the Egyptian symbol for "100" as "1000," thereby multiplying all figures tenfold. Eliminate that extra zero and the destruction took place 900 years before Solon—in the 15th century B.C., which coincides with the destruction of Santorini. Atlantis's size, then, would have been 80,000 square miles, which accords nicely with the dimensions of the eastern Mediterranean islands. Galanopoulos notes, too, that there are two promontories on the coast of Greece near Crete also called "Pillars of Hercules."

From Plato's descriptions, the plain on which the "Royal City of Atlantis" was located closely resembles the plain on Crete where the Minoan city of Phaistos stood. And the description of the part of the kingdom which was sacred to the sea god Poseidon, with its steam fissures, hot springs and concentric circular canals, "fits perfectly the features, shape and size of the island of Santorini," says Galanopoulos. "Traces of the canals and harbors are discernible even now on the floor of the caldera, or undersea crater." These and other parallels have induced at least one distinguished historian to note, "It seems that the riddle of Atlantis has finally been solved."

A second great historic consequence of the Santorini cataclysm is the effect it may have had on northern

A quarry's strata show the layers of ash that buried a flourishing civilization on Santorini.

Egypt, 450 miles away, where the children of Israel labored as slaves at the time. Historians have long noted the resemblance between the Ten Plagues, as recorded in the Bible, and disasters that have accompanied volcanic eruptions. The surrounding waters may turn a rusty red, fish may be poisoned, and the accompanying meteorological disturbances frequently create whirlwinds, swamps and red rain.

The Ten Plagues produced similar phenomena. The waters of Egypt turned red as blood, killing fish and driving frogs on shore. Darkness covered the land for three days. The heavens roared and poured down a fiery volcanic hail. Strong winds brought locusts, which destroyed what crops remained. Insects, which bred in the rotting bodies and swamps, brought disease to cattle and humans. Death was so rampant as to amount to the killing of the "firstborn" of every family.

Egyptian documents confirm the disaster. "The land is utterly perished . . . the sun is veiled and shines not," says one papyrus. "O that the earth would cease from noise, and tumult be no more!" laments another. "The towns are destroyed . . . no fruit nor herbs are found . . . plague is throughout the land."

Did the enslaved Israelites take advantage of the confusion and begin their epic migration to the Promised Land? As evidence, some biblical scholars cite I Kings 6:1: "And it came to pass, in the 480th year after the children of Israel were come out of the land of Egypt in the fourth year of Solomon's reign over Israel. . . ." Since Solomon reigned from 970-930 B.C., that puts the Exodus right around the time that Santorini exploded.

The Bible relates that Pharaoh pursued the Israelites and drowned in the sea with his army. Egypt inscriptions also refer to this event. Galanopoulos attributes the disaster to the tidal waves created when the cone of Santorini dropped into the sea—which could have occurred weeks after the eruptions.

He points out that the Hebrew words *yam suf* can mean either "Red Sea" or "Reed Sea," and declares that many scholars believe it was the latter that the Bible refers to. He identifies the location as Sirbonis Lake, a body of brackish water in northern Sinai between the Nile and Palestine, which is separated from the Mediterranean by a narrow barrier of sand. He believes that the Israelites fled across this dry bridge, with the waters "on their right hand and on their left," during the interval when the sea was drawn back toward the Aegean, and that the Egyptians were caught in the huge returning tidal wave. The interval would have been about 20 minutes.

These theories about the Exodus stand on shakier ground than those concerning the destruction of Minoan civilization and the disappearance of Atlantis. Nevertheless, they seem to have occurred too closely together in time to be ascribed to mere chance. They fit together like parts of an incomplete jigsaw puzzle. Today scientists and historians are working hard to find the missing pieces that will prove the contention that Western civilization was born in the flame and ashes of a volcanic eruption in the Aegean, during a summer day 3400 years ago.

Where Was Atlantis?

Atlantis—the lost civilization! Has there ever been a myth so tantalizing to man's romantic fancy? Over the years the land of the noble and artistic people who disappeared suddenly beneath the sea has been located by overactive imaginations in several places: the Bahamas, the East Indies, the North Sea, South America, Spain. The most enduring version of the legend can be traced to the Greek philosopher Plato, who described Atlantis as a sizable country and placed it beyond the "Pillars of Hercules"— present-day Gibraltar—as indicated in the map below. New investigations show, however, that Plato misread data in Egyptian manuscripts and multiplied both the age and the dimensions of Atlantis by a factor of ten. A drastically smaller and more recent Atlantis could have been located easily at the site of the cataclysmic 15th century B.C. Santorini explosion, and many scientists are convinced that this is the answer to the mystery. Indeed, recent surveys have established that the size of Santorini before the explosion matches the Egyptian specifications, and excavations have turned up remains of an advanced civilization on the last rocky remnants of the island.

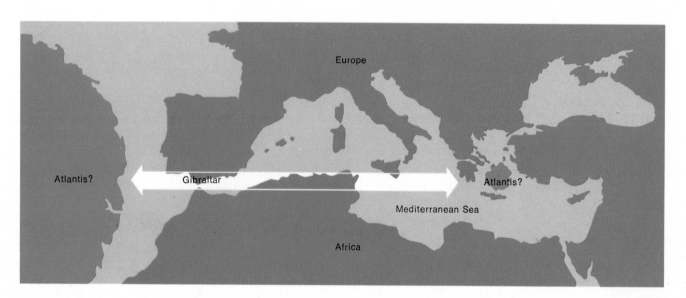

Europe

Atlantis?   Gibraltar   Atlantis?

Mediterranean Sea

Africa

Mount Etna Attacks a Town

*After years of silent brooding over the landscape of eastern Sicily, the giant black
peak blew up and the people on its lower slopes saw a molten river creeping toward them*

For almost a generation Mount Etna was quiet, a thin wisp of smoke from its eastern crater the only hint of its potential for terror and destruction. Then, in April 1971, it erupted for the 136th time since man began counting. Under enormous pressure from below, new fissures burst open on the mountain's flanks, and fiery lava began welling up.

The scene was a familiar one to the people in the area. Etna is Europe's oldest continuously active volcano—the Greeks took note of it as far back as 475 B.C.—and it has exploded three times in the 20th century alone. But the rich volcanic soil has lured farmers to plant their crops higher and higher up the mountain, and a score of towns and villages nestle uneasily on the lower slopes. Thanks to an observatory located near the volcano's rim, there were warnings about the eruption and time for threatened families to gather their possessions and get out of the area. But there was no way to tell what would be left of their farms and homes when the lava stopped.

Etna's 1971 eruption began with new vents opening on the slope of its main cone. Spewing lava, smoke and ash, this one widened to 450 feet. Later, under great pressure, lava gushed forth from three more big vents.

While sparks shower from newly formed vents on Etna's flanks, an incandescent ribbon of lava surges down a river bed toward villages below.

Continuing along the old river bed, the lava front —a mix of glowing cinders and molten rock—begins to engulf a bridge on the road between the villages of Fornazzo and St. Alfio. The lava paused when it reached this obstruction, and the villagers could watch in safety. That evening the lava overwhelmed the bridge and resumed its all-consuming march.

In early April 1971 the constantly active spout of gases at Etna's eastern crater suddenly became quiet. To an international team of volcanologists operating the observatory near the summit, this was a warning signal. Lava had solidified like a plug in the volcano's throat; with Etna's escape valve closed, the pressure of its deep-seated magma chamber would build up to explosive levels. On April 5 the feared explosion came. It ripped two vents in Etna's flanks about 1000 feet beneath the summit. One of these was only some 80 yards from the observatory and smothered it with molten lava at a searing temperature of about 2000° F. (The scientists got away safely.)

Soon after, two more vents appeared slightly lower on the mountain. Thus four lava flows were moving down Etna's flanks at about 25 feet per hour. By May 7, cooling off along the way, the lava had reached two miles downslope to an altitude of 7000 feet. Here the initial flows halted. Meanwhile molten rock poured from the main vent into a subsurface fault.

Invaded but not destroyed, this home in St. Alfio was abandoned by its occupants but withstood the relentless crunch of the advancing lava front.

The intense heat and toxic gases at the edge of the lava flow called for special asbestos suits whenever a scientist wanted to make close-up tests.

Painstakingly nurtured plants became flaming torches as the probing fingers of the lava front reached downslope over hundreds of acres of farmland.

This hidden channel brought more lava downslope.

During the next week this subsurface lava broke into the open. From thin fissures bright red liquid rock bubbled up, emptying into an old river bed that carried it rapidly along a course between the villages of Fornazzo and St. Alfio. Cutting a swath through the cultivated slopes located at 3000 feet and lower on the mountain, the lava piled up as high as five stories, burying acres of vineyards, fields and orchards, engulfing two stone bridges and scores of homes in St. Alfio. Fortunately the lava front —tumbling blocks of solid, incandescent lava, glowing clinkers and smoking cinders—ground to a halt before anyone was killed. The lava front ended its march at the edge of a dry waterfall only two miles upstream from the town of Macchia—whose 6000 inhabitants have built their homes astride the dry river bed. Disaster on a larger scale was thus narrowly avoided. But no one living in the shadow of the great mountain believes it has spoken its last word.

The Great Killer Quakes

By Ann and Myron Sutton

All Europe was earthquake-conscious in the 18th century. And for good reason. A spate of tremors left the people jittery, ripe for prophets who used past shocks to predict future ones. In London, dire warnings of an earthquake of doomsday proportions created enough panic in April 1750 to cause thousands of persons to leave the city in fear of a catastrophe that never materialized.

Then came the calamity of the century. It was All Saints' Day—Saturday, November 1, 1755—in Lisbon, Portugal, and much of the population was crowded into temples and cathedrals. At 9:30 A.M. there was a rumble and buildings trembled.

A second shock struck, lasting two full minutes. This brought down roofs, walls, churches, homes and shops in a roar of destruction and death. Then a third tremor quickly followed, after which a suffocating cloud of dust settled like fog over the city.

Day turned into night. Fires broke out. Aftershocks slammed again and again into the stricken city, and marble churches swayed like ships on the sea. The waters of the Tagus River roared into town in three great towering waves.

The whole southwest corner of Portugal had been jolted. So had parts of Spain. North Africa experienced violent shocks and much loss of life. Reverberations from the seismic wave raised tides along the European coastline and as far away as the West Indies.

Some of the magnificent and monumental buildings of Lisbon survived the earthquake only to be ruined by the fire that followed. This fire burned for a week,

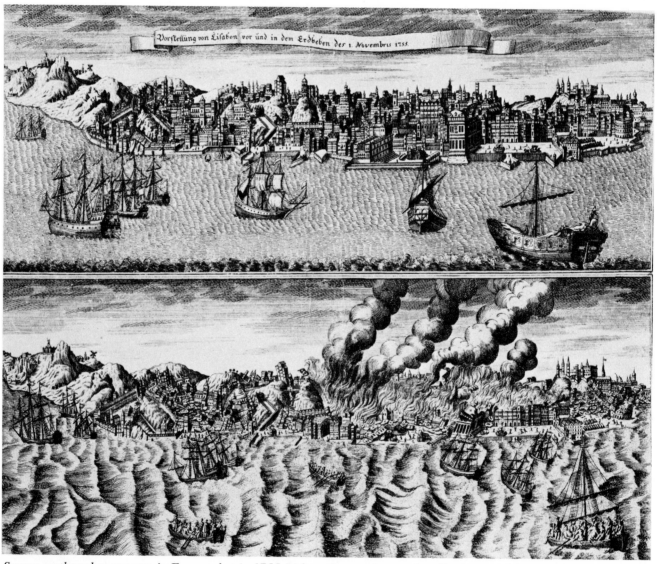

Severe earthquakes are rare in Europe, but in 1755 Lisbon, Portugal, was virtually destroyed by shocks of unusual violence, as recorded in these remarkable before-and-after engravings by an anonymous artist.

gutting the low parts of the city. Much of the treasure of Lisbon was consumed by flames: tapestries, libraries, textiles, paintings, furniture. One report lists the loss in the palace of the Marques de Lourical at 200 pictures, including works of Titian, Correggio and Rubens, 18,000 printed books and a thousand rare manuscripts, plus priceless archives, maps and charts, some relating to Portuguese discovery and colonization in the New World.

In the Hospital Real, hundreds of patients burned to death. Six hundred persons were wiped out in one church, 400 in another, 300 in another. The total number killed may have reached 60,000.

Month after month the smaller shocks continued. "Will your Earth never be quiet?" a Spanish diplomat asked an envoy from Lisbon.

The Lisbon earthquake was one of the earliest great seismic events for which we have accurate descriptions. But for centuries learned men had tried to account for these terrifying visitations. Aristotle, the philosopher who established scientific thinking in ancient Greece, reasoned that wind imprisoned inside the Earth caused earthquakes. "We must suppose the action of the wind in the earth," he said, "to be analogous to the tremors and throbbings caused in us by the force of the wind contained in our bodies. Thus some earthquakes are a sort of tremor, others a sort of throbbing." This was the beginning of a theory that would last for centuries and be taken as scientific gospel for generations to come.

In Renaissance times one common belief was that the airs imprisoned beneath the earth were "diseased." When the Earth ruptured, these were supposedly released, producing the cholera and other dire pestilences that commonly raged in disaster areas. Shakespeare dramatized this notion in *Henry IV, Part I*, when he wrote:

Diseased Nature oftentimes breaks forth
In strange eruptions; oft the teeming Earth
Is with a kind of colic pinched and vexed
By the imprisoning of unruly wind
Within her womb. . . .

The change in scientific thought from darkness to enlightenment came with men like Robert Mallet, a British civil engineer and scientific investigator, who published the first general study of earthquakes in 1846. Mallet suggested that the only way cliffs could be sheared off and landscape features displaced was with the passage of a "wave of elastic compression, or of a succession of these, in parallel or in intersecting lines, through the solid substance and surface of the disturbed country."

This was a novel notion—an earthquake wave passing through the crust of the Earth. Mallet believed that earthquakes were primarily volcanic in origin, and that some subterranean explosion such as might result from

Downtown Tokyo was a burnt-out ruin after the catastrophic earthquake of 1923, perhaps the world's most costly: killed and missing, more than 140,000 people.

superheated steam or lava entering a fissure would cause the jarring of the Earth. He supported this contention by pointing out that earthquakes frequently occurred in volcanic areas.

Mallet once estimated that for the Earth as a whole, over a period of 4000 years there had been at least 13 million lives lost through earthquakes. This may not be an unreasonable estimate, for the crust of the Earth has always been heaving.

A tremor may start simply enough: by the sudden release of strain in rocks along a fault in the Earth's crust. As the rocks along a fault snap into new positions, they create seismic shock waves that travel great distances through the solid outer layers of the Earth. Most earthquakes appear to originate within a depth of 40 miles, but some are known to occur as deep as 450 miles. However deep an earthquake may originate, seismologists speak of the spot on the land surface directly over the point of origin as the epicenter. This location is established by means of triangulation from several seismic reporting stations.

Damage is not always severe in the area around the epicenter. Seismic waves may cause more serious trouble far from the point of origin if they strike faults in the Earth's crust that are more sensitive than those around the epicenter.

To understand fully an earthquake and what it does, we need to know the point of origin and the power

Within a space of five minutes in 1964, one of the most powerful earthquakes ever recorded ripped apart the business district of Anchorage, Alaska, dropping some city blocks 20 feet below their normal level.

released by the rock slippage at the epicenter. Several years ago the American seismologist Charles Richter fashioned a scale on which the smallest earthquakes identifiable by instruments would have magnitudes of just above zero; the smallest earthquakes felt by man would have magnitudes of about three, and for the largest earthquakes there would theoretically be no upper limit. Earthquakes are assigned values on the Richter scale by correlating and averaging data from several reporting stations.

So far, the largest earthquakes measured on this scale have had magnitudes of 8.9, and chances are that this is about as high as they will go. If there is ever an earthquake of magnitude 10, the whole world will know it at once. Though magnitude scales have come into use only within the past three decades, reliable measurements with various types of instruments since the turn of the century allow us to estimate magnitudes of

earthquakes that occurred prior to the development of scales. Only two earthquakes have reached a magnitude of 8.9, highest of all. These were the Colombia-Ecuador earthquake of January 31, 1906, and the Sanriku, Japan, earthquake of March 2, 1933.

Each year, besides the countless number of small tremors, some good-sized shocks occur. Seismologists consider any shock having a magnitude of 7.0, or above, a major shock. How many major shocks occur annually? Richter indicates about 25.

How strong can an earthquake get? Richter says: "It is evident that there must be some upper limit to earthquake magnitude. . . . A physical upper limit must be set by the strength of the crustal rocks, in terms of the maximum strain which they are competent to support without yielding. . . . There is no historical seismic event to which we are inclined to assign magnitude above 8.9."

If that is true, then the Lisbon earthquake, which proved so devastating, was no more powerful than some others in recent years.

Even the most celebrated United States earthquake—San Francisco, 1906—does not rank at the top. Its magnitude is estimated at 8.3. But this was sufficiently powerful and near enough to a heavily populated area to cause tremendous destruction.

While earthquakes occur all over the world, they are especially abundant in well-defined tracts called seismic belts. One of these is the well-known "Ring of Fire" bordering the Pacific Ocean on the north, east and west. Another stretches from the Mediterranean through the Alps, Caucasus and from the Himalayas to the East Indies.

Seismic disturbances are transmitted through water as well as land. There are hundreds of accounts of "seaquakes" such as this report from a seaman aboard a British ship in 1855:

"While standing near the wheel I heard a sound as of distant thunder; on walking over to port side to look to the southward, I experienced a tremendous and grating motion of the ship, as if grazing a coral reef. It caused everything to shake for about a minute after the sound had ceased. The whole lasted two minutes. I tried for soundings, but had no bottom with 120 fathoms line. There was not the least ripple on the surface of the water, but the sound seemed to come from the ship's bottom, and the motion was not unlike letting go the anchor in deep water when the chain runs out quickly."

Back on land, some rare and freakish things take place. If there is sufficient groundwater, an earthquake may disrupt the crust enough to force subsurface water out in fountains and geysers. These are sometimes called sand spouts or mud volcanoes. A man caught in an Indian earthquake in 1934 vividly described this phenomenon:

"My car suddenly began to rock. . . . As the rocking ceased, mud huts in the village, on either side of the road, began to fall. To my right a lone dried palm trunk without a top was vigorously shaken, as an irate man might shake his stick. Then waterspouts, hundreds of them throwing up water and sand, were to be observed on the whole face of the country, the sand forming miniature volcanoes, whilst the water spouted out of the craters. Some of the spouts were quite six feet high.

"In a few minutes, on both sides of the road as far as the eye could see, was a vast expanse of sand and water, water and sand. The road spouted water, and wide openings were to be seen across it ahead of me. My car sank, while the water and sand bubbled, and spat and sucked till my axles were covered. 'Abandon ship' was quickly obeyed, and my man and I stepped into knee-deep hot water and sand and made for shore.

It was a particularly cold afternoon, and to step into water of such temperature was surprising."

Of all the manifestations of earthquakes, none seems to be more dreaded than the opening and closing of fissures in the ground, even though such events are extremely rare. In a 1797 earthquake at Riobamba, Ecuador, fissures opened, men fell in and the fissures closed again, pinning the hapless citizens by their legs. The British geologist Sir Charles Lyell tells of an earthquake in Jamaica in 1692 in which the ground heaved like a rolling sea, cracking open and closing rapidly. Many people were swallowed in the cracks, some caught by the middle and squeezed to death, some pinned with their heads protruding; others were engulfed and then cast up with great quantities of water.

If man had possessed the ability to foresee earthquakes in the past, some awful calamities might have been avoided. Possibly much of the horror of what happened in the heart of Japan on September 1, 1923, could have been eliminated.

The day dawned with a strong wind accompanied by rain, and when the shower ended, the wind abated a little, the sky cleared, and a brilliant morning sun came out with great intensity. Tokyo and Yokohama, together with other communities in the seven surrounding prefectures, sweltered in the heat. Midsummer was past, but the lingering heat seemed almost unbearable, especially on this morning.

The disaster struck at lunchtime just as coal, charcoal and gas fires had been lighted to prepare the midday meal. Down went the houses and up went the fires, hundreds of them. Thick smoke columns swirled up to join clouds of yellow dust that billowed skyward with the collapse of buildings. The air turned from yellow to brown to black.

Sampans in the bay caught fire and broke loose. Floating unattended and driven by the mounting gale, they spread the flames to other vessels in the harbor.

At that moment a terrifying seismic tidal wave—a "tsunami"—rose out of the bay and swept into the shoreline, carrying bodies and flaming debris into the city.

On the southern part of the Bo-So peninsula, the hills gave way in thousands of landslides. One of them rolled over the entire village of Nebukawa and buried all of its inhabitants.

In Tokyo, a special correspondent for the *Japan Chronicle* wrote: "Some had gained the street and escaped the direct flames. Their bodies lay almost entire, but with cruel blisters. Theirs must have been a far greater agony than that of others whose remains are but charred fragments—perhaps just a bit of blackened skull visible amid piled-up bricks and twisted wire and shop goods. How many are buried in that tangled mass? No one will ever know. But already the stench is high, especially around the canal to which scores had run in vain hope of safety."

The wind shifted and the fire burned on. High winds coupled with intense heat from the fires stirred the atmosphere into devastating violence. A tornado appeared and led to a flood on the upper reaches of the Sumida River about four in the afternoon. It swept downstream, lifting small boats out of the water. It roared over the flaming Higher Polytechnic School at Kuramaye, scooped up a huge mass of burning debris and transferred it to the opposite shore, where the buildings immediately caught fire. Flames promptly spread to the open grounds of the Military Clothing Depot, in Honjo Ward, to which an estimated 40,000 persons had fled from falling and burning buildings.

Smoke and fire quickly enveloped the whole scene. Refugees and their belongings were seared by the flames. Immediately the place became a veritable sea of fire, spinning and roaring as the tornado swept down upon the luckless refugees. In all the annals of human tragedy it was one of the closest to hell on earth. Only a handful of the 40,000 survived.

For three days, conflagrations raged through Tokyo. To stave off destruction, buildings were blown up in the path of the fire. Earth shocks kept on, relentless, terrifying. On the day of the earthquake there were 224 aftershocks. The next day there were 245, the next 95, the next 83, and so on in decreasing numbers until they finally stopped.

Tokyo was almost totally destroyed. So was Yokohama. The number of killed (99,331) and missing (43,476): 142,807.

Later there were heavy rains and floods, and what had not slid during the major shocks slid then. A precipice overhanging a mountain stream collapsed and blocked the stream, subsequently creating a flood that carried houses down the mountain. A hamlet containing some 170 families was buried under a layer of earth more than a hundred feet deep. A train bound for Manzuru with over 200 passengers was carried away by floodwaters and sent to the bottom of Sagami Bay with every soul on board. It was perhaps the only time in history that a single area was visited at one time by so many elements on the rampage, with such an enormous loss of life and property.

If the victims needed an epitaph, a newspaper reporter provided it: "What is to be said? A boy kicks an anthill and leaves only a few meaningless and shapeless pillars of earth standing above the leveled mass. Such is Yokohama. The ants get busy and build again, there or elsewhere. Men build in the same place, because nature bids them use the opportunities she gives, whatever the risk."

Perils of the San Andreas Fault

By Sandra Blakeslee

In February 1971, a sudden, violent earthquake along a branch of California's San Andreas Fault killed 64 people, injured hundreds and caused more than one billion dollars' worth of damage. But today you would hardly know it. All along the 600-mile fracture in the Earth's crust, which has been generating severe earthquakes for eons, Californians are living as if there would never be another tremor.

It is not that evidence of the danger is lacking. Going to California after the quake was over, I saw a winery near San Juan Bautista being pulled apart as the result of having been built across the trace of the Fault many years ago. In Daly City, houses built a decade previously were starting to fall into the sea. In other areas along the Fault, I observed fences, roads, bridges and structures of every kind that will remain in constant need of repair as the ground underneath them imperceptibly grinds and creeps.

Judging by dozens of conversations I had during a 450-mile journey along the course of the Fault, the people who are buying land and raising families there realize they are playing a sort of "earthquake roulette," but are unwilling or unable to think about it—at least for long. "Ignorance is bliss," one woman told me, smiling. But when pressed about the risk, she and many others faced up to it with an unsmiling fatalism, or a slightly rakish attitude of "well, you only live once; besides, the scientists may be wrong."

In this case, however, the scientists are sure they

Landslides triggered by the 1964 Alaska earthquake destroyed this home. Others in the same Anchorage suburb slid into chasms opened by the upheaval.

are right. The fact is that Californians live atop one of the most active earthquake zones in the world. It is a region where violent upheavals are certain to recur. Although efforts at predicting and even controlling earthquakes have begun in the last decade, at the moment scientists cannot do anything to remove the threat.

Since the middle of the 19th century more than 20 severe earthquakes have been recorded in California. With today's highly concentrated living conditions, even a small-to-moderate shock—one which the Indians here might have dismissed as merely a giant rolling over in his sleep beneath the Earth—could have disastrous consequences.

The San Andreas Fault begins in the Gulf of California, runs north through most of the state, then turns out to sea north of San Francisco. It is the "master fault" in a zone of continually shifting, straining rock and rock fractures. The Fault extends 20 to 30 miles down into the crust, and at various depths ranges from a few feet to a mile in width. At the surface, its presence is usually disclosed by a narrow trace—either an earth-filled depression in the terrain or a crack with raised lips where rock grinds together.

During an earthquake, rock on either side of the Fault slips horizontally in parallel but opposite directions. Total slippage along the Fault since movement began about 30 million years ago is put by geologists at a minimum of 350 miles, while some individual rock formations that once were in adjoining locations on either side of the Fault have been displaced as much as 450 miles.

An earthquake relieves the stress that builds up in rocks locked together across such a fault. A major fault, geologists say, divides two plates of the crust that are in motion relative to each other. Recent studies suggest that such plate movement along the San Andreas Fault has averaged a little more than two inches per year for the last several millennia. Because this constant movement results in almost daily slippage somewhere along the Fault, nearly every Californian has experienced a small earthquake at one time or another. In 1970, seismologists recorded 130 tremors strong enough to be felt in various parts of the state.

Some authorities hold that numerous small earthquakes—gradual slippages—reduce the chances for a major quake in the same region. Where rock remains locked across a section of a fault the dangers of a major quake increase with time, they say. But not all scientists concur. Some studies show that small tremors often precede big earthquakes.

All the experts agree, however, that the ceaseless movement of plates along the San Andreas Fault will lead to a large earthquake in some part of the network in the future. Dr. Robert Wallace of the United States Geological Survey says: "The San Andreas Fault sys-

The San Andreas Fault slices out to sea at Daly City, California. As the Fault moves, backyards fall away from homes perched along the crack.

tem has been moving and creating great earthquakes for more than 20 million years. Just because we're here now is no reason to believe it will stop."

My journey up the San Andreas began at a point just west of Los Angeles. The terrain is hilly, rough and rocky. But the Fault is easy to spot. Its course is marked by ground breakage, long, straight escarpments, narrow ridges, pulverized rocks and sag ponds (which form in the depressions created by the sagging of underlying rock within the fault zone).

Wrightwood, a small town situated 6000 feet up in the San Gabriel Mountains outside Los Angeles, is bisected by the Fault. There, a middle-aged woman talked to me as she waited to meet friends.

"What causes an earthquake and the movement along the Fault?" I asked.

"I haven't the slightest idea," she answered. "But I do know that I'm here, and you just can't run away from things. If you go someplace else, something may happen to you there just as easily as here. We've been here six years and I'm learning to adapt to this way of life, including earthquakes."

Only two people interviewed throughout the journey

had heard of the new theories that explain earthquakes along the San Andreas Fault: the concepts of sea floor spreading, continental drift and plate tectonics.

These concepts, which have gained broad scientific acceptance in recent years, have revolutionized the science of geology. Previously, scientists had no adequate explanation of the cause of movement along the Fault.

According to the new theories, the Earth continually builds up new crust at the mid-ocean ridges. At the same time, old crust is being destroyed in the deep ocean trenches. The crust is divided into plates. These plates form the sea floor and the underpinnings of the continents. As the new crustal material—molten rock welling up in the rifts of the mid-ocean ridges—is added to one edge of a plate, the plate moves slowly outward across the ocean floor on either side of the ocean ridge at a rate of a few inches a year. Ultimately, the oldest edge of the crustal plate plunges back into the Earth's mantle in deep ocean trenches and melts back into the molten rock of the mantle.

Geologists believe that wherever two plates meet or a plate plunges into a trench earthquakes are

Farewell, California— But Not for a While

On the first map—showing the west coast of North America as it is today—the San Andreas Fault (the red line) divides a narrow segment of southwestern California from the rest of North America, whose true edge, the continental shelf, is shaded light blue. Embedded in the Pacific plate, this sliver of California is known to be drifting northwest at a few inches a year. If this motion continues, five million years from now (second panel) the fragment consisting of Los Angeles, San Diego and Baja California will be some 300 miles farther northwest, blocking off San Francisco harbor. In 20 million years (third panel) the narrow land mass will project out into the Pacific as a peninsula off Seattle. Within 50 million years it may be an island (final panel). But scientists cannot be sure that the same northwest motion will continue so far into the future.

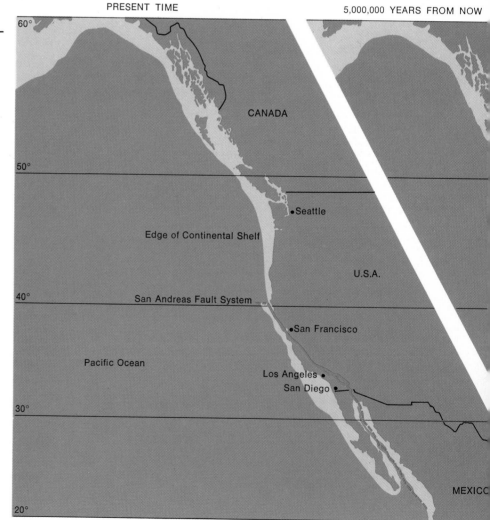

PRESENT TIME   5,000,000 YEARS FROM NOW

CANADA

Seattle

Edge of Continental Shelf

U.S.A.

San Andreas Fault System

San Francisco

Pacific Ocean

Los Angeles

San Diego

MEXICO

touched off. In fact, plate boundaries are best mapped through the detection of such seismic activity. A good example is the western coast of South America. There, a huge oceanic plate is plunging into a deep trench beneath a continental plate. As this ocean plate dives beneath the continental plate, its action is responsible for pushing up the Andes Mountains and triggering the deep earthquakes that frequently rock Chile and Peru.

The picture on the western coast of North America is believed to be somewhat different. Here, two plates are indeed bordering one another. One, the Pacific Plate, comprises the whole northern Pacific Ocean, including a coastal sliver of California. The other includes all of North America and the western half of the Atlantic Ocean. Their dividing line is the San Andreas Fault. But their driving forces (which are not fully understood) are such that these two plates are crunching slowly past one another at the rate of over two inches a year. The Pacific plate is moving in a northwesterly direction; at its remote northwestern edges it is plunging into the Aleutian trench off Alaska, pushing up the Aleutian Islands. The North American plate is heading in a southeasterly direction.

The San Andreas Fault, with its related fractures, is the weak, somewhat ragged boundary between these two massive, moving plates in the Earth's crust. Given such tremendous forces at work, how safe is it to live in the vicinity of the San Andreas Fault? Experts say that it does not really matter how close one lives to the fault—as long as one's house is built on sturdy ground. Houses built on loose or weak soil will probably not withstand severe earthquakes, no matter whether they are only 50 feet or 50 miles distant from the fault zone.

People gave many reasons for deliberately building their homes in the fault zone. For one thing, the Fault often causes sags, which fill up with sediment to make nice flat depressions in otherwise hilly ground—perfect spots for building.

But why take the chance? An electrician and his wife were in the process of building a house in Wrightwood, just a few hundred yards from the Fault. "We know the Fault runs through here," the woman said bluntly, "but we don't care. The living conditions are not so crowded. The air's clean and there's no smog. That's reason enough to be here."

20,000,000 YEARS FROM NOW

50,000,000 YEARS FROM NOW

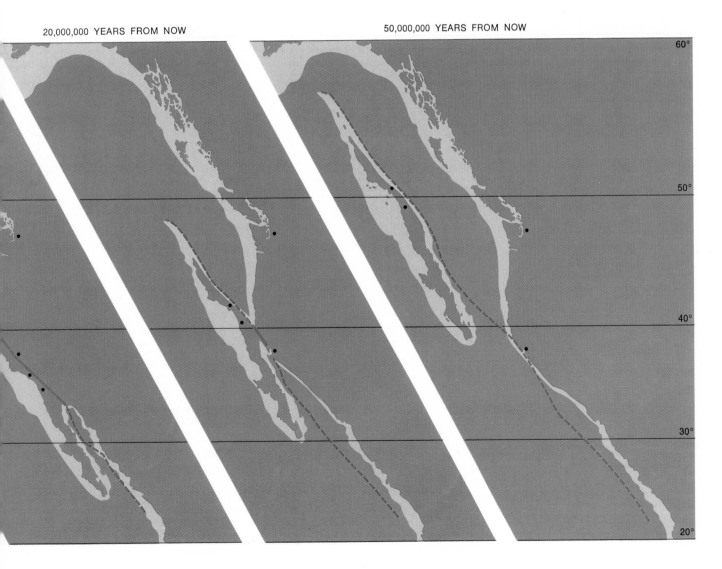

60°

50°

40°

30°

20°

In the desert hills east of San Diego, California, the San Andreas Fault shows up as a straight, well-defined break in the Earth's surface that stretches to the northwest almost as far as the eye can see.

Her husband agreed. He commutes 50 miles one way every day to his job for the advantages of living in fresh, mountain air. "You hear those scientists talk," he said, "but when you come right down to it they really don't know. And when have you ever had an earthquake in Wrightwood?"

In 1857, Wrightwood was near the epicenter (or central point) of the worst earthquake in California history. In that tremor, in one minute's time, some of the rock along the Fault moved 30 feet laterally. There has been no movement along this segment of the Fault since then, however.

"Well, I'm inclined to believe that if we did have a quake here once, maybe the strain has lessened and it won't happen again," the wife of the electrician said. "Just like they say, lightning never strikes twice . . . only I'm not sure that's true."

Lightning may or may not strike again in the same place, but scientists insist that earthquakes will almost certainly continue to occur along the active traces of faults. The only question is when will the next big one occur?

As yet, geologists have no way to predict earthquakes. They can—and do—measure strain building up in rocks along the Fault, hoping through such research as well as other investigations to develop a method of forecasting quakes. Using the short historical record available to them (California has been settled only within the last 200 years and most of its population growth has occurred within the last 50 years), geologists have worked out rough cycles for major California earthquakes. Severe earthquakes seem to occur

along the fault system approximately every 100 years.

Driving north from Wrightwood, one encounters dozens of recreational parks that are in wide year-round use. The common features of these parks are small lakes for fishing and boating. Actually, they are not lakes but sag ponds that trace the route of the San Andreas Fault ever northward. Northwest of this park land, the Fault moves through a region where the terrain is flat and open. The mountain ranges on either side of this plain are dun-colored and covered with a gossamer haze the year round. A few ranchers graze herds on the semi-arid land.

This region is a favorite of California's geologists because it so clearly illustrates the lateral slipping motion along the Fault. Although the fault zone has been dormant on this plain since 1857, the land is vividly scarred with a record of older rock movements.

The most graphic view is from the air. From a plane, the observer sees line upon line of streambeds—well over 150—crossing over the Fault. Only the very youngest streams cross the Fault in a straight line. The courses of the others have been offset by as much as 1000 feet. Offsets occur during earthquakes when the rocks on either side of the Fault suddenly snap into new positions. Streams that were once straight are thus pulled apart into two channels that no longer meet, one on each side of the Fault.

Farther north, the Fault exhibits another kind of behavior. For some unknown reason—perhaps the rocks are extra tough in this region—the rocks along the Fault slip only infrequently. Stresses build up for decades before there is an earthquake, usually a severe one, relieving the strain. Geologists speak of this portion of the San Andreas Fault as "locked." Some believe that such locked sections of the Fault are likely to give rise to major earthquakes in the future.

Approaching the San Francisco Bay the Fault again behaves differently. In this area, rocks on either side of the Fault are creeping an inch or two a year. No one knows why the Fault locks in one place and creeps in another. But geologists have noted that earthquakes tend to be more frequent and less severe in creep zones than in locked zones.

All along this restless stretch of the Fault I saw fresh evidence of creep. Fences, roads and bridges must be continually patched up as surface features are displaced by slowly moving rocks. Sometimes the foundations of buildings are distorted by the displacement. The Almaden Wine Company makes all its red wines in California (five million gallons of it at a time) just south of San Juan Bautista. One of their buildings straddles the San Andreas Fault, and here, for 15 years now, geologists have been measuring the fault creep. They report an average of one-half inch of movement a year. Posts have shifted positions and the ceiling structure is pulling apart at the seams as one half of the building creeps southeast with the North American plate while the other half of the winery moves northwest with the Pacific plate. Mrs. Alan Cullumber, who sells wine in Almaden's tasting room, admitted that she had never heard of fault creep. "I'm not too curious about earthquakes," she said. "I guess I'd rather not know. Kids ask me about it a lot and I tell them it's gases in the earth."

North of San Juan Bautista, the San Andreas Fault takes a slight westward swing into the rugged Santa Cruz Mountains that lie along the Pacific coast. At this point the Fault locks again. Thereafter, it remains locked all the way to its northern end. The last violent break along this locked segment of the Fault occurred in 1906, causing the great San Francisco earthquake of that year, the worst disaster in California's history.

Before disappearing beneath the ocean west of San Francisco Bay, the Fault cuts through an area that is sprawling with new construction. During the past ten years, one company alone has built 30,000 houses here on a small part of the fault zone around Daly City. Oddly enough, few people who live on top of the Fault in the Daly City area know of its presence.

What can be done to protect man against the violence of earthquakes? Scientists are concentrating their efforts on three main areas of research—prediction, modification and mitigation. Dr. Jerry P. Eaton, of the National Center for Earthquake Research at Menlo Park, California, a specialist in earthquake prediction, says the most promising areas of investigation seek to understand how earthquakes are generated and to identify warning signals in the Earth's rocks.

Dr. Wallace, of the United States Geological Survey, says: "Earthquake prediction is now a credible and respected venture. We hope in the next few years or decades it will become a reality. But there's a long way to go. Just five years ago, prediction was science fiction."

One idea on modifying and controlling earthquakes is particularly intriguing. Some researchers hope to learn how to safely trigger small earthquakes in stressed rock, thereby relieving the growing strain that can lead to major upheavals. But such projects would be extremely expensive. Thus many scientists feel that it would be more realistic to focus on minimizing the impact of earthquakes.

"We must work out in better detail what happens at the time of an earthquake," comments Dr. Bruce Bolt, a seismologist at the University of California. "We must give more information to engineers on such things as the size of earthquake waves, their frequency, how different soils stand up under shaking and more about the duration of waves."

Such studies, Dr. Bolt believes, could supply structural engineers, soil engineers and city and town planners with the information to avert the worst effects of earthquakes.

Just now, however, in man's continuing battle against earthquakes, nature still has the upper hand.

Lakes, such as this one near San Francisco, form when land subsides along the San Andreas Fault, creating depressions that stay filled with water.

Chapter Two

ICE ON THE MARCH

*Glaciers and avalanches scour the Earth with enormous power, but they are
mere tokens of the great ice sheets that swept over the globe during past ice ages*

The Power and Glory of Glaciers

By Paul Friggens

Although few people have ever seen one, glaciers are in many ways as important to our future on this planet as the seas we sail and the air we breathe. For if the world's climate were to cool substantially, as it has in the past, the glaciers of Antarctica and Greenland would greatly expand, lowering sea levels with disastrous consequences. On the other hand, if these ice caps should melt further, and thus *raise* the level of the sea, the destruction of coastal areas throughout the world would be severe.

But if glaciers pose a threat to man, they also offer unique and prodigal gifts. Consider: Most of the world's great river systems, from the Amazon to the Ganges, the Rhone to the Columbia, originate in glaciers. About three fourths of all the fresh water in the world—some seven million cubic miles—is stored in the form of glacial ice. Scientists estimate that this reserve is equivalent to about 60 years' rainfall over the entire globe.

How are glaciers born? Simply put, glaciers form whenever winter snowfall exceeds summer melt. The excess accumulates, and is gradually transformed into ice. The first change is from snowflakes to minute ice crystals called "névé" or "firn." As storm follows storm, and the snowfall accumulates, the delicately wrought crystals are compacted and recrystallized into nearly spherical granules of solid ice. With each year's accretion of snow, the ice mass grows, becoming tougher and harder all the time.

What makes this rock-like glacial mass flow? Glaciologists debate several theories, but generally agree that at a thickness of 100 to 150 feet the ice undergoes a further transformation. The crystalline ice, under heavy pressure deep within a glacier, becomes a quasi-plastic substance and, initiated by the force of gravity, starts to flow. Nobody really knows, however, where this happens or at what temperature. Tunneling into the ice in Greenland, scientists found the glacier solidly frozen to the ground—yet in motion from one to 30 feet above the ground.

Whatever the precise mechanism, most glaciers move only an inch or two, or maybe a foot or two per day (some do not move at all). But some record breakers have shown astonishing speeds. In 1966 a pilot flying over Mount Steele in the Yukon Territory spotted a spectacular glacier that galloped two feet per hour—nearly 50 feet a day! Pushing ahead in great pulsating surges, the river of ice, 22 miles long and over a mile wide, sheared through everything in its path including the stagnant ice of a previous advance.

Over the centuries, glaciers have dramatically changed the face of the Earth. Scouring with tremendous erosive power, they helped scoop out the Great Lakes of North America and Norway's fiords, chiseled the mighty Matterhorn and helped dig the breathtaking valleys of the Rocky Mountains.

How do glaciers sculpt these masterpieces? As they advance, they literally quarry rock and soil from the sides and floor of the valleys through which they move—everything from finely ground stones to boulders as big as a house. Shoved along by the ice, this glacial debris abrades and rasps away the underlying bedrock, widening and deepening the trough.

During the last great advance of the Ice Age, which reached a maximum 18,000 years ago, ice sheets covered nearly 30 per cent of the Earth's land. Of these only

The melting edge of a glacier in Canada's Banff National Park creates a river that washes down glacial debris as it flows into dazzling Lake Peyto.

two remnants exist today—in Greenland and Antarctica—and they account for about 97 percent of the world's glaciers. Antarctica covers some 5.5 million square miles, an area larger than Europe. The entire continent is covered with a mantle of ice, pierced in many places by mountain ranges as high as the Alps, while in places the ice sheet is up to 12,000 feet thick.

The world's second largest ice cap, with 708,000 square miles, is in Greenland. From the forbidding Greenland plateau, huge icebergs "calve" into the open ocean to travel hundreds of miles. One such iceberg, moving out into the Atlantic on the night of April 14, 1912, sank the "unsinkable" British luxury liner *Titanic*.

The rest of the world's glaciers, important but tiny by comparison with Greenland and Antarctica, are located in North and South America, Europe, Asia, Africa and New Zealand. In all, about ten percent of the land surface is presently covered by glaciers.

Modern glacial research stems from the curiosity of a Swiss scientist, Louis Agassiz, who set out a century ago to measure the precise movement of glaciers high in the European Alps. Today an entire army of glaciologists is probing the world's glaciers, studying, among other things, the causes of advance and retreat, the flow mechanism and the probability and timing of a new ice age.

The Antarctic is a major theater for this glacial study. There, scientists from nearly a dozen nations, using an imposing array of new tools and techniques, are making extensive surveys and investigations. In 1969 a team of scientists and engineers drilled more than 7000 feet into the ice sheet at Byrd Station in western Antarctica. Their specially designed rotary drill brought up a continuous core, four inches in diameter, which provides a clear profile of polar history—and an exciting look into the Earth's past.

By analyzing the oxygen in ice formed 25,000 to 100,000 years ago, for instance, scientists were able to study the composition of the Earth's atmosphere back to that early time. In ice 10,000 to 14,000 years old, the scientists found layers of volcanic ash, possibly deposited by an ancient, major cataclysm. At approximately 850 feet, they recovered ice that had fallen as snow at the time of Christ, and in the topmost layers they identified ice contaminated with radioactive fallout from our Atomic Age. The drillers discovered water at the bottom of their 7100-foot hole-in-the-ice, indicating melting from pressure, assisted by the heat of the Earth itself.

Why are scientists so interested in the Antarctic glacier? "The Antarctic ice cap," says the U.S. National Science Foundation, which sponsored the drilling, "has a powerful influence on the world's weather. Any significant change in this great freshwater reser-

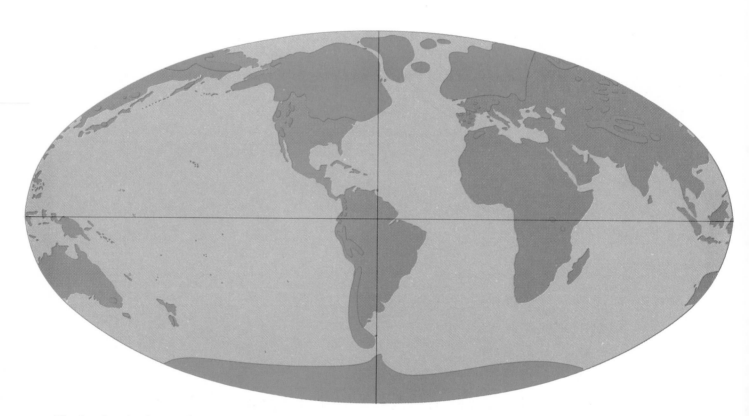

The last Ice Age began about 25,000 years ago. Ice increased in polar regions and advanced over the land, reaching a maximum—the bright blue areas on the map—about 18,000 years ago. Over the next 12,000 years the ice withdrew, leaving only the smaller polar caps and mountain glaciers existing today.

*Rock and soil gouged from the Swiss Alps form
dark stripes of moraine as the Rhone Glacier
flows down to the headwaters of the Rhone River.*

voir would affect man's environment by causing
changes in sea level, rainfall, river flow and lake levels."

"In fact," says William O. Field, noted glaciologist
of the American Geographical Society, "with about
90 percent of the world's ice, Antarctica is a super
refrigeration system. Should the ice cap wholly melt,
the level of the oceans might rise as much as 200
feet, flooding the present coastal regions and engulfing
the world's great port cities."

What caused the last Ice Age? Scientists don't agree.
All they are sure of is that several million years ago,
the earth began to grow cold. One after another, four
mammoth ice sheets invaded the Northern Hemisphere. In Europe the ice piled more than 8000 feet
high on the Scandinavian Peninsula. Pushing south, it
covered northern England and Germany, and, to the
east, penetrated almost to Moscow. In North America
the ice spread out until it covered more than half the
continent, mile-deep, as far south as the Ohio and
Missouri river valleys. The last ice began to retreat
rapidly here about 13,000 years ago, and the melting
raised the oceans 400 feet or so to their present level.

Scientists vigorously debate several hypotheses on
what brought about this climatic pendulum. According to the "solar-radiation" theory, variations in the
amount of energy radiated by the Sun produce climatic
changes on Earth. Thus, during periods of less intense
solar radiation, the Earth could be cooled sufficiently
to trigger an ice age.

Closely linked with radiation is a second theory
which holds that unexplained changes in the composi-

*Driven by the pressure of an 8000-foot-thick mass
of continental ice, Antarctica's Beardmore Glacier
squeezes out of a mountain pass and onto a plateau.*

*Where the Antarctic ice sheet meets the sea, its
sheer edge melts and refreezes over and over again
to produce brilliant tiers of sculptured forms.*

85

tion of the Earth's atmosphere—increased cloud cover, for example—might black out some of the Sun's radiation, and so lower the Earth's temperature. A similar effect could be produced by air pollution, volcanic dust and, possibly, concentrations of meteoric debris or other matter in space between us and the Sun. Whatever the reason, the whole subject is profoundly complex, with enough variables to defy a computer, and it will be a long time before man unravels the true story.

Meanwhile, where do we stand now? Are the world's glaciers melting sufficiently to raise the sea level and submerge our great coastal cities? Or is the Earth cooling and heading for another ice age?

"Very probably we are now in the middle or perhaps toward the end of an interglacial period lasting thousands of years," says Richard P. Goldthwait, founder of the Institute of Polar Studies in the United States. "The Earth has swung back and forth between ice ages and interglacial periods for a million or two years, and we can expect it to continue. But don't worry about the return of an ice age soon."

At Greenland's Iron Mountain, conveyor belts of ice converge, their dark streams of eroded material blending into a pattern that shows how glaciers transport large amounts of debris over long distances.

How Icebergs Are Born

Icebergs are chunks broken from glaciers. Nearly all of the world's huge icebergs are "calved" by the great ice sheets of Greenland and Antarctica, which average 5000 and 7750 feet in thickness, respectively. As polar snows add new ice to these glacial masses, they descend from mountains and plateaus at speeds varying from a few inches to perhaps 100 feet per day. When they reach the coast, they push out beyond it, forming ice shelves in the seas offshore. As these shelves extend farther from land, they crack under their own weight, calving icebergs. Greenland's bergs may rise several hundred feet out of the water and weigh as much as a million tons. Yet Antarctica's flatter bergs are thicker, longer—some measure as much as 100 miles—and far heavier. Because ice is lighter than sea water, icebergs float with about one seventh to one eighth of their mass above the surface. In polar waters, a berg can last indefinitely. But if it drifts into seas with temperatures above 40° Fahrenheit, the iceberg melts rapidly.

As many as 1400 icebergs a year break off from glaciers in Greenland's Jacobshaven Fiord and float out to sea.

Mountain Sculpting with a Chisel of Ice

By John A. Shimer

One of the best known and most impressive relics of the Ice Age is the Yosemite Valley in California. This dramatic gorge on the western slope of the Sierra Nevada Mountains was born millions of years ago as a mere stream-cut trench on the slope of a rising mountain mass. The magnificent shapes we find here today are results of the awesome power of a glacier that came along much later.

At one time a thick continental ice sheet covered almost all of Canada and extended down into the central and eastern two-thirds of the United States. In the west the ice which covered the mountainous regions consisted of many separate individual masses. They were valley glaciers which were merely large and extensive versions of those found today in high mountains on every continent except Australia.

In the Sierra Nevada area surrounding Yosemite there are still about 60 small glaciers. They appear very insignificant in comparison with Alaskan or Alpine glaciers and, in fact, many look much more like large piles of snow that have not quite melted than

like true glaciers. However, they persist throughout the year, they have many layers and they move. These characteristics qualify them as true glaciers. Today there is nothing left of the original glacier that carved out Yosemite Valley. But the effects of this extinct glacier become noticeable very quickly as you head up into the mountains.

The western approach to Yosemite is gentle at first as it follows the Merced River upstream from the central valley of California. As far as the little town of El Portal, the canyon is narrow and winding, and there is barely room for the road as well as the stream. Here the traveler is always conscious of the steeply sloping rock walls as they descend right down to the road. The slopes of the valley up to this point have the rough V-shaped profile characteristic the world over of youthful stream-cut valleys.

A few miles above El Portal a change occurs and everything seems to open out. The valley floor becomes wider and the steep cliffs retreat. The road and stream both wander at will over a wide floor, and the valley profile is obviously U-shaped. The feeling of scale here is curiously unreal. It is almost impossible to realize the height of the immense cliffs soaring upward. Some of the early explorers in the region experienced this same sensation of unreality, and they made estimates of the height of El Capitan Mountain that varied from 400 to "at least 1500 feet." Actually El Capitan

Walled in by granite cliffs that rise as high as 4800 feet above its floor, California's Yosemite Valley today is a breathtaking seven-mile-long, mile-wide gorge in the Sierra Nevada range.

Before the onset of the last Ice Age, Yosemite was a shallow upland valley bounded by rocky bluffs. Following a winding course, the Merced River slowly eroded the slopes and built up a bed of gravel.

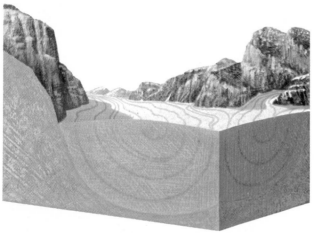

During the last Ice Age, which reached its peak 18,000 years ago, a glacier filled the valley, gouging a deep U shape. Several tides of ice moved down the valley during the Ice Age, each deepening it further.

With the glaciers gone, the valley was left slightly deeper than today. Now it has been carpeted with river sediments. When the main ice retreated, waterfalls formed where side glaciers had joined the flow.

rears fully 3600 feet above the valley floor, or about two-and-a-half times the height of the Rock of Gibraltar.

There are two especially fine vantage points from which the major features can be seen. The first is on the lower road as it climbs up the steep south side of the valley. From the road the view looks east up the valley. The massive cliff of El Capitan towers almost vertically from the valley floor to the left, while on the right, Bridal Veil Falls makes an exquisite leap of 620 feet from the end of a small hanging valley perched high up over the main floor. Such hanging valleys are common all over the world in areas of mountain glaciation. They owe their origin to the relatively rapid and deep carving of the main valley by a glacier

which lowered the valley floor and increased its width so that it lies well below the level of tributary valleys entering the main valley.

Yosemite Valley has a U-shaped cross section here, with the higher sides of the U standing almost vertical. In the distance, nine miles away, appearing just over the shoulder of the valley wall is Half Dome Mountain which soars 4800 feet above the valley floor.

Farther up the road, and 3000 feet higher than the first vantage point, is Glacier Point, the second good place to stop and look. From here the view is directly down on the wide, flat floor of the main part of the valley with the meandering Merced River 3200 feet below.

Across the valley and slightly to the left, Yosemite

Yosemite's Half Dome Mountain was dramatically sliced by a glacier. As the ice swept down the valley it undermined the mountain, causing it to split. The loose half was then carried away by the moving ice.

Creek makes a breathtaking drop of 2425 feet in two major leaps and a series of very steep cascades from its hanging valley. In the spring when the snows melt in the uplands and the rivers are flooding, this thunderous cataract, falling from a height of almost half a mile, is a most impressive sight. Later in the season the volume of water is greatly reduced in this and all the other waterfalls of the valley, and in summer many may cease altogether. It is thus in early summer when its flow is diminishing that Bridal Veil Falls lives up to its name as the winds in the valley blow the thin stream of water first one way and then another in a slowly cascading, filmy veil.

To the east, Half Dome Mountain dominates the scene with its insistent presence. Farther to the east,

as if to put these beautiful and impressive valleys into their proper place, the high Sierra Nevadas rise peak on peak to a jagged snow-covered horizon 20 miles away. This range is located on a block of the Earth's crust which has been tilted westward as the result of an uplift of many thousands of feet along a crack or fault on its eastern margin.

In preglacial times the Merced River flowed down the westward-facing slope of the range in a V-shaped valley for its complete length. Yosemite and Bridal Veil Creeks joined it as ordinary tributaries and did not fall as they do now into the valley of the Merced. Later the mountain block rose and the westward tilt increased. The Merced River, flowing directly down the slope, started to cut a much deeper channel for itself. Its tributaries, however, flowed across rather than directly down the increased slope and did not cut their channels as rapidly, and thus eventually were left behind in such a fashion that they had to complete their journey to the Merced via a series of rapids and small falls.

Next came the Ice Age. Glaciers developed in the higher reaches of the Sierra Nevadas and flowed down any available·valleys. The Merced was such a valley, and the ice which flowed through it eroded both sides and the floor of the valley with the result that it was widened and carved into the characteristic U shape. At the time of maximum glaciation we can picture Yosemite Valley as almost abrim with ice which extended down just about to El Portal before it melted away. We know that El Portal marked the limit of the glacier because here the U shape of the valley changes into the unglaciated V shape.

At that time in its history the main part of the Yosemite Valley had a filling of ice at least 5000 feet

Plunging almost half a mile in two steep cascades, Yosemite Falls are the highest in North America, and throughout the world are exceeded only by Venezuela's Angel Fall and Tugela Falls in South Africa.

thick. The only landmark that would have been visible above the icy waste was probably the peak of Half Dome. The ice streams would have been an impressive sight where they met to form one mighty ice river. Each glacier would have had a load of rock debris riding on its surface or frozen solidly into its icy mass. Such material, being especially thick near the sides, formed what are called "lateral moraines." Below Half Dome, where two of these lateral moraines joined to form a single line of debris at the center of the main stream, a "medial moraine" was formed. It would have appeared as a black streak on the lighter-colored ice, delicately following each turn of the glacier and thus emphasizing its flow pattern. Air views of some of the modern Alaskan and Greenland glaciers show medial moraines to perfection, and they help us to visualize what Yosemite Valley must have looked like.

When the ice in Yosemite Valley eventually melted away it uncovered Bridal Veil and Yosemite Creeks which, instead of completing their journey to the Merced via the series of preglacial cascades, were forced to descend in flying leaps.

After studying Yosemite Valley, the origin of the fiords of the Alaskan and Scandinavian coasts can be more easily understood. A fiord is simply a drowned, glacially produced U-shaped valley. If one imagines the ocean rising and covering California until there is a thousand feet of water in Yosemite Valley, the result would be a long, narrow estuary facing westward, with deep water and precipitous cliffs—a perfect fiord.

In addition to glacial features the Yosemite region is also noted for the impressive forms into which the granite has been shaped. There are steep, very extensive cliff faces with no cracks for great distances, and there is a sweeping landscape of rounded knobs and domes of rock obviously shaped into their present configuration by the weathering away of concentric layers of rock.

Half Dome Mountain illustrates perfectly the development of both the cliff face and the rounded dome. The cliff on its northern side resulted from the collapse of half of the dome when glacial erosion in the valley below undermined it. The rounded part of Half Dome was shaped by the weathering and peeling of layers from the mountain's granite rock formations.

Will Yosemite Valley ever be filled with ice again? Already there have been four major and many minor advances of the ice in this part of North America. Predictions are difficult and depend a great deal on which theory one favors for the cause of glaciation, but perhaps in the future there may be other advances, and we may only be in an interglacial time. At any rate, glaciation is the most dramatic of recent geologic events of major importance to occur on the globe, and provides the basis, as we have seen, for understanding some of our greatest scenic wonders.

White Terror of the High Mountains

By James H. Winchester

On December 24, 1969, high above the Alpine village of Zinal in southwestern Switzerland, 14 mountaineers moved across a snowy slope to prepare a ski run for holiday vacationers. Suddenly, the unstable snow began to slide from beneath their feet in a wild surging mass. Three helpless men in the group were swept downward to their deaths in tons of churning snow.

That Christmas Eve tragedy signaled the start of an unprecedented chain of white death. Some of the most destructive avalanches of the century cascaded during that terrible winter like awesome tidal waves down steep slopes from Europe's Alps eastward to Iran's Elburz Mountains. In the first three-and-a-half months of 1970, these violent cataclysms killed more than 200 people and injured several hundred more. They ripped away homes, hotels, hospitals, ski lifts, sections of bridges and stretches of highways and railroads.

France was particularly hard hit. In mid-April an earth slide, triggered by an avalanche of melting snow, sent hundreds of thousands of tons of debris crashing at high speed into three dormitories of a tuberculosis sanatorium at Plateau d'Assy, across the valley from Mont Blanc. Seventy-two were killed. Most of the victims were less than 15 years of age. Two months earlier at Val d'Isère, 39 young skiers died when a massive snowslide burst on a youth hostel.

During the same period Switzerland also experienced one of the worst avalanche disasters in her entire history. On February 24, 1970, 30 died at Reckinger, a village near the Italian border that hadn't seen an avalanche in 220 years. More died in continuing small accidents.

Avalanches are common in high mountains everywhere; perhaps as many as 250,000 major snowslides occur every year, though not all of them destroy life or property. Avalanches and glaciers are entirely different phenomena. While glaciers are composed of thick, slow-moving masses of rock-hard ice, avalanches are fast-moving masses of loose snow. Snow on a slope is always in delicate balance. With successive falls, tiers are built up over the ground, something like the layers of a cake. As temperature and weather conditions change, these layers settle, pack down, melt and refreeze. New snow, for instance, may fall atop a previous layer that has become a crystalized mass of ice. The bond between the two tiers is tenuous. Any small disturbance can start the top layer of snow sliding over the icy crystals underneath.

Mountaineers divide avalanches into two categories: "ground" and "dust." A dust avalanche commonly

A wave of snow surges down a flank of 9000-foot Weissfluhjoch near the town of Davos, Switzerland. This type of snowslide—attaining speeds of 200 miles per hour—is called a "dust" avalanche. It occurs when a mass of newly fallen powder begins sliding over a hard base of older snow.

occurs when fresh heavy snow fails to cling firmly to older layers. The danger is considered greatest when ten or more inches of new snow have piled up over a period of 24 to 48 hours. Pulled downhill by the force of gravity, the top layer starts sliding. The heavy mass picks up speed quickly and gathers more snow and weight as it moves. In just a few seconds, down a long slope, it reaches fantastic speeds. One such slide in Switzerland was timed with a stopwatch at more than 200 miles per hour.

Dust avalanches also create enormous air pressures in front of the fast-moving wall of snow. Such a powerful "wind" once uprooted 250 acres of sturdy, 100-year-old trees in the Alps. Another blew a tourist bus off an Austrian bridge, killing 23 skiers. Snow from the avalanche itself never touched the vehicle.

On the other hand, ground avalanches—slides of wet snow—seldom achieve more than 60 miles per hour, but also can be destructive. This type of avalanche is common in the spring when snow starts to melt. The snow in ground avalanches tends to roll up into balls as it moves, gathering up debris such as earth, uprooted trees and boulders along the way in a mass that may weigh a million tons. Forces of as much as 100 tons per square yard have been measured on obstacles in the paths of such avalanches.

While avalanches are an inevitable characteristic of mountainous areas, during the past few years they have been causing more death and destruction than ever before. Here, according to experts, are the reasons: Skiing and other winter sports are booming. As winter homes and resorts proliferate on land where avalanches have been crashing down for centuries, danger increases. Says one Swiss authority: "People just don't seem to care whether the land is safe or not. Financial self-interest seems to come first."

But unrestricted development is not the only cause of avalanche deaths. Careless skiers start more than half of all reported avalanches and account for many of the casualties. "Too many die each year through ignorance or because they underestimate the dangers," points out one leader of a Swiss rescue service. "They don't think of the possibility of being buried alive, crushed or suffocated." (Swiss records show that after two hours, only 19 out of every 100 persons buried in avalanches are found alive; after three hours, only nine out of 100.)

Switzerland, which had an estimated 20,000 major snowslides in the 1969-70 season, had proportionately fewer casualties than her neighbors—an achievement made possible primarily by a well-established early warning system. A component of the Federal Institute for Snow and Avalanche Research, this system is already being imitated in Italy, Austria and Scotland and to some extent in France. Once a week in the snow season, and more often in particularly threatening conditions, the possibilities of serious Alpine slides are

calculated. Thousands of individual pieces of intelligence are analyzed before an avalanche warning bulletin is released. Institute and outside observers, including Alpine guides, ski instructors, schoolteachers, monks and cable-car operators, report every morning by phone and telegraph on every aspect of snow and weather conditions from 52 strategic Swiss sites. Similar information comes in on an exchange basis from Austria and Italy.

Newspapers, radio and television feature the Institute's bulletins. They are also recorded by the Swiss Federal Telephone Service and are available to anyone who calls. Their reliability guides police, rescue services and resorts on limiting transportation, closing slopes or evacuating people.

Founded in 1942, the Institute is located in a modern, four-story building that clings to the top of the rugged Weissfluhjoch, a steep 9000-foot peak towering above

When the spring thaw came to this Swiss slope, melting weakened the bond between the snow and the ground, and the entire accumulation slid downhill.

the town of Davos. A two-mile-long funicular railroad, which also serves skiers going up the mountain, links the Institute to the valley.

Each of the two dozen staffers at the Institute is a specialist in some aspect of snow control. They are meteorologists, hydrologists, engineers, foresters, geologists and physicists. As this team brings into clearer focus the basic causes of avalanches, and how to protect against them, snow experts from around the world flock to the Weissfluhjoch for new information.

Another Institute activity is a week-long avalanche academy, held every other January. The indoor and outdoor classes are open to anyone paying a small fee and many of the students are Alpine rescuemen and ski resort instructors from abroad. One of the things they are taught is that for locating a human body buried in

an avalanche it is hard to beat a dog's acute sense of smell. Dogs' educated noses are unbelievably sharp.

For example, one night in 1968 when a series of avalanches in Davos buried 11 people, avalanche dogs located and unearthed every one of them—four were alive—within a couple of hours. These canines can sniff out victims buried beneath as much as 16 feet of dry snow. A well-trained avalanche dog, the Institute reports, can search a two-and-a-half-acre snow field in 20 to 30 minutes; it takes 20 men, poking rods down into the snow at closely spaced intervals, 20 hours to do the same job.

It is said that in olden days, coachmen coming through an Alpine pass would crack a whip to loosen snow and get it to drop harmlessly. Today, the Institute seeks to eliminate potential trouble by starting

An avalanche in 1968 crashed into the town of Davos, Switzerland. Rescue teams used trained dogs to sniff out victims buried in the rubble-strewn snow.

Another avalanche in 1970 piled snow so heavily around Davos that villagers had to cut a new road to reach the outside world. Splinters of a tree ground up in the slide protrude from wall at left.

harmless slides artificially. The best avalanche busters, they find, are explosives. Mortars, anti-tank bazookas, remote-fired land mines, hand grenades and just old tin cans packed with dynamite and set off by match-lit fuses, are all employed. In the Swiss Alps alone, as many as 10,000 explosions are set off each year to create artificial avalanches. Says one resort owner: "We may be a peaceful nation, but you'd never know it from all the shooting." Similar defenses are employed in the U.S. and Canadian Rocky Mountains to protect railroads and highways.

Snow scientists still have great gaps in their knowledge of avalanches and how to protect against them. Thousands of baffling questions remain to be answered. How can snow be artificially stabilized? What specific snow surfaces are sensitive to specific weather conditions? How can protective trees best be grown in avalanche areas?

"If only a fraction of these and other pressing problems are solved," says the Institute's director, Dr. Marcel de Quervain, "avalanche deaths and damages can be cut considerably—perhaps by as much as half."

With skiing continuing to grow in popularity, there is an imperative need for skiers in all countries to be more aware of avalanche dangers and to act sensibly. For best avalanche protection, everyone should heed these common-sense rules:

• Even without a warm-up of temperature or wind, a foot of fresh snow presents a likely peril on any ski slope steeper than 25 degrees. Eighty percent of all avalanches start during or immediately after snowstorms.

• As temperatures go up, avalanche odds increase. Snow begins to melt, becomes loose and slides start easily and frequently.

• Whatever the weather—fair or foul—check with local authorities before beginning any ski excursion and *believe them*.

• Whenever possible, when skiing in unknown territory, choose a route through heavily timbered slopes and long ridges. It is best, too, to avoid crossing steep slopes. If it has to be done, go as close to the top as possible. When crossing a slope in a group, *never bunch together*. Form a single line and space out so that only one person at a time is in any possible danger.

• Always carry an avalanche cord: a 100-foot red nylon string. When skiing across an avalanche area, it should be tied around the waist and trailed behind. If you are caught in a slide, there's a chance that part of the cord will remain on the surface to guide rescuers.

• When someone in your group *is* caught in an avalanche, mark the spot where he was last seen. Use a ski or ski pole. Start searching immediately and, if your party is large enough, send someone for help. Keep up the search, if possible, until trained aid and dogs arrive. Time is life.

Antarctic Glaciers— in the Sahara Desert!

By Rhodes W. Fairbridge

The central Sahara Desert is one of the most desolate places on Earth. At noon in summer, the rocks and sand become so hot they can blister your hands. During the hot months, the Arabs travel only after dark. Oases are far apart, and only a mile or two from the life-giving springs there are no signs of life—not a shrub, bush or tree. At midday, even in winter, the Sun's rays burn the skin unmercifully and there is rarely a cloud. Rain usually falls for only a few hours a year and then disappears instantly into the sand. Quite a few hardy tribesmen have lived all their lives in the Sahara and have never seen rain.

Several years ago in the central Sahara, French geologists found traces of what looked like evidence of ancient glaciers: long parallel scratches and grooves on rock surfaces. Along with these formations, they found fossils just below and just above the glacial layers. By determining the age of the fossils that bracketed the glacial evidence, the French geologists concluded that glaciers were present in the Sahara about 450 million years ago.

The Sahara is the world's hottest place. Recently a world record high temperature of 137 degrees Fahrenheit in the shade was registered there. How could there have been glaciers in the Sahara?

By chance I met one of the French geologists in the summer of 1967. He showed me pictures of the glacial evidence he and his colleagues had found. They certainly looked like many other glacial formations I had seen. From what I knew of the way the continents had drifted after the Sahara glaciation 450 million years ago, it dawned on me that the Sahara—here in what we like to call the Northern Hemisphere—might have been located near the South Pole at one time.

The problem now was to prove it. Glaciers, after all, can be formed in high mountains, such as the Himalayas, which are closer to the Equator than to the Poles. And rocks can be scratched by landslides as well as glaciers.

In January 1970, at the invitation of the Petroleum Institute of Algeria, I joined a group of specialists in Algiers to undertake an expedition that would enable us to see all the glacial evidence in the central Sahara and bring back information on the drilling potential of several oil reservoir areas.

Over the years, many geological expeditions have been sent into the Sahara by the French government and by private oil companies. Reports of glacial formations have come in from all across North Africa— from Morocco, Mauritania, Algeria, Niger and Chad,

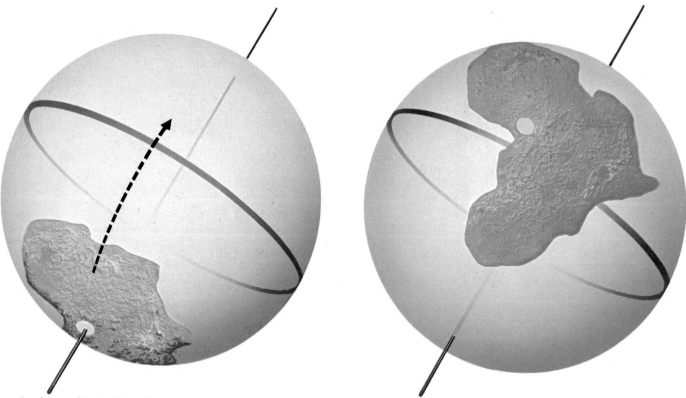

In the world of 450 million years ago (left), Africa lay at the South Pole under a sheet of ice. The continent slowly drifted north, and today the former South Pole is located in a desert near the Equator.

a spread of 3000 miles. Glacial evidence over such a wide area could hardly be the result of local phenomena like mountain glaciers; only a polar ice cap the size of that in Antarctica today could account for an ice sheet that left its traces over a continent-sized area.

Moreover, traditional geological evidence also argued for a South Pole position in the Sahara 400 to 500 million years ago. A decade ago a Norwegian scientist, Nils Spjeldnes, concluded that northern European fossil animal life of this period was characteristic of warm water, while in southern Europe the fossils were characteristic of cooler water. Careful plots of all the known fossils of the same period suggested that the South Pole of the time lay somewhere south of Europe, probably in west Africa. Subsequent independent work by two other scientists based on different fossil data also suggested something very similar.

One more piece of evidence virtually predicted that we would be able to locate an ancient South Pole somewhere in the Sahara Desert. When rocks are formed they become slightly magnetized because of the Earth's powerful magnetic field. The magnetism in the rocks is always aligned with the Earth's magnetic poles at the time the rocks are formed . No matter how much time has passed, nor how much the rocks have shifted since their formation, they retain their original magnetic orientation. By plotting the magnetic orientation of ancient rocks it is thus possible to determine where the poles lay at the time the rocks were formed.

Working independently of the French geologists in the 1960s, a number of British and Australian geophysicists were exploring South America as well as other parts of Africa at about the same time, measuring the magnetic orientations of ancient rocks. Their studies indicated that if South America were brought side by side with Africa, according to generally accepted theories of continental drift, the polar orientations of South America's ancient rocks would coincide almost exactly with those of African rocks and pointed to a pole position somewhere in or near northwest Africa.

Our 1970 expedition centered around the Ahaggar region of Algeria, some 950 miles south of the city of Algiers. The Ahaggar is a wilderness of jagged, rocky mountains ranging in age from two billion to 600 million years. Around the edges of this ancient area lies the Tassili Plateau, a rough, barren terrain of younger sandstone rocks whose structure could be read like chapters in a book.

Beneath the sandstone layers were the remnants of peaks in an ancient mountain range that had weathered away to form a plain. We found traces of a clay soil in the products of this weathering, indicating that the mountains had been worn down under warm, wet tropical conditions. Therefore, the climate here had been tropical before any glaciation occurred.

The bottom layer of the overlying sandstone contained pebbles with a distinctive wedge shape due to abrasion by desert sand. These pebbles were also very ancient and indicated that the earlier wet tropical climate had been followed by an arid climate.

Masses of ice moving out from the ancient South Pole leveled this pavementlike rock in the Sahara and scored long, straight grooves that helped geologists determine the former location of the Pole.

Several hundred feet above the base of the sandstone, we came across subtle changes in the composition of the rocks. Here and there the surface of the rocks showed ripple marks exactly like those that can be seen on sandbanks at low tide today. In some places the rocks were imprinted with the shapes of seaweeds; in others the rocks contained the tracks of trilobites, extinct crablike crustaceans characteristic of that period. Later on we found complete specimens of fossil trilobites together with many other creatures. We were seeing evidence of an ancient sea that had pushed in over the surface of a primordial continent.

The underlying sediments in other areas were deeply scoured into channels as if tremendous currents had swept over the sea floor; then the channels were re-filled and covered with sand containing pebbles and boulders of a strange type, evidently transported from a considerable distance.

Those boulders told the story. Some of them were ancient granite and quartzite and were far too big to have been rolled along the sea floor by marine currents. Some of the boulders were smoothly scraped or polished on one or more sides. Only glaciers do this. And only ice floes or icebergs can transport large boulders far from their source and out to sea.

The curious-looking sandstone that enclosed and lay over these boulder beds contained higgledy-piggledy mixtures of all sorts of rocks. They were typical of glacial deposits, particularly the sandy ones of the Baltic. Similar rocks are known, however, to occasionally result from submarine landslides. We knew they could well be "tillites," rock debris that is the product of melting ice, but we needed more evidence to be sure.

In sandstone lying above and to the north of the glacial deposits, there was an extraordinary feature: a belt where the sands had been swept by powerful currents into giant ripples ten feet or more from crest to crest, in a formation up to 100 feet thick that stretched out for several hundred miles. Giant ripples like these are formed by tidal currents in certain constricted areas today, such as in the southern North Sea, the Irish Sea and the Strait of Malacca. But the ripples in the Sahara are of vast dimensions. Could they have resulted from the outpourings of millions of tons of meltwater from the edge of a huge glacier? A similar spillover occurred in the Baltic after the last Ice Age and likewise in Hudson Bay.

Traces of subglacial volcanoes were perhaps the most peculiar features we found that are not normally associated with sandstone. Eruptions commonly occur under the glaciers of Iceland today, so we have some living models. The lava rises up but is quenched by sudden contact with the ice. In the Sahara, we first saw these circular structures (from 500 to 3000 feet across) in aerial photographs and thought they might be meteorite craters. But at ground level, we found that each had a core of basaltic lava surrounded by a curious belt of fused sand and volcanic rock, all within a ring of glacial sandstone forced up into a vertical position. In some places the glacial rocks and these intruding volcanic necks were overlapped by beautifully layered shale dating from about 430 million years ago, which showed no traces of ice or volcanoes. By that time the Sahara Ice Age was definitely past.

Pierre Rognon, a French specialist in modern glacial processes, showed us many details of the Sahara glaciation. Some outcrops revealed a maze of vertical cracks, a sign of former frost action. Spring mounds were found such as could have been produced by meltwater being forced up by the pressure of nearby ice. Somewhat similar "tundra craters" are found in northern Canada, Greenland and Siberia. Meltwaters from the Sahara glaciation carried out and laid down sand as river or tidal flat deposits. Because the Tassili formations lie flat, it was still possible to follow these old meanders across the country; they were spectacular from the air. The sediments are weathered into strange shapes: from a distance the ridges look like ruined castles.

At this stage of investigation we had abundant evidence of former ice sheets in the Sahara, but nothing to prove it was polar and not mountain ice. The ice could have come from immense mountain glaciers, as in southern Alaska today. Then the great day came. In the eastern part of the Ahaggar, in the valley of Wadi Taffassesset, we came across mile upon mile of pavementlike rocks with long, deep parallel grooves. They were running from south-southeast to north-northwest, just like the scrapings of some giant bulldozer. We traced the scored pavements on the ground for miles. Later, from the air, we followed them across hundreds of miles. Others have confirmed their continuation right across northwest Africa. We had proof! Grooved rock pavements of such size are found

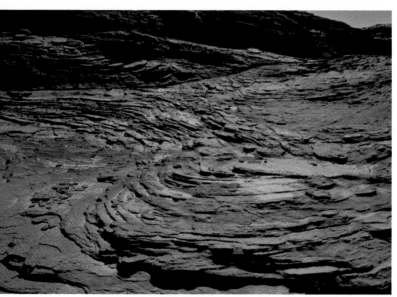

As Africa moved north, its continental ice sheet melted and released swirling currents that rippled a sea floor and left traces in solidified sediment.

These delicate seaweed impressions in the rippled Saharan rock confirm the presence of an ancient sea in what is now the world's hottest desert.

only in the regions of the Antarctic ice of today, or in the Canadian or Scandinavian glaciated areas of the last Ice Age.

The debate was over; we were unanimous in certainty. Only a giant continental glacier could have done this. Our expedition to North Africa had confirmed that 450 million years ago there had been a mighty glaciation. An ancient South Pole had indeed existed somewhere near what is now the Sahara.

The discovery in the Sahara has not revolutionized our ideas of geology, but it has been a triumph of international scientific research and it has started a chain reaction of ideas about shifting continents, migrating poles and the changing nature of the planet.

What is the globe doing at the present moment?

The opposite shores of the Atlantic Ocean are known to be spreading apart at one or two inches per year. The entire crust seems to be wheeling around, so that the Aleutian Islands and Alaska are pressing southward toward the Pacific, while Europe seems to be setting northward.

The crust of the Earth appears hard enough, but it is really thin skin on the massive inner mantle and core. At depths of 50 to 100 miles there is a zone of potential melting, and over this layer the crust can slide, albeit sluggishly. It now appears that the whole crust can slide, sometimes moving equatorial lands toward the poles, sometimes moving polar land masses of former times into warmer latitudes as we found in the Sahara.

Rising above the desert floor like ruined battlements, these jagged rocks were sculpted by thousands of years of wind and sand long after North Africa's immense polar ice cap had disappeared.

Chapter Three

MONUMENTS TO TIME

Though the land changes so slowly it seems eternal, each of its most ancient features has its own story to tell of the forces that have altered it down through the ages

Why We Have Mountains

By Lowell Thomas

Everyone knows what a mountain is—or do they? Because mountains mean different things to different people, definitions are likely to vary widely. Everyone admits, for example, that Everest is a mountain, the titan of them all, thrusting its snowcapped peak above the clouds to an altitude of almost 30,000 feet above sea level. If this is not a mountain, nothing is. But what of Snowdon, the loftiest peak in Wales, rising just over 3500 feet above sea level? Yet we speak of *Mount* Snowdon, just as we speak of *Mount* Everest. The term seems incongruous when applied to such unequal contenders. Still, Snowdon and its slightly smaller sister peaks are giants compared with the surrounding terrain, which descends by degrees until it merges with country you wouldn't even call hilly.

Geographers generally agree that a land mass must reach an elevation of at least 3000 feet before it can be considered a topographic mountain. When Mount Everest is said to be 29,028 feet high, for instance, that means it is 29,028 feet above the level of the sea. Actually it lies many miles inland and only rises some 15,000 feet above the surface of the neighboring Tibetan Plateau.

Geologists refine the meaning of the word "mountain." They define mountains by their structure—a mountain is a mountain by virtue of its geological structure and not because of its elevation above sea level. Some rugged highlands in the form of plains and plateaus, such as those in Tibet, are certainly "mountainous," but they are not really products of mountain building in the geological sense. On the other hand, there are flat, low-lying rock surfaces in Canada and elsewhere which are true mountains. They are low now because they have been worn down to near base

level, but they are still called mountains because of their underlying geological structure. And there are true mountains—the Mid-Atlantic Ridge, for example—under the sea.

Why are there any mountains at all? And why are they situated where they are? To understand completely what we see on the Earth's surface then, we must also understand the nature of the Earth's interior and the power of the forces at work thousands of miles beneath us.

For many years it was customary to think of our planet as consisting of a thin crust wrapped around a molten interior. If this were so, the Earth's rotation would produce internal tides so violent they might eventually cause the crust to break up; or if the crust were sufficiently strong to withstand this motion, the internal friction might cause the Earth's rotation to stop. You can easily see why when you visualize a favorite demonstration of 19th-century British scientist Lord Kelvin. He spun two eggs, one raw and the other hard-boiled. The raw egg, with its liquid interior, would spin for a much shorter time.

Let's assume, then, that the planet on which we live is an oversized hard-boiled egg, essentially a solid body, although not of a uniform structure throughout. Rather, it is made up of concentric "shells" or zones. The boundary between each zone is sharp and distinct.

The deeper the rocks, the hotter they become. Temperature increases with depth at an average rate of one degree Fahrenheit for every 60 or 70 feet. Since pressure usually raises the temperature at which a substance melts, deep rocks remain solid at temperatures that would be high enough to melt them on the surface of the Earth where the pressure is much less.

Generally speaking, the Earth is composed of two major outer layers—the crust and the mantle—and a core composed of inner and outer sections. The crust is 20 to 30 miles thick under the continents and about three to four miles under the oceans.

The Cauca River in Colombia runs in a natural aqueduct among "fold" mountains in the northwestern Andes. The entire Andes system was created by buckling in the Earth's crust and volcanic eruptions.

| BLOCK | FOLD | DOME |

"Block" mountains form when solid blocks of the Earth's crust thrust up and then tilt over; "fold" mountains are compression wrinkles in the crust; "domes" occur where crustal material pushes up from below and bulges the surface. The broken red line shows the contours of the hills prior to erosion.

The Earth's crust amounts to hardly more than a film surrounding the rest of the planet. Yet it has a unique importance. The crust is relatively stable because of its rigid rock structure. With its upper side turned to the atmosphere, the surface of the crust provides the conditions of life. Here the drama of evolution has been played, leading from unicellular creatures swimming in primeval seas to man—who, among other things, climbs mountains. Mountains are strictly crustal phenomena, whether above the land surface or under the oceans.

The mantle, or intermediate zone, is about 1800 miles thick. It is thought to be composed of rocks heavier than those on the surface which are not only hot but plastic as well, owing to their depth.

Beneath the mantle is the dense outer core of the Earth, 1310 miles thick. Heat and pressure are both fantastic at such a depth, so fantastic that the iron and nickel of which this core is believed to be composed is certainly in a molten, almost liquid state. The 850-mile-thick inner core lies at the Earth's center. The inner core also seems to be composed principally of iron and nickel, and is much denser and more of a true solid. The temperature at this depth has been estimated to range from 6000 to 7200 degrees Fahrenheit. The pressure is more than 3.5 million times greater than on the surface of the Earth. At any rate, we cannot imagine the subcrustal interior of the Earth as a solid chunk analogous to a block of concrete.

Mountain building involves twisting and buckling, collapse and eruption of rock at or near the surface of the Earth's crust. It is probably a fairly continuous process. The Alps, Himalayas, Andes, Rockies and other great mountain ranges of the world did not rise at a single moment of geological history in lone, shattering episodes. They evolved by long, slow, grinding movements within the Earth's crust. The name given to this movement is "diastrophism," or deformation of the Earth's crust. The peaks of mountains and the depths of the seas, low-lying swamps and rolling prairies, cliffs and canyons—all of these diverse landforms are as they are largely because of diastrophism.

Many localities in the highest mountains reveal that parts of the ocean bottom have been raised to form land. Attesting to this are the fossils found in limestones, sandstones and shaley rocks in the Alps, Andes, Himalayas and many other ranges. What caused the ocean bottom to rise to these heights? The process must have started when tremendous amounts of mud and sand were washed into low-lying troughs in the oceans. As the plates that compose the Earth's crust shifted, these sediments became folded and gradually rose above the level of the sea. In other regions the sea floor was uplifted by upwellings of hot magma from below. These mountain-building processes go on for millions of years, until parts of the original sea floors become highlands. Then various agents of weathering and erosion tear them down and wash the debris into the sea once more. Again the sediments fill in the basins until they too are uplifted in the never-ending cycle of the birth and death of mountains.

What are the causes of this uplift? One explanation is the principle of equilibrium. Knock over a tower of blocks and they will end up in a heap, some on top of the others, the rest scattered around, depending on their size, shape, position, weight, angle and speed of fall. They do not stop rearranging themselves until gravity brings them into balance and a state of equilibrium is achieved.

Different parts of the Earth's crust try to attain equilibrium in different ways. In some places the crust is lifted and folded by the extreme converging pressure generated when two crustal plates meet and grind against each other. The Himalayas, Alps and Andes are all "fold" mountains created by plate pressure.

Sometimes the plate pressure is divergent. When crustal plates move away from each other, the rocks are pulled apart into huge blocks, some moving sideways, others moving up or down. This breaking of the Earth's crust is called "faulting." Mountains formed in this way, such as the Sierra Nevadas in California, are "fault block" or "block" mountains.

In regions of the world where crustal activity is intense, violent movement may occur along fault lines, or zones of weakness, at the Earth's surface. Notable examples of this geological phenomenon are the San Andreas Fault in California and the Great Rift Valley that begins near the Dead Sea and slices across east Africa.

Sometimes the Earth's crust does not fault or fold intensely but folds more gently to form what appear to be giant blisters. These are known as "dome" mountains. Britain's Lake District mountains and the Black Hills of South Dakota are examples of mountains of this type.

As we have seen, the temperature rises as you go below the surface of the Earth. Owing to the extreme

These mountains in Britain's Lake District are stumps of a seven-million-year-old dome which was deeply scoured by erosion and polished by glaciers.

The jumbled, irregular formations in the American Rocky Mountains near Vernal, Utah, resulted from uplift and tilting of a huge crustal block which then weathered and was grooved by Ice Age glaciers.

Exposed by erosion, the skeleton of central Australia's Krichauff Range reveals the folded structure that once supported mountains many thousands of feet high.

pressure exerted upon the rocks in the lower crust and the mantle zone, these rocks will not melt. They do not become molten unless they are brought to the surface. Where the pressure is relieved by faults or fractures in the crust, rocks near the surface become molten magma, a viscous silicate melt that may flow like molasses to the surface and at the surface.

In places of crustal weakness, where fault and fracture are present, volcanoes may erupt, sending lava (surface magma) over the side in a torrent, a cloud of smoke and gas towering into the sky. On the other hand, the heat may seep up into the ocean and be dissipated in thousands of cubic miles of sea water. Or it may cause faulting or a gradual pushing up of rock layers over a wide area.

The present belts of major earthquake and volcanic activity also happen to be areas where young mountain ranges occur. Some of the present young titans are less than 50 million years old. The Alps and the mighty ranges that fan out from the Pamir knot in Central Asia rose as recently as 40 million years ago. If we could telescope the history of the Earth, we would see a kaleidoscope of geological patterns, a convulsive billowing of mountain ranges as contorted as the waves of a stormy sea.

A Mountain Higher Than Everest?

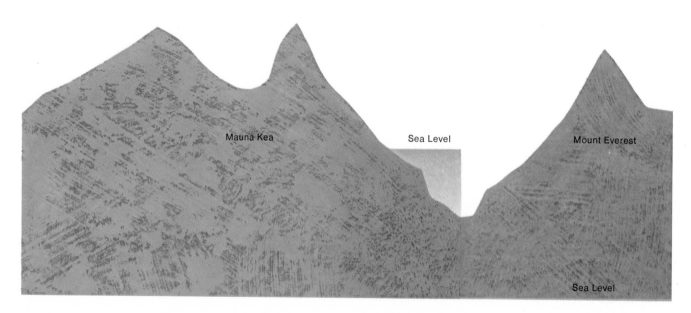

Is Mount Everest really the highest mountain in the world? It depends on how you measure a mountain. From sea level to summit, Everest is 29,028 feet, unquestionably the highest point on Earth. But Hawaii's Mauna Kea is 33,476 feet from its base on the floor of the ocean to its highest peak,

looming larger than Everest if the level of the ocean floor beneath Mauna Kea is aligned with sea level beneath Everest, as in the diagram above. Moreover, Everest's height above its true base is only half the rise of Mauna Kea from the sea floor. But Everest is still growing (see page 132).

Wastelands of Wind and Sand

By David I. Blumenstock

Beyond the dry, dusty Dasht-E-Lut tableland of eastern Iran begins the desert region of Seistan, a 7000-square-mile area that straddles the Iranian-Afghan border. It is a place of desolation, a land of naked rocks and searing sand. Except on rare occasions in winter and early spring, its riverbeds lie dry and baking in the constant rays of the Sun.

Today, Seistan is inhospitable to life and only sparsely inhabited by man. Yet less than six centuries ago Seistan was a populous locale—a center of commerce with a thriving agriculture based on elaborate irrigation systems. Then Mongols, riding under the banners of Tamerlane, swept down from the Asian steppes into Seistan, decimating the population, razing buildings and destroying the dams and weirs of the irrigation systems. A culture that during previous centuries had managed to wrest prosperity from a dry land was wiped out by one stroke of the sword.

In other places on the Earth, destruction on the scale of Tamerlane's has seldom led to the abandonment of the land by mankind for hundreds of years. True, no desert is an ideal environment in which to survive while starting again from scratch. But irrigation works can be repaired; houses can be rebuilt. A patient people, even though greatly reduced in numbers, can with resolution often regain its wealth.

Seistan was unique because wind and sand quickly completed the desolation that Tamerlane began. After the Mongol horde rode out, as if on cue the sand marched in, driven by the *bad-i-sad-o-bist roz*—the "wind of 120 days."

This continental wind sets in every year toward the end of May. Throughout the summer it blows with appalling violence. There is little or no relief until it dies down around the end of September. It always blows from a heading a little west of north, a dry wind sweeping down from the arid heartland of western Asia. With velocities sometimes exceeding 70 miles per hour, the wind brings a pandemonium of noise, dust and sand. The sand arrives in echelons. First come "skirmish lines" in the form of advancing dunes. Steadily they roll forward. Where they encounter obstacles too large to bury quickly, they mark time as if awaiting reinforcements. As more sand comes in from behind, each new wave of dunes climbs up the windward slopes of those ahead—building massive, towering ridges that swallow up whatever has given them momentary pause.

The late Sir Henry McMahon, who led a party that surveyed the Baluchistan-Afghan-Persian border early in the 20th century, reported that in a single summer the Seistan village of Kila-i-Nau was totally buried by

Driven by a high wind, dunes creep across a desert, forming a relentless tide of sand that neither the land nor the works of man can resist.

sand. The village lay in the leeward shelter of a ridge, where windblown sand would tend to collect. As with snowdrifts and river silts, deposition occurs where the currents propelling the grains quiet down, allowing particles of dust and sand to settle. In a few weeks, McMahon witnessed a large pond filled in and transformed into a ten-foot high mound by the action of Seistan's wind and sand.

Not only does the Seistan wind pile up sand in dunes in some places, it also excavates the ground in others. Hence, ruins, buried for centuries, are abruptly exposed through the scouring action of the wind in a few weeks. Even in clay, huge depressions are excavated by the Seistan wind; these depressions may go 200 feet deep into the Earth.

There is little doubt that in Seistan the strong, persistent summer winds have blown from the same direction and with great violence for seven centuries or more. Present-day houses of sun-dried brick are built with an extra thick, solid wall—a wall without windows—facing in the direction from which the wind comes. Those of six centuries ago are similarly constructed and of similar orientation.

Perhaps more intriguing than the question of Seistan's failure to recover after Tamerlane is the problem of how during Seistan's prosperity the people had ever managed to cope with the great summer winds and the encroaching dunes. Certainly they could not control the movements of the wind and sand. Probably they waited out sandstorms in their houses and then

Curving sweeps of fine sand give this magnificent Arabian dune the classic crescent shape of a "barchan," a type that forms when there is ample distance between dunes to allow their tips to stream out.

dug themselves out. It must have taken a high-spirited and politically united people to maintain a civilization in Seistan.

The summer winds of Seistan are related to the tropical monsoon circulation in the broader South Asian region. Seistan lies just to the northwest of the great atmospheric low-pressure area that develops over central Asia in the heat of summer. Cold, dry arctic air is drawn to this low from the north at the same time that cool, moist air moves in from the Indian Ocean. The resulting counterclockwise circulation about this immense central Asian low accounts for the dry summer winds of Seistan. There are other contributing factors, such as the funneling effects created by the great Asian mountain chains farther to the northwest and to the northeast, as well as the effects of local heating. When autumn arrives and the central Asiatic low disappears, the Seistan wind slackens and by wintertime is replaced by winds from other directions.

The sand dune is the most impressive land form created by the wind. Dunes are formed wherever there is an abundant supply of sand and a prevailing wind—a wind that generally blows from one dominant direction. The shores of oceans and lakes and desert terrains are the sites where dune development is commonly quite extensive. More rarely, dunes are found along the margins of large rivers—in flood plains where ample sediments have been deposited in times of flood.

Dunes are of three main types. The first, the "transverse" dune, common both to beaches and deserts, is simply a ridge of sand that collects along lines generally perpendicular to the heading of the prevailing wind. Thus beach dunes, where the prevailing wind blows from the sea, form in a series of inland waves parallel to the shoreline.

The second type is best known from the Sahara, where the Arabs call it the *seif*, or "scimitar." These dunes run parallel to the wind direction and only develop under very high velocity winds and plentiful supplies of sand.

The third type, the "barchan," is shaped like a crescent moon or horseshoe, with long curving horns that trail off downwind from the edges of the dune, decreasing in size away from the crest. Powerful winds in barren regions with relatively little sand produce the barchan form. If the wind were not powerful, the sand could not be swept around the sides of the dune to form the curving horns; and if the sand were too plentiful, the dunes would be so densely crowded that the

waterline is partly covered with vegetation and so moves more slowly before the wind. Generally, the fourth or fifth row—if not the third—is entirely overgrown with grasses, or some combination of grasses, brush, vines and scrub trees. It has become stable. This zonation of vegetation and of dunes is the product of a developing balance between the natural forces of wind and sand supply, on the one hand, and the ability, on the other hand, of various plants to take root and maintain themselves upon dune surfaces. Such growth, of course, is only possible in non-desert areas. From the study of such sandy areas, land managers have been able to develop conservation control measures for seashores, fastening down dunes whose unimpeded movement would threaten coastal highways, houses and other facilities. For example, kudzu, a hardy, fast-growing desert vine, has been found to take hold readily on fresh dune sand, pinning it down to suppress migration.

Many years ago someone observed that sand grains found in dunes along the seashore are uniformly rounder than on the adjacent beaches from which the dunes are formed. The reason for this difference at first appeared obvious enough: when the sand particles were swept across the beach and onto the dune slope, they became more rounded through having their corners worn down. But sand put into a vibrating machine can be agitated to create friction effects like those that come from being blown upslope by the wind. Microscopic examination of the grains before and after such treatment discloses no rounding down of the grains' edges.

What, then, explains the presence of rounder sand grains in the dunes? The answer must lie in the action of the wind, which picks up and drives smoothly rounded grains that roll most easily. Hence the wind performs a sorting action, carrying off primarily spherical grains toward the inland dune rows.

Sand dunes are not the sole art of the wind. In desert areas the winds remove fine sediments from basin floors, leaving behind only rocks and rocky surfaces. Often these form a pavement, giving the terrain the look of a rough, irregular flagstone terrace. When rock pedestals, rock outcroppings and cliff faces are composed of strata of differing hardness, the wind, armed only with abrasive sand particles as a cutting edge, bites most deeply into the softer layers. Like an abstract sculptor, wind and sand produce strangely chiseled forms of intriguing and often beautiful complexity. Moreover, sand polishing develops a hard sheen often observed on desert floors and boulders. By sweeping away all the finer soils washed down from the mountains in flash-flood waters, the scouring winds may keep basin floors in a state of perpetual luster.

Thus, in addition to its climatic role, the wind is everywhere a force in the molding of the surface of the planet's major land masses.

wind would not have full play to curve the dune's corners back. The dunes of Seistan are barchans. There are many other desert areas characterized by barchan dunes, including parts of the Sahara and its smaller American counterpart, Death Valley.

In its broad contours, the form of the dune reveals the action of the wind. To windward, dune slopes are streamlined—long and relatively gentle. Here the sand grains are rolled upward over the dune as they are driven before the wind. The dune's steeper side is to leeward, down which the grains roll after they have been blown over the crest and reach the comparative shelter of the lee. The larger the grains of sand in the dune, the steeper is the dune's leeward slope; this is because coarse grains will maintain a steeper slope than fine sand. The constant movement of sand grains from the windward side across the crest causes the dune to migrate. This migration ceases only when the dune encounters damp soil or where grasses, scrub and other vegetation have taken hold upon the dune surface.

At seashores unaltered by the activities of man, row upon row of dunes often reach far inland from the coast. The newest, most seaward dunes, almost free of vegetation, are blown inland toward their older, leeward neighbors. The second row back from the

In California's Mono Lake region, rock columns resembling stumps of petrified trees were carved by frost and wind from vertical plugs of lava that remained after a deep blanket of volcanic ash eroded away.

The Great Pageant of Erosion

The duel is unending—what the Earth's mightiest forces build up, the subtle powers of wind and water wear away, leaving weirdly sculptured forms scattered about the landscape

"Ceaseless dripping of water will wear away the hardest stone," observed the Roman poet Lucretius. While his insight remains just as valid today as ever, scientists now know that such action by water alone is just one of several forces that constantly erode the land. Streams and rivers do have the power to carry away loose soil. But they could not carve so quickly if the rock particles that give them their abrasive cutting edge were not loosened beforehand by the constant assault of weathering and chemical decomposition.

Weathering depends very much on climatic conditions—especially temperature and rainfall. Repeated rise and fall in temperature, as in deserts where nights are markedly cooler than days, cause alternating expansion and contraction of rock—and this stress makes rock crumble.

Moreover, water that has seeped into pores and crevices in rocks expands when it freezes, acting like a wedge that breaks up the rock, producing rounded corners from jagged edges and splitting big masses into smaller pieces. Plant roots also enter the cracks and pores of rocks, creating further splitting pressures as they grow. And where rocks on a cliff or hillside are weakened by the tools of abrasion, heating and cooling and plant growth, gravity exerts its inexorable force to pull debris down into fans of "scree" or "talus" at the base of a slope. As snows melt and rains fall, the debris manufactured by these processes is washed away into streams and rivers. Now its role changes, and the eroded material becomes a flowing abrasive agent that can scour canyons and cut river beds deep in solid rock.

Wave Rock sweeps skyward like a line of breaking surf, its granite wall shaped by moisture and polished by the sand of a western Australia desert.

In the foothills of North Africa's Atlas Mountains, water coursing down strips the soil and creates ranks of impressive gullies along the main stream.

The solid rock domes of Australia's Mount Olga were rounded by millions of years of weathering. Alternating heat and cold, aided by occasional rain, caused thin layers of rock to flake off.

109

Besides its role in wearing away land, water is
also a crucial factor in chemically decomposing
rocks. Almost pure as it leaves the clouds, rain
dissolves carbon dioxide as it falls and washes
out particles of salt and other solid matter. As
it runs along the ground, it dissolves other chemicals that make it more corrosive. The strength
of the acid or alkali varies according to the
soil chemistry. But even slightly acidic water
can dissolve marble and any other rocks held together by a cement of limestone, iron oxide or
other acid-soluble minerals. As the acidic water
attacks, it pits and etches the rock surface.
Then it percolates through pores and the seams
between strata—or into cracks caused by faulting—and can, in time, hollow out enormous caves.

Spectacular rock structures result from chemical
weathering and erosion. For example, the arches
found in North America's southwestern desert
(next page) are remnants of sandstones whose
cement weathered chemically, releasing the rock's
sand grains to be carried away by wind and water.

*Giant basaltic columns, formed when lava cooled,
rise 75 feet up a Mexican cliff exposed by erosion.*

*Blowing sand, cutting with great abrasive power
near the ground, sculpted these huge, mushroom-
shaped rocks in the Enchanted City of Spain.*

In arid lands, wind is a powerful erosion agent. It carries away loose bits of earth in dust storms, driving before it abrasive grains that have the power to sandblast whatever they strike. Usually, though, the wind raises blowing sand no more than a few feet off the ground. Thus wind-eroded structures are nearly always distinguishable from other types. Their upper portions, which lie above the sandblasting air flow, are generally bigger and less worn away than their bases.

Wind, water, ice—these tools of erosion tell but half the Earth's unceasing story of change. For whatever is worn away from a mountain crest or a canyon floor must be deposited elsewhere—as mud in the mouths of great rivers, sediment on the continental shelves, gravel on a valley's bottom, silt in a coastal plain, majestic sand dunes in desert areas. For countless ages, when the Earth's deep-seated forces of upheaval have erected new highlands, the forces of erosion have immediately set to work leveling the surface again—a self-renewing cycle that goes on and on.

Like a solitary face looking out to sea, this rock on a Faeroe Islands beach has been battered by the combined erosive forces of water, ice and weather.

Bourke's Luck potholes in South Africa's Transvaal were cut out by sand swirling in river currents.

Sandstones in the Arches National Monument, Utah, were weathered by frost, heat and cold, wind and rain.

Grand Canyon—
Rock Calendar of the Ages

By Wolfgang Langewiesche

The gigantic thing is hidden, so that you come on it suddenly. You are in high country on the Colorado Plateaus, but on a flat plain covered with fragrant pine woods. You might camp in these woods only a few hundred feet away from it and never suspect a thing. And then one evening you walk in that direction, and there . . . the earth opens up, the Grand Canyon.

Right at your feet is a gash a mile deep, four to 18 miles across, 217 miles long. And this is one of the peculiar things about the Canyon: its bigness shows. With a mountain, if you get close enough to make it seem big, you also have to look up at it slantwise and that makes it seem small. With the Canyon, when you stand on one rim and look squarely across, you see a mile-high cliff of rock and it looks gigantic.

More than two million people a year visit the Canyon. Some of them look for a while, but you can see it doesn't take. "Terrific, yes. But what does it *mean?*" They know they're missing something.

Other people—I am one—try to find the missing thing by legwork. I have walked down into it and climbed up out of it again. The main thing I found down there: it's hot! I've flown over the canyon and down into it. Main finding: scary. In the end, I found that all this poking around is unnecessary. The best way to see the Canyon is to do exactly what most people do: walk over to the rim and look.

Look at what? At layers of the Earth's crust that are normally hidden. When you stand there at the rim looking down, you see two things. One—the hole, which is an odd, terrific, scenic thing, unique. The other—what you see through the hole. And this, by contrast, is not unique. It is a typical cross section of the uppermost part of the Earth's crust. If you could cut the Earth open in France or England, Texas or Arabia, you would see much the same thing.

But the Canyon is such an odd and violent sight that you look at the hole, rather than at what it shows. At best, one asks, "What happened? What made this hole?" Well, let's see what it shows.

The Earth's crust is built up of layers of different rocks. At the Canyon one can clearly see 12 major layers, some red, some gray, some brown. Many of them are so regular, clear-cut and neat that they look almost artificial.

Toroweap Overlook offers one of the most breath-taking views of the Canyon. Here the almost sheer walls plunge 3000 feet down to the Colorado River.

The Colorado Plateaus in the vicinity of the Grand Canyon are etched by a network of tributary gorges, such as National Canyon, that become deeper and more precipitous as they approach the main chasm.

These layers are the stuff that settled out of the water of ancient seas. If you let muddy water stand in a glass, the water clears and the mud collects at the bottom. The same thing happens in a sea. The rivers bring sand and mud, which settle and form thick layers. In deeper seas the skeletons of fish and the shells of tiny sea animals sink to the bottom and, mixed with mud, clay and sand, form deep layers. These layers, through millions of years under their own weight, turn into rock. Sandstone is sand cemented grain-to-grain. Shale is former mud. Limestone is former seashell and other fine-grained lime material.

These Grand Canyon layers prove that this region, now a mile and a half above sea level, was under the sea not only once but several times. At one time the river brought red mud. At another the sea was deep, and limestone formed. At still another time it was a river mouth, and sand bars and beaches formed.

If you look down into the deepest part of the Canyon where the Colorado River is cutting a V-shaped gorge, you see that the rock is dark and quite different in texture from all the other layers. That's the oldest layer of land that went under the sea. Geologists call it the "basement."

The interesting thing is that this is not a local Arizona circumstance. It is worldwide. Most parts of the world have been under water at least once and have come up again encrusted with marine deposits. This thing happens slowly, over millions of years.

And the process is going on right now. The west coast of North America is slowly rising, the southeast coast sinking. Scandinavia is tilting, the northern part going up, the southern part going down. The Netherlands is sinking an inch a century.

Submerged in the Gulf of Mexico south of Texas is the site of the mysterious lost country of "Llanoria," as the geologists call it. The sea floor, which the Mississippi is now covering with mud, was once dry land. At some future time Llanoria may be high country again. Somebody may then point to a layer of rock and say, "This was once mud. It must have been brought by some river from a land that is supposed to have existed north of here."

As you stand at the Canyon rim looking at the

A New Gorge With Old Roots

Although the actual cutting of the Grand Canyon required only a few million years—a mere moment in the Earth's 4.6-billion-year life span—the orderly descent of the Canyon's rock layers to the Colorado River far below opens a unique vista into distant geologic history.

Over the eons, the Earth's crust in this area has seen some astonishing changes. Geologists read the Canyon's rock strata like the pages in a diary going back 1.7 billion years—the age of the oldest exposed rocks as determined by radioactive dating. They have found that at least two great cycles of mountain building, erosion and subsidence predate the Canyon's formation. At least twice before the cutting of the chasm, volcanic eruptions and crustal upheavals pushed up the North American Southwest into a mountainous plateau. Then erosion wore away the land, producing plains. Twice too, the region sank profoundly and was covered by ocean. When mountain building drove back the sea, the area was left thickly blanketed with marine sediments, including telltale plant and animal fossils.

The Colorado plateau of today was formed during a third period of uplift which proceeded slowly at first, then accelerated over a period of several million years. Meandering river systems started to cut deeply into the sedimentary rock beneath, carving the curved terraces that we see today. Averaging 300 feet in breadth, the Colorado River eroded only a narrow part of the Canyon, which is up to 18 miles wide. Weathering and the river's tributaries did the rest, with the Colorado River carrying away the debris. Today its turbulent, reddish yellow waters wash downstream an estimated million tons per day of mud, sand and gravel.

1. About three billion years ago the site of the Grand Canyon was washed by an ocean that deposited more than 25,000 feet of sediment. Volcanic activity produced dark bands of rock between sediment layers.

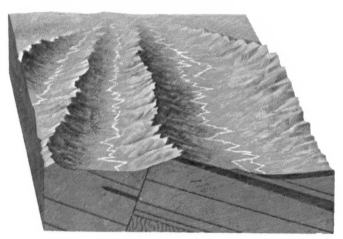

4. In a new cycle of upheaval about one billion years ago, sediments deposited by the second invasion of the sea were tilted by faulting in the crust and formed new ranges of block mountains.

stack of rock layers formed in the sea, there's a big idea waiting to be seen. It is this: when each layer was formed, the layer underneath it *was already there*. Of two such layers, the lower one is older. It's so obvious it's hardly worth saying—but 300 years ago it was an astonishing idea. It makes the stack of rock layers into a calendar. As you go down into the Earth you go down into the past. Look at the "basement" rock of the Inner Gorge and you're looking back through at least two billion years.

And now another vast perspective opens: buried in the rocks of each time are signs of the life of that time. Most of them are marine life, but you find also ferns, trees, insect wings. Covered more and more deeply through the years, these things petrify; the organic matter is replaced, cell by cell, with mineral matter. So fossils remain for us to study—a rock image of the once-living thing. An expert can look at a chip of rock under the microscope and date it approximately by certain characteristic fossils, the way you might date an old photograph by the models of the automobiles you see in it.

So the rock calendar is also an illustrated history of life on this planet! The basement rock down there

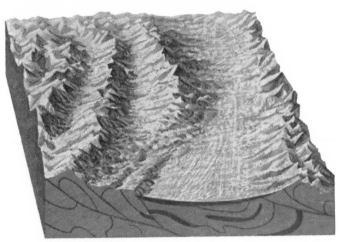

2. Crustal folding 1.7 billion years ago produced mountains whose deep-seated cores pressed upon layers of sedimentary rock, changing them into the hard "schist" now found at the bottom of the Canyon.

3. Over the next several hundred million years, the forces of erosion wore away the mountains and created low-lying plains. The ocean rolled in, as in illustration No. 1, laying down new sediments.

5. A long period of erosion leveled the second mountain formation and added new layers of sediment. The land rose again, speeding up in the last few million years, and the Colorado eroded its Canyon.

6. Today the land around Grand Canyon has risen more than 8000 feet. The river's ability to eat into the hard "basement" rock matches the continuing uplift; thus the Canyon is still getting deeper.

contains no signs of life. Was there no life in this region then? Have all its traces been destroyed? Not all the answers are in yet about the basement rock; in a few places some obscure traces have been found that seem to have been formed by bacteria.

Next layer up, scientists recognize algae, one of the simplest forms of life. (The green scum that forms on a pond consists partly of algae.) Next layer up, a leading citizen of the world was the trilobite. Next, strange fish with rigid shells instead of skins. Then, in most of the Canyon, there comes a big gap in the calendar; millions of years are unaccounted for.

When the story resumes, something big has happened: life has come out of the sea onto the land. The living creatures have now learned how to breathe air. They are still only lizardlike amphibians, but their tracks have been found, complete with marks that show where the little fellow dragged his tail on a mudbank. At the Canyon, that's where the story stops, because even the topmost layer is very old.

To the experts, then, the Canyon's layers of rock are like the pages of a picture book, depicting the past.

But the Canyon itself, this astounding piece of scenery; how was it made?

A farmer recognized the cause right away when he saw the Canyon. "Golly," he said, "what a gully!" It's erosion, all right. But there's a difference. On eroded farmland a stream forms and digs its way down, down, down. Here on the Colorado Plateaus the river was there first. And then the country rose up, up, up.

John Wesley Powell, ex-U.S. Army major and the first scientific explorer of the Canyon, was the first to understand this process. He went through the Canyon in a boat in 1869 and again in 1870. In the intervals between almost drowning and almost starving he looked at the stony puzzle and solved it.

The puzzle was of the kind, "What's wrong with this picture?" The river seems to have picked its course with complete disregard for the terrain. The Colorado Plateaus region is a plain. But this plain is really the flat top of a huge dome, a mountainlike upland raised thousands of feet above the surrounding country. The river comes out of the north, flows in a narrow cut right through the high plateau country and moves out into low country beyond. It should have flowed *around* the high plateau country. Powell's explanation was that the river must have been there first when all this was lowland. The country must have risen later, and the river stayed in its groove and sawed its way down as the land rose.

In ranks that match almost perfectly across the yawning gorges of the Canyon, layers of brick-red rock are sediments several hundred million years old.

One night, down in the granite gorge, I heard the river working. You hear a peculiar sound through the hiss and rush of the water—a sound like the clinking of marbles. That's boulders and pebbles, the river's cutting tools, rolling along the river floor with the current. That way the river has cut through a mile of rock. The time it must have taken! Yet this cutting was a quick job compared to what went before—the laying down of all that rock in the first place, a grain at a time.

That's the big thing you see at the Grand Canyon: time, how much time there is. The thing we feel we have the least of, there's the most of. It calms you down.

The awesome scale of even the smallest tributary chasm dwarfs a group of people winding their way along a ledge near the center of the picture.

Part Three

WONDERS OF THE LAND

The core of an old volcano, Ship Rock, New Mexico, endures like a monument in the American desert.

NATURE'S LAST FRONTIERS

Man explores but does not tarry long in the planet's more forbidding regions—thus their beauty and mystery remain while more hospitable lands are altered by man's activities

Earth's Awesome Belt of Green

By Marston Bates

The word most often used to describe the tropical rain forest is "cathedral-like." The comparison to a cathedral is inevitable: the cool, dim light, the utter stillness, the massive grandeur of the giant trees combine to justify this metaphor. The gothic details of the thick, woody lianas plastered against the trunks, or looping down from the canopy of leaves, add to the illusion. Awe and wonder come readily in the forest, and sometimes exultation—or, for a man alone, fear. Man is out of scale: the forest is too vast, too impersonal, too deeply shadowed. Here man needs his fellow man for reassurance. Alone he loses all significance.

Tropical rain forest occurs in a globe-circling belt extending roughly 20° N and 20° S of the Equator. There are three major areas: the South American, the African and the Indo-Malayan. It covers most of the land area in Equatorial regions, with the notable exception of a part of East Africa that is desert or semiarid savanna. The South American forest is by far the largest and most continuous, covering much of the Amazon drainage basin in north-central Brazil. It extends southward along the eastern foothills of the Andes in Bolivia to the drainage area of the River Plate and reaches north into Colombia to the drainage area of the Orinoco. This vast forest expanse is separated by the northern Andes from a Pacific coastal strip of forest that follows the shoreline north from Ecuador into Panama and continues, in Central America, along the Caribbean coast almost to the line of the Tropic of Cancer in Mexico. There is also an isolated stretch along the southern coast of Brazil. While rain forest once covered much of the West Indies, now there are only scattered remnants on the islands of the Caribbean.

The African rain forest is the smallest of the three, and there is considerable debate among naturalists as to its present limits and its former extension. It covers, essentially, the central drainage basin of the Congo, with a north and west extension along the Gulf of Guinea as far as Liberia.

The Indo-Malayan rain forest is the most fragmented of the three main areas. It covers the large islands of the East Indies—Sumatra, Borneo, Celebes (Sulawesi), New Guinea—as well as the Philippines and the Malay Peninsula. Outlying areas range across the west coast of India, Burma, the coast of South Vietnam and Cambodia and along the coast of northern Queensland and the Northern Territory of Australia. These are more properly called monsoon forests, as their regions are subject to a wet and a dry season.

In structure and appearance the forests of the three major areas are generally similar. The taller trees reach a height averaging about 150 feet, though individual specimens more than 200 feet in height are not uncommon. The tallest reported rain forest trees are somewhat less than 300 feet. Rain forest trees, in general, are taller than trees in the temperate forests of Europe or North America, where the average height for trees in the least disturbed forests is around 100 feet, with 150 feet an exceptional height. But trees in the tropical forest do not reach the gigantic proportions of the California sequoias or the Australian eucalyptuses. California sequoias frequently attain heights of more than 350 feet while the tallest Australian eucalyptus on record is said to have reached 375 feet.

The rain forest everywhere has a multistoried can-

The fierce light of the tropical Sun is all but lost as it filters down through mist and foliage to the floor of a monsoon forest in northern Australia.

opy. The layers of the canopy, like the depth zones of the sea, are hard to define since they are not sharply separated. It is customary, though, to refer to the upper, the middle and the lower zones of the forest canopy. This multiplicity of foliage layers is due to the great variety of trees that make up the rain forest. In contrast, most Temperate Zone forests consist of just a few species of trees, achieving a maximum in the Appalachian forests of the eastern United States of perhaps 25 kinds. On the other hand, 50 or so species of trees is the minimum for a rain forest. And in most places there are many hundreds of species.

The British naturalist Alfred Russel Wallace (1823–1913) spent many years exploring both the Amazonian and Malayan regions. In his book *Tropical Nature* he commented on the surprising distribution of trees within the tropical rain forest: "If the traveler notices a particular species and wishes to find more like it, he may often turn his eyes in vain in every direction. Trees of varied forms, dimensions and colors are around him, but he rarely sees any of them repeated. Time after time he goes towards a tree which looks like the one he seeks, but a closer examination proves it to be distinct. He may at length, perhaps, meet with a second specimen a half a mile off, or he may fail altogether, till on another occasion he stumbles on one by accident."

Because of this immense variety, the scientist's knowledge of rain forest trees is still far from complete. Plant families that in temperate zones are known only as herbs—the *Compositae*, or daisy family, for instance—are represented by trees in the rain forest. Even grass takes on the tree form in bamboo; and ferns, in the tree ferns. There is a great development of woody vines—lianas. A considerable proportion of the foliage in the canopy, sometimes nearly half of it, issues from the great vines that are supported by the trees, and these woody vines, like the trees themselves, belong to a variety of families and species.

Epiphytes, or "air plants," are characteristic life forms in the tropical forest. These plants perch high up on the branches and trunks of trees to obtain the light they need to grow. While lichens and mosses are familiar epiphytes in temperate forests, in the rain forest the branches and trunks of the trees are crowded with a bewildering variety: ferns, orchids, cactuses and bromeliads, including hanging mosses. In all, something like 33 types of seed plants and ferns are represented by epiphytic forms in the rain forest flora.

Without access to the soil, the epiphytes are faced with a water problem. Their special niche on dry branches high in the trees is, in a way, a sort of microdesert. Many of them, the epiphytic cactuses, for instance, are in fact relatives of desert plants and have their fleshy look. There is a good reason for this, since in both situations it is important for the plant to be able to conserve water; hence, the frequency of succulent bulbs and leaves, as in the orchids. The bromeliads have

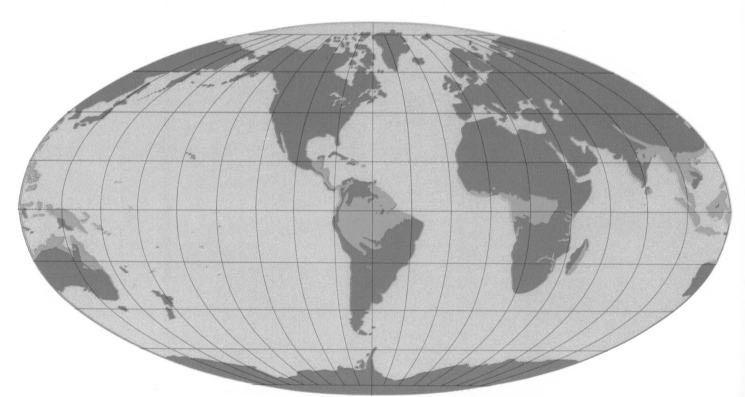

Rain forests and monsoon forests girdle the Earth in broad, green belts wherever extremely heavy rainfall combines with the year-round high temperatures of tropical and equatorial land areas.

122

In the never-ending struggle for a place in the sun, a rain forest tree sends up a branchless trunk and spreads a leafy parasol in the light.

Slit only by a jungle river's narrow, curving channel, the lush foliage of a South American rain forest forms a dense, impenetrable-looking canopy.

solved the water problem by forming from their overlapping leaves watertight bowls in which they can collect their own supply, and where they may also gather up rotting organic matter as food. The bromeliads, as someone has remarked, are often so numerous as to suggest a "marsh in the treetops."

The epiphytes are also faced with the problem of obtaining minerals. They must get the salts they need from the extremely dilute solutions in the rain that washes down through the canopy, or from the humus and debris that collect in the cracks of the bark of the host trees and in the tangle of their own roots. Somehow they receive sufficient minerals from these sources. Frequently the epiphytes live in a close, mutually advantageous "symbiotic" partnership with fungi; the weblike fungal growths form masses within the root hairs of the epiphytes, helping both partners in food collection.

The roots of epiphytes are used, with surprising frequency, as nesting sites by the ants that abound in the forest; it has been suggested that the epiphytes have an additional source of food in the material that the ants are constantly hauling into their nests, forming a

sort of soil around the epiphyte's roots. This too would be a symbiotic relationship—the ants receiving a well-protected home, built by the plant, and providing food to the plant's roots by way of rent.

The rain forest crawls with ants of many different kinds, occupied with many sorts of business. Close associations between particular kinds of plants and particular species of ants are quite frequent. Sometimes the ants clearly serve to protect the plants that provide them with nesting sites in hollow stems. One learns to avoid brushing against the trunks or foliage of certain trees, like Cecropia, with the same care that one learns to avoid poison ivy or stinging nettles in temperate climates. With the ant-protected trees the fiery consequences of transgression are immediate as well as painful.

The ants are one example of the incredible abundance of insects in the rain forest. It is an abundance of kinds, rather than of individuals. Within a range of about 10 miles of a laboratory in Colombia, my colleagues and I once found 150 different species of mosquito. Although you may get more mosquito bites in northern woods than in tropical forests, in the former,

A fern finds enough nutrients to flourish high off the ground in the crotch of a tree where organic debris has collected and formed a soil.

Rising from the rain forest floor where the dimness discourages the growth of leaves, a tangle of tree roots creates a grotesque, impassable thicket.

Roots descending from a strangler fig plant wind around a jungle tree trunk in a deadly embrace that will eventually enclose and kill the host tree.

the mosquitoes are apt to be from a large population all of the same kind, while in the rain forest, almost every bite will be from a different species of mosquito —if that is any comfort.

The largest known insects are found in the tropical rain forest: with respect to wingspread, butterflies and moths; for bulk, rhinoceros beetles; for length, walking sticks. But there are relative giants among almost all insect groups: cockroaches that look like small turtles, big flies, big wasps, monstrous grasshoppers.

Generally the mammals and birds represented in the forest are not particularly big. Frequently, in fact, for-

est mammals—forest deer and the forest-dwelling pigmy elephants of equatorial Africa, for instance— tend to be somewhat smaller than their cousins in the savanna or outside the tropics. Perhaps only the insects and the cold-blooded animals—anacondas and boas and pythons, for example—find a special opportunity for bigness in the rain-forest environment.

The color that is seen in the rain forest is mostly high in the canopy; in the flowers of the trees, lianas and epiphytes; in the birds and butterflies. There is, it seems to me, a special tendency for flying creatures, especially the birds and butterflies, to be colored in metallic blues and greens. The wings of the great *Morpho rhetenor* butterfly, native to the American rain forest, are dazzlingly brilliant when flashing in the sunlight. Many other kinds of day-flying insects have metallic colors, even mosquitoes! However, species that fly at night and species of the forest floor are apt to be dull. With the butterflies, as with the birds, the brilliance tends to be a male characteristic.

Among the rain-forest fauna are many relics of the past. In the South American rain forest, for example, there are dozens of species of marsupials. They are not as spectacular as the kangaroos and their other Australian relatives, but they are nonetheless interesting. There is a sleek water opossum living on the margins of the forest streams; a bright-eyed woolly opossum with lovely, thick fur; and many different species of tiny, mouselike opossums. The sloths are another archaic group, now confined to the trees of the American rain forest. In the recent geological past, the sloths were not restricted to the forest canopy, but ranged over the ground as well.

The catalog of ancient animal types surviving only in the rain forest is a long one. Because of this, the rain forest has sometimes been regarded as a sort of biological backwater, a refuge from the more strenuous and progressive evolutionary pressures of the biosphere. But the rain forest is, in fact, anything but a backwater; rather, it is the place where evolutionary change is most active.

It is in the rain forest that the struggle for existence has produced some of the world's most extreme adaptations. The strangler fig, for example, starts out as an epiphyte, a seedling growing high on some tree. It sends down roots which reach the ground and grow until finally the host tree is smothered by the encircling fig, which then stands alone. And I can think of nothing more devastatingly fierce, more irresistible or more efficient than a horde of army ants on the move, killing and picking clean the bones of every animal they encounter that cannot fly away or outrun them. We had a snake pit in the laboratory in Honduras where I spent some time, in which we kept many kinds of vipers for their venom, which was used to make serums to treat snakebite. Once, I remember, we were invaded by army ants. We tried everything we could think of to stop them or at least to make them change their course, but to no avail. The ants poured in by the tens of thousands, swept through our snake pit, and left us with a collection of bare skeletons.

In one important sense, I suppose, the tropical rain forest does represent a refuge of sorts. One gets the feeling that there is so much constant warmth, so much light (at least in the canopy), so much moisture that almost any organism can survive. Organisms must always cope with other organisms; but outside the rain forest, they must increasingly cope with climate, too.

How Life Defies the Desert

By Edwin Muller

What is a desert? Looking into my dictionary, I found this definition: "a place left unoccupied . . . a region lacking moisture to support vegetation." Another source in speaking of deserts uses the terms "barren," "uninhabited" and "forsaken." Yet in every desert are living things that defy the dictionary as well as the desert itself. In fact, if you would look for supreme examples of the way life can endure, go to the desert and you will be rewarded.

Picture a sandy stretch of the Arabian Desert where no rain has fallen for five years. The terrain appears desolate and devoid of all life. Then one day storm clouds gather and a tropical downpour follows. Afterward the water drains off, leaving the surface of the sand moist. Now, as if by magic, the desert breaks out in a sheen of green grass. Millions of tiny flowers also appear, painting the dunes with bright patches of color. The plants grow with remarkable swiftness—from seed to blossom to seed in one week. When the new seed falls to earth, it lies in the sand—to wait perhaps another five years for rain, and the brief repetition of its life cycle.

Flowering annuals survive in the desert because their seeds lie dormant in sand and dust through long periods

Occupying a windswept plateau in the Andes, the rocky Atacama Desert of northern Chile is as much a no-man's-land as deserts made of searing sands.

of drought, sprouting only when a rain sets in motion the chemistry of germination. By this device, they avoid the whole problem of growth in dry seasons. Grasses and other desert annuals cling to survival in this same way. The seed of the *ashab*, the favorite grass of the camel, may lie dormant ten years, finally sprouting when a rain comes. The "belly plant" of the deserts of the western United States (so small you have to lie on your belly to see it) also has seeds that lie dormant for years, sprouting after a rain drenches the sands.

But there are other desert flora—the perennials—that do not avoid the long season of desert drought; instead, they have developed special ways of obtaining moisture and conserving it. Such plants make extraordinary efforts to reach water. The low, shrublike, flowering tamarisk tree of Mediterranean and Asiatic desert regions may send its taproot down 100 feet. The giant cactus or saguaro of the western United States and Mexico may have a taproot only three feet deep, while sending out other roots horizontally 90 feet in all directions. Still other plants derive sufficient moisture from dew. They absorb the moisture through their leaves, excrete the surplus through their roots, and reabsorb it from the root soil as needed.

Cactus plants have an efficient water-storage system in their thick stems; one species of tree cactus may contain hundreds of gallons of moisture. Cactus plants have lived six years without rain. One cactus native to the southwestern United States and Mexico, the bisnaga, has been called the "well of the desert" because of the ready accessibility of its stored water. The center of the spiny green globes and barrels can be cut out, forming a bowl that quickly fills with water from the plant's tissue. Desert travelers who have not known about the bisnaga have died of thirst with these bowls of water available all around them.

Like the plants, animals have demonstrated a remarkable capacity for adapting to the harsh environment of the desert. The camel, that familiar symbol of desert lands, can, after drinking its fill, exist on the water in its body for seven to nine days. It has a good instinct for finding water. A nomad lost in the desert may give his camel free rein, letting the beast have the lead until it reaches a water hole.

A species of frog in the Australian desert stores water in its abdominal cavity, bloating itself into a spherical shape. During a drought the frog burrows a foot deep into the soil and there it stays, baked in by the Sun, until the next rain. Among the insects, certain desert ants store water in the stomachs of specialized worker ants. The bellies of the latter become enormously distended and supply the colony in times of drought.

Man, too, has learned how to adapt himself to desert life. When in their wanderings Bedouins come to a spot where a recent rainfall has caused grass to spring up, they stay until their grazing animals have eaten every

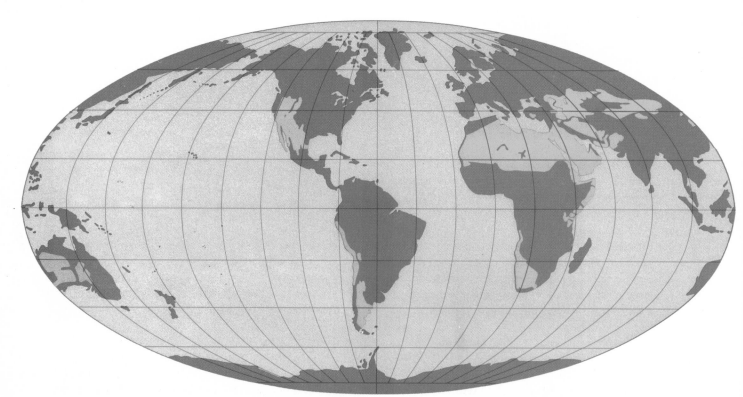

About a seventh of the Earth's land is desert (yellow areas). Not all deserts are hot; Asia's Gobi extends far north into Mongolia, while the coastal deserts of Peru and Chile rise into cold plateaus.

126

Giving no quarter, massive drifts of sand overrun an oasis in the Sahara. Stretching from the Atlantic to the Red Sea, this desert covers some 3.5 million square miles—almost the size of the United States.

blade, then move on. Over thousands of trackless square miles these nomads have memorized nearly all of the permanent landmarks. A Bedouin may have seen a certain hill only once in his life, yet remember its contours well enough to recognize it in the future— and the distance and direction from it to the nearest lifesaving water hole.

Summertime temperatures of 120° F are common in the Sahara. The world's highest known air temperature was recorded in the Libyan sector of the Sahara, 136.4° F in the shade. The sun-baked surface of the sands may reach temperatures 50 degrees hotter than the air. The Bedouin has devised the best clothing for protection against this heat—thick, loose, voluminous garments which trap air that insulates the body.

It would be incorrect to picture deserts mostly as sand. Less than one sixth of the Sahara is sandy. A more typical surface is gravelly hardpan, hard as iron, covered with fist-sized rocks. Nor, obviously, are desert terrains always flat or smoothly modeled with great, undulating dunes. In its central regions, for example,

the Sahara piles up with a jumble of fantastic rock peaks.

In the Sahara today there is much travel by car, but this is dangerous. Water holes are often 100 miles apart, and men lost in the sun without water for just one day from dawn to dusk have sometimes died of dehydration and sunstroke. In some locales, if one's arrival at the next water hole is not reported by radio within 24 hours of departure from the last stopping place, a car is sent out on a search. Desert regulations require the tourist to carry in his car spare auto parts, shovels, wire netting, pails and rope. The length of the rope can be a matter of life or death, as evidenced by the fact that men have died of thirst at the edge of a well where their ropes proved too short to permit their pails to reach water.

There are deserts in five of the six inhabited continents (only Europe has no desert). To rate scientifically as true desert, the land must receive less than 10 inches of annual rainfall. About 14 percent of the Earth's land surface is covered by true desert. (An-

After a brief rain, parched desert earth cracks into a harsh mosaic under the blazing rays of the Sun.

other 14 percent is covered by semidesert and steppes receiving no more than 20 inches of rain per year.) Some desert regions are located in the hearts of continents where the winds that reach them are dry—already having given up their moisture as rain in coastal regions. Deserts also form on the leeward sides of mountains that cause air currents to rise and cool, shedding most of their moisture on windward slopes. Other deserts lie in tropical belts where the prevailing winds are dry winds.

Man, too, has contributed to the formation of deserts—or at least to their expansion. Most true desert regions are surrounded by a broad fringe of semiarid land, where there is enough rainfall to support grass, bushes and some trees. Frequently these semiarid desert margins are invaded by nomadic herdsmen who cut the trees and scrub growth for fuel while their flocks overgraze the land; hence vegetation disappears. In time, the soil bakes dry and the wind carries it away. Thus the fringe becomes true desert, and the nomads move on unwittingly to destroy the next fringe.

Some nations are now experimenting to find ways of reclaiming desert lands for forestry and agriculture. Several species of tough desert plants may prove helpful to man's efforts. These plants hold moisture in the ground and slowly build up the soil by adding organic matter to the sand. The roots hold the newly forming soil, protecting it from erosion by winds and flash floods. Later, belts of trees can be established. They act further to build soil and retain moisture. In addition, modern technology promises new approaches to desert reclamation. Experimental, petrochemical surface-coatings that block the evaporation of water from the soil are now being tested in desert regions.

Graveyard of Vanished Water

To some, the "Pinnacles" (right) along the coast of Western Australia north of Perth are reminiscent of church spires. But local residents call them the "tombstones," and, in a geological sense, this is what they are. Once this was an arid coastline with large dunes. Then the climate changed and heavy rains fell. As the rainwater washed through the dunes, it dissolved calcium in the sand. The mineral-laden water collected in low spots and drained away, forming underground "pipes" as the calcium was redeposited along the walls of these drainage conduits. In time, the land became arid again and the wind stripped away the sand around the limestone-lined pipes, leaving them as weather-sharpened memorials to the water that once was here.

Shrinking in perspective against the stark backdrop of Afghanistan's Hindu Kush Mountains, a walled oasis offers refuge from the winds in a cold desert where temperatures plummet far below freezing.

Of course, the essential need in fighting the desert is water. Today, the oldest method of obtaining water is still the most successful method—bringing it from a river through canals or irrigation ditches. This is how Egypt has lived for 5000 years. From the time of the Pharaohs the floodwaters of the Nile have been drawn off through a network of canals and ditches and spread over the fields.

Engineers in Mesopotamia (present-day Iraq) criss-crossed canals back and forth between the Tigris and Euphrates rivers, turning the desert into lush wheat-growing country that supported many more people than live there today. Recently Iraq began devoting a large part of its oil revenues to restoring and improving this ancient irrigation system that has lain in disuse since its destruction by the hordes of Genghis Khan 800 years ago. And in Libya more than 200 ancient storage cisterns have been cleared out and are today functioning as well as they did 2000 years ago.

Modern man is developing a new method of getting water in the desert—tapping underground reservoirs.

In the course of centuries, desert rains seep down, perhaps thousands of feet, until the water reaches a layer of porous rock, where it is held as groundwater. In the Sahara there are two enormous underground reservoirs formed by groundwater. At Zelfana, in the northwestern Sahara, one was reached by drilling a well 4000 feet deep, creating a flourishing oasis.

One of the most remarkable underground bodies of water is a subterranean "river" flowing under the Nile. This water exists in strata of porous sediment of varying thickness that lie under the river bed. At depths ranging between 300 and 900 feet, the subterranean river follows the Nile's course for 560 miles, from near Luxor to the Delta. Attaining a width of up to six miles, its storage capacity is much greater than that of the river above it.

More than 20 years ago spectacular use was made of this "hidden river" by an Egyptian statesman and financier, Hafez Afifi. North of Cairo, Afifi acquired 300 acres of desolate, sandy desert, then bored two wells down to the underground water and installed powerful pumps. He plowed manure and other fertilizer into the sand. Around his villa he soon had green lawns with fountains; his orchards and vineyards produced oranges, lemons and grapes; his new fields of clover supported a herd of cattle; bumper crops of grain and vegetables were forthcoming.

But to make the desert fertile by pumping up subsurface water, of course, requires power. Fuel for conventional power is expensive to transport to remote areas, so man is experimenting with other means of obtaining energy—among them the use of windmills and solar power generators. And eventually one of the most important uses of atomic energy may be to make desert areas fertile through the production of electrical energy for pumping.

So in time the hot, dry deserts—which occupy one square mile out of every seven of the world's land surface—may help solve one of mankind's most urgent problems: how to increase food production. When this happens, the deserts of the world will no longer represent a peril to man but a new cornucopia of agriculture, an ally in time of need.

Australia's Lake Eyre is really a desert most of the time. The rare flash-flood waters that reach it evaporate quickly, leaving a harsh salt crust.

The Rooftop of the World

By Barry C. Bishop

Tibetan Buddhists call her *Chomolungma*, "goddess mother of the world." In Nepal, the Hindus refer to her as *Sagarmatha*, "summit above the oceans." Both seem more fitting for the Earth's highest peak than Mount Everest, the name given her by British officials in honor of Sir George Everest, the first Surveyor General of India. For this mountain is the greatest jewel in the crown of the most spectacular mountain range in the world—the Himalayas.

Pushing skyward to 29,028 feet, Mount Everest's summit is so lofty that it penetrates the jet stream. Winds that sometimes reach 200 miles an hour blow a perennial plume of snow from her peak, while lower down, expanses of glittering glacial ice and bare rock give Everest a beauty that is both tantalizing and forbidding.

For thousands of years various peoples have lived in the shadow of Everest and her sister peaks in the Himalayas—ancient Sanskrit for "abode of snow." The mountains permeate the mythology, religion, literature, politics and economy of the region. Travelers, explorers, mountaineers and scientists like myself have been equally fascinated by the Himalayas. Like all men before us, we have pondered such mind-boggling questions as: Why are the mountains here? What was here before them? How old are they? What tremendous forces produced them? And just why are they so high?

Towering above the low alluvial plains of the Indus, Ganges and Brahmaputra river systems, the Himalayas stretch in a broad arc for 1500 miles and separate the subcontinent of India from the Tibetan Plateau. In breadth the Himalayas span 100 to 250 miles. Within this rugged belt the mighty range builds through successive zones. The Siwalik Hills, the first wrinkles north of the Indian plains, rise 3000 to 4000 feet. The Lower Himalayas contain successive ridges that rise 12,000 to 15,000 feet. The Great Himalayas themselves culminate in ice-carved crests four to five and one half miles high. Beyond are the lower Tibetan Marginal Mountains which descend and merge with the 15,000-foot Tibetan Plateau. In northern Kashmir the Himalayas merge with the towering Karakoram, a great trans-Himalayan range with the world's second highest summit, the 28,250-foot K2.

The stupendous arc of this mighty mountain chain is so great that it could completely encircle the European Alps. Its crest averages more than 19,000 feet. Moreover, the Himalaya and Karakoram together boast over 500 peaks above 20,000 feet. And more than 100 of these exceed 24,000 feet! North America can claim only one 20,000-foot peak, Alaska's Mount McKinley, while the highest summit in Western Europe, Mont

In this view into Nepal from shadowed Singalila Ridge in India, 28,208-foot Kanchenjunga, the world's third tallest peak, appears as the apex of a colossal Himalayan crown gilded by a sinking Sun.

Blanc, in the French Alps, rises up a mere 15,781 feet.

For the few geologists who have actually roamed and researched the Himalayas, working out the details of this mountain-building epic has not been easy. Both political restrictions and the rugged environment have impeded their work and contributed to an uneven understanding. Immense scales of space and time also have conspired to present a multi-dimensional picture puzzle, for compression, uplift and erosion have varied from place to place and from time to time. Ferreting out the correct sequence of events is further hampered by a dearth of age-indicating fossils and complicated by confusing rock structures.

Although geologists often have different interpretations of different sites and continue to argue over particular points, most have long recognized that the series of Eurasian mountains that range from the Alps to the mountains of Southeast Asia, including the Himalayas, were formed by forces that reached a climax within the last 65 million years. All the ranges in this system are products of titanic crustal upheavals that uplifted thick layers of marine-deposited rocks from a deep, ancient oceanic trench that geologists call the Tethys Sea.

But what primeval forces could produce such a mammoth upheaval as the Himalayas? Today most geologists look to continental drifting, an idea first propounded by German geologist Alfred Wegener more than 50 years ago.

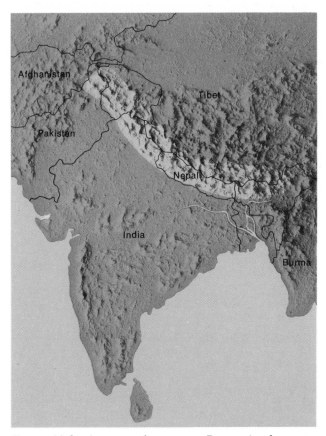

From Afghanistan on the west to Burma in the east, the Himalaya Mountains form a great wall separating the Indian subcontinent from the continent of Asia.

131

About 180 million years ago the Tethys trench bordered the entire southern fringe of Eurasia. The old southern supercontinent of Gondwana was breaking up and its fragmented, platelike segments were on the move. After splitting away from southern Africa, the Indian subcontinent pursued a northward collision course during the ensuing 100 million years, and the Tethys trench became gradually confined in a giant pincer with India to the south and the rest of Asia to the north.

As the relentless crunch continued with increasing intensity, compressive forces built up, filling the former seaway with crumpled sedimentary rock forced up from the ocean bottom.

The critical collision of the Indian plate with the Eurasian plate occurred 65 to 70 million years ago. Despite its immense power the Indian plate could not overcome the resistance of the Eurasian plate. And so the Indian plate sheared downward into the Tethys trench at an ever-increasing pitch.

During the next 30 million years shallow parts of the Tethys Sea gradually drained away as the sea bottom was pushed up by the plunging Indian plate. Ultimately parts of the Tethys Sea became the Tibetan Plateau. The Marginal Mountains on the southern edge of the Plateau became the region's first major watershed. The mountains rose high enough to become a climatic barrier which caused heavier and heavier rains to fall on the steepening southern slopes. With increasing energy the headwaters of the major rivers eroded along old fracture lines and fold structures, capturing the streams flowing onto the Plateau and laying the foundations for today's drainage patterns.

To the south old estuaries of the Arabian Sea and the Bay of Bengal rapidly filled with debris carried down by the ancestral Indus, Ganges and Brahmaputra rivers.

70–65 million years ago: What is now Mount Everest (yellow symbol) was sediment in Tethys Sea between advancing Indian plate (left) and Asian mainland.

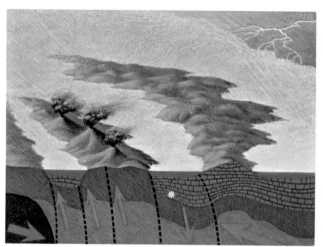

60–30 million years ago: The area in the yellow square, enlarged above, begins to thrust up under pressure from the still advancing Indian plate.

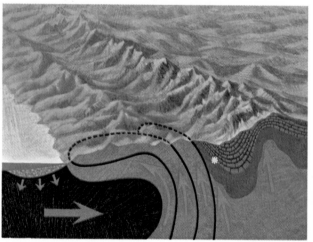

30–2 million years ago: The Himalayas emerge as more rock squeezes up and erodes (dotted lines). The Tethys Sea remnant becomes an isolated basin.

Today: Everest still rises slowly as pressures continue in underlying crustal plates. Tethys sediments have now thrust up 29,028 feet above sea level.

A dazzling plume of snow trails off in the fierce, jet-stream winds that sweep the naked summit of Mount Everest. The wavy layers of rock testify to Everest's birth beneath the sea—they are ancient sediments laid down in the ocean before continental drifting drove the Indian subcontinent into Asia.

Completely surrounded by peaks that tower 20,000 feet or more, the sun-dappled Buddhist monastery of Thyangboche nestles serenely before the white horn of Ama Dablam in the Everest region of Nepal.

Extensive erosion and deposition continue even now as these rivers carry down more than three million tons of material each day.

In a crescendo reached some 30 million years ago, the tempo of mountain building sharply increased and the Himalayas began rising up in earnest. The former seaway closed altogether. As the Indian plate continued to push down into the Tethys trench, its topmost layers of old rocks peeled back over themselves for long horizontal distances to the south. Such overthrust waves of rocks are called nappes.

Wave after wave of nappes were thrust back over the Indian land mass as far as 60 miles to the south. Each new nappe contained rocks older than the last. In time these nappes became folded like pleats in an accordion, closing up the former trench by 250 miles.

All the while, downcutting by the rivers almost matched the rate of uplift. A tremendous quantity of weathered material was eroded from the rising Himalayas and carried to the plains where it was dumped by the Indus, Ganges and Brahmaputra rivers. The weight of this sediment created a depression which in turn could hold more sediment. In some places the sediment under the plains of the Ganges is now 25,000 feet thick.

The early Himalayas were probably about 12,000 feet high, like the Alps today. Just when did the Himalayas become the highest mountains on Earth? Only within the last 600,000 years, when glacial ice was sometimes much more extensive.

The Himalayas reached maturity in a burst of fantastic uplift. Along the core of the northernmost nappes—and just beyond—young crystalline rock, formed from molten magma below the Earth's crust, was forced up to produce the staggering granite crests we see today. On some of the peaks, such as Mount Everest, fossil-bearing sediments from the old Tethys Sea were carried piggyback to the summits by the new crystalline rocks.

Toni Hagen, a Swiss geologist who surveyed Nepal in the 1950s, believes that the recent rise of the giant crystalline backbone of the Himalayas was due to the continued compressive crunch of the Indian crustal plate which forced up rock in this core area "like squeezing toothpaste from a tube." Other geologists feel that the spectacular rise of the crystalline peaks was more a response to the Earth's never-ending tendency toward "isostatic" balance: If one area of the Earth's crust sinks down, another tends to rise up. Augusto Gansser of the University of Zurich suggests that this is why the highest peaks occur opposite the greatest depth and weight of Ganges' alluvium.

Are the Himalayas still growing? We cannot be certain because surveying techniques have not yet provided accurate enough measurements. However, we do know that the Earth's crust is still active and moving. Violent earthquakes in the Lower Himalayas and Ganges Basin bear witness to deep-seated crustal activity. The terrible earthquake of 1934 that killed 10,000 in Nepal and Bihar had its epicenter below the Ganges oceanic trench due south of Everest.

As the highest—and most fascinating—mountain in the world, Everest has had an irresistible appeal to all mountaineers. Since 1921 more than two dozen expeditions have striven to reach her pinnacle—British, Swiss, Indian, Chinese, American, Japanese and international teams. Before 1950, when the Kingdom of Nepal finally opened its doors to the world, the way to Everest lay through Tibet. On the British expedition of 1924, geologist Noel Odell, climbing above 27,000 feet on the treacherous north face, first discovered that *Chomolungma's* summit pyramid was made of Tethyan fossil-bearing limestone, 350 million years old.

Except for the Chinese expeditions, all attempts to reach the summit since 1950 have been through Nepal. Because of the obstacles imposed by her forbidding height, Everest repulsed all assaults until the 1953 British expedition in which New Zealander Edmund Hillary and Sherpa Tenzing Norkay reached the top. Since then Swiss, American, Indian and Japanese teams have been successful. As a member of the American effort in 1963, I was fortunate enough to stand for a moment on her snowcapped crest.

The 180-mile march from Katmandu, Nepal, to the foot of Everest is either up or down, hot or cold, wet or dry, exhausting or exhilarating. The painful process of getting into shape sometimes makes you forget that this is one of the most pleasant periods on the expedition. Traversing the weathered nappe surface of the Lower Himalayas, you soon sense that man is a modern modifier of the landscape. Most slopes are terraced with farmland. But where deforested hillsides have not been cultivated, erosion runs rampant. To the north the Great Himalayas appear deceptively close, their ridges and buttresses, valleys and glaciers often opalescent in the sunlight. Somehow they seem to tower suspended in space.

However, it is only after reaching Khumbu Valley, home of the Sherpa mountain men, that you can get the first good view of Everest. From the commanding position of Thyangboche Lamasery at 13,000 feet on a great glacial moraine above the Imja Khola gorge, Mount Everest's black wind-swept summit pyramid rears ominously behind the stupendous wall of sister peaks, Nuptse and Lhotse. It is difficult to grasp the significance of Everest's height at this point, for the scene is dominated by the magnificent blunt spear of a lower and closer mountain, Ama Dablam.

On our 1963 expedition we let our bodies acclimatize for a few days at the idyllic location of the lamasery. We moved on when signs of mountain sickness—nausea, vomiting, headaches, loss of appetite, insomnia—disappeared. By the time we reached the rubble-strewn snout of the Khumbu Glacier at 15,000 feet, trees and habitation were beneath us. Now we entered a sterile environment of snow, ice, rock and wind that would be our home for the rest of the expedition. Walking up-

stream on the natural glacial highway against the gentle current (the glacier moves only a sluggish 150 feet each year), our route, called Phantom Alley, passed beneath giant pinnacles of ice that sometimes rose 85 feet above the rest of the glacial surface. This eerie icescape is formed by intense melting and evaporation.

The head of the Khumbu Glacier lies in a high "cirque" or circular valley surrounded on three sides by Everest, Lhotse and Nuptse. Called by the British the Western Cwm (pronounced "coom"), a Welsh word meaning high mountain valley, this great cirque has been eroded from an old weakness in the geological structure. At 20,000 feet the glacier tumbles from the Cwm as a 2000-foot icefall that moves about three feet a day.

Just below at 18,000 feet, we established base camp. This is probably the highest altitude to which a healthy man can adapt permanently. Here atmospheric pressure is only half that at sea level. On the summit of Everest it is only one-third. Above 18,000 feet, fatigue, weight loss and general physical deterioration become inevitable from the lack of oxygen. This plus cold and wind are the chief stresses when working or climbing at extreme altitudes.

The Swiss nicknamed the Western Cwm the "Valley of Silence." True, its precipitous flanks protect it from the wind. But quiet it is not. At night a cacophonous symphony of screeching summit gales harmonize with the roar of thundering avalanches and make sleep all but impossible. Thoughts of Everest's harshness and hostility come easily at night on the Western Cwm. George Mallory, who lost his life in 1924 on Everest's

Dwarfed by the monumental Himalayan summits that loom everywhere, a party of climbers picks its way carefully along a treacherous cornice of ice and snow.

north face, once wrote a friend. . . . "It's an infernal mountain, cold and treacherous. Frankly the game is not good enough. The risks of getting caught are too great; the margin of strength when men are at great heights is too small. Perhaps it is mere folly to go up again. But how can I be out of the hunt?"

At an altitude of 23,000 feet we began using supplementary oxygen. Our success and safety depended on how well we reduced the effects of oxygen starvation. Soon our crampons scratched a landmark of tawny rock, the Yellow Band, one of Everest's ancient Tethys sediments. Now we climb higher into a world where man cannot remain for long.

Finally the white snow cornice of the summit ridge pushes up to meet a blue-black sky. The wind whips at us. We push back. It abates for a moment and we fall, then struggle on. Eight breaths and then one step. The body cries out for and wants relief, but the mind has locked on to that point in space like radar. At last there is nothing higher.

To the north stretches the purple-brown Tibetan Plateau. To the south we look down now on the "abode of snow." Beyond, the plains of India lie somewhere below the haze. With the euphoria of altitude, my mind jumps from thought to thought: "I'm standing on the same latitude as Tampa, Florida. . . . Somewhere beneath my feet are the highest fossils on Earth, and they have traveled much longer and farther than I. . . . How small and insignificant—and tired—I feel!"

Two climbers are photographed by author Barry C. Bishop as they reconnoiter a route to Everest's summit during the successful 1963 U.S. expedition.

The Arctic—Unspoiled, But Not for Long

By John P. Milton

Between July 22 and August 29, 1967, three conservationists—Kenneth Brower, Steven Pearson and myself—walked from the southern foothills of Alaska's Brooks Range across the high mountains of the Arctic Divide, then down across the open tundra to Barter Island on the shores of the Arctic Ocean. Much of our route had no record of earlier exploration.

On this 300-mile walk, my own purpose was to blend into the soul of these Arctic mountains and plains, to sense the moods and manifestations of the land. Because of the recent, expanding exploitation of Alaska's North Slope oil deposits, we may be among the last to see this Arctic wilderness in all its unspoiled majesty.

These excerpts from my Arctic journal open on the day prior to the beginning of the 39-day trek through the Alaskan wilds.

FRIDAY, JULY 21, 1967

After three rainy days in Fairbanks, we are about to leave for Last Lake near the upper Sheenjek River.

Our 250 pounds of gear are now loaded in the little plane that will carry us to our remote starting point, Last Lake. For the next month or so—hopefully—we will see no other men. We will be alone, on our 300-mile hike, to take in the wilderness of the Arctic Wildlife Range. To our knowledge, no other men have made this south-north journey through the high mountain passes of the Brooks Range. From now on, our life will center on the immediate horizon.

SATURDAY, JULY 22

We are here. My sense of space and distance is greatly expanded by this country. It is incredibly beautiful. The broad valley of the Sheenjek curves through a jumbled series of steep side valleys and slate-gray mountains practically barren of visible vegetation. The valley floor is green and lush, with numerous patches of open forest breaking up the grassland and wet muskeg. To the north, at the head of the Sheenjek, a band of snowy mountains reach skyward 6000 to 7000 feet. (At Last Lake the valley floor has an elevation of 2500 feet.) Over all this, a wonderfully clear atmosphere lets the eye wander on and on, drawing it deeper into the hidden mystery of far valleys and mountain passes. The Sun, wheeling slowly in a great circle, casts a deep, warm glow over the wilderness. For only a few hours does the Sun disappear behind the northern mountains, casting a fiery sunset all "night" long. I cannot walk far in any direction without wanting to stop and look about me. We will camp here for several days.

A sudden shift of mood and weather is in the wind as a midsummer thunderstorm builds above Alaska's Brooks Range, whose bare peaks brood over a vast, still virgin expanse of North American tundra.

SATURDAY, JULY 29
Although it's nearly midnight, I can still write by the evening light. We find abundant signs of wolf, lynx, Dall sheep, moose, grizzly, caribou and smaller animals all around us. Ken reported seeing a yellow-haired porcupine that looked just like a small bear. "When I talked to him," Ken told us, "he faced me and wasn't afraid. When I came back for a second look, he grew suspicious and waved his spiny tail at me." Ken and I sat late photographing a colony of beavers.

Yesterday I walked a mile south from Last Lake to frozen "Ice Lake." The lake is cut across by a flow of clear water which slices a serpentine course through bluish-white ice walls eight to ten feet in height. A small ice cave sheltered me when a storm blew up. The ice walls reached out over me, not quite meeting. The melting inner walls, drizzling steadily, were neatly scalloped—as though by some fine craftsman's chisel. After the storm passed, I left and plodded back across the brownish moorland to camp.

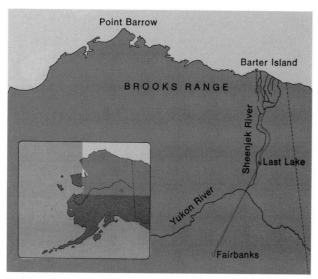

The author and his companions flew to Last Lake. Then they trekked 300 miles, through mountain passes and empty tundra, to the Arctic Ocean's shore.

FRIDAY, AUGUST 4

Today, exploring side valleys, we came to a high cliff with a talus, or sloping deposit of fallen rocks, at its base. This slope dropped steeply into a tributary of the Sheenjek River, which flows south from the Arctic Divide to feed Last Lake. We climbed 800 to 1000 feet to the top. Below, the valley, bounded by impressive ridges and peaks, curved gently to the west. The river wound along the center of the valley, its meandering bed marked by large gravel bars.

Arctic mountain valleys bear strong personalities that stick in the mind. And so it is here, in Nameless Valley. Arctic mountains are distinguished by an ever-present mood of impending if not actual darkness, cold and snow. The short duration of the Arctic summer only intensifies an awareness that here the gift of warmth is only a brief respite from cold, that light will soon be followed by a deep and long-lasting darkness. This mood seems to dominate the land and every living thing in it. The mountains also know the cold of glaciers; ice wedges carve their slopes and chip away the cliffs, causing great cones of talus to pile up below. In these mountain valleys, summer seems but a fragile moment of hesitation before the onset of another dark, cold cycle.

Walking back to camp along a high ridge of rock in the valley center, I was struck by the dual qualities of this wilderness. On the one hand, I marveled over tiny and numerous details: red mosses, wild flowers and gurgling mountain rivulets. But these delicate details were set against the larger scale of grizzled mountain slopes; the vast expanses of wide valley; far, snow-flecked peaks, clouds and sky—all contributing to an intoxicating sense of the monumental in nature.

TUESDAY, AUGUST 8

The hour is 10 P.M. and it is snowing steadily. We are camped at the junction of the two streams that combine to form the headwaters of the upper Aichilik River.

By five today we were all chilled, with freezing-cold feet and hands. We set up camp in record time. After stripping off my wet trousers inside my tent, I changed into warm socks and crawled into my sleeping bag. Dinner—frozen pork chops, sliced potatoes and vegetables—was cooked by Steve. He had set up our portable stove in his tent. That little stove has been doing very well by us since we crossed the Arctic Divide; recently we've had no opportunity to build a wood fire (either there has been no fuel or the weather has been too wet).

My ankle, injured the other day, feels a little better,

The Handwriting in the Arctic

Conservationists warn that oil-company construction may irrevocably ravage the Alaskan tundra. Its vulnerability can be seen in the photograph at left. To mark the spot for a supply drop, a bulldozer operator scraped the letters "GSI" into the permafrost. His blade exposed the ground only to a shallow depth. But in a chain reaction of melting and erosion, the summer heat penetrated deeper and deeper, destroying the soil's rigid cement of ice. Soon the 150-yard-long letters had subsided into deep, watery trenches that may not fill in again, say some ecologists, for 500 years.

Primroses blooming in the gravel along a tundra stream grace the gay palette of Arctic summer.

Unlike other species, both male and female Barren Ground caribou grow antlers during mating season.

although I was in almost constant pain for the last two hours of hiking. I've noticed I prefer to walk either last or first and to have our group well spaced out. When in a tight group, I always tend to feel distracted by the sounds and movements of the others. Much better to wander as individuals, collecting our own feelings and experiences to be shared in the evening or at rest stops.

SATURDAY, AUGUST 12

In spite of my ankle problems, we managed to cover another eight miles today between 2 and 7 P.M. The northern border of the Brooks Range is now quite near —and beyond that, open tundra.

In our thoughts the tundra now represents the major unknown. We worry about crossing the Jago River, but much more about traversing the tundra beyond it. Will blizzards hit? High winds? Will we find enough scrub growth to fuel our fire? Can we be sure of finding the right stream to follow in all that flatness? How will we know precisely when to cut cross-country to Barter Island—lacking landmarks? Will the walking be all muskeg, tussock grass and bog—or gravel bar and level grassland? Will our feet, legs and general health hold up long enough to get us there?

An hour ago Ken strode into camp after some hiking. From an overlook he had spotted a medium-sized grizzly that came within 70 yards of camp, rose on its hind legs, then apparently caught our scent. The bear bolted up into a side valley, still running as it disappeared from sight. What makes such a powerful beast run away? Fear of the unknown? We are probably the first men whose scent the bear has ever smelled; but the bear has never had grounds to fear other animals in its domain, so why run away from a new smell? No other wild animal can conquer the grizzly in its Arctic

home, and it should be outside its experience to take flight. But the bear ran, nonetheless.

It is now nearly midnight, and I write by candlelight; the horizon to the north outside my tent is afire with a deep, reddish glow. Outside, the thermometer stands at 23° F. Inside the tent it is 28 degrees.

MONDAY, AUGUST 14

Today has been magnificent. We've walked about eight miles, over two passes, and camped atop a wide, two- to three-thousand-foot-high plateau (which we call Emerald Basin). The plateau lies at the northern edge of the Brooks Range, near the junction of the mountains and the barren tundra stretching north to the Arctic Ocean.

On a mud bar in Green Willow Creek I saw a line of wolf tracks, unusually large in size. In one spot they met and mingled with caribou tracks of about the same freshness in the earth. An encounter?

Earlier in the evening a red Sun poised above Lone Mountain, and fog drifted thickly from the tundra through a pass and sent out numerous fingers of white across the valley floor. The Sun turned the top of the fog to gold and bathed the green valley in amber light. Sheer, snowy mountains looming over the scene turned rosy-colored from flank to summit, while ice formed in thin sheets of transparent crystal on the small pools between tussocks. The torment of leg and foot injuries and the punishment of blizzard are made bearable by these few hours of evening.

This land seems to be forever unfolding new surprises, in both weather and terrain. These two elements combine and recombine in an infinite variety of form and light. Who can say that he has seen these Arctic mountains as they really are? They are too varied in

Under the shadowed wall of the Brooks Range on an Arctic summer night, the Sheenjek River cuts a path through a year-round Alaskan ice field that reflects the haunting light of the "midnight sun."

form and too changing in appearance. Each of us who has come here has lived but a part of the truth; tomorrow will not be the same.

TUESDAY, AUGUST 15
About an hour after we had risen to a clear, bright dawn, Steve yelled "Caribou!" I grabbed my binoculars, my camera, the two telephoto lenses and ran from the tent. About a half mile away a pair of bulls lay contentedly in the snow. While Ken watched, Steve and I began to stalk them. We got to within one quarter mile of the animals, then sat down to watch and wait.

Eventually, the bulls got up from their resting place and grazed between snow patches. Then, as though curious to see what we were, they began to trot directly toward us. Our cameras clicked madly. They passed within 30 yards, their heads lifted, their antlers still encased in their coat of velvet. After passing us they nearly went through camp—coming to within 20 feet of Ken—then on out into Emerald Basin.

FRIDAY, AUGUST 18
I now feel little desire to exchange this Arctic wilderness environment for civilization. All of us have now been here long enough to settle completely into our new pattern of life. Although a difficult existence, it carries its compensatory comforts and the sharp joys of wilderness experience. At each of our three main base camps, Last Lake, Nameless Valley and now at

the mouth of the Jago mountain valley near the edge of the tundra, this sense of belonging to the Earth, of the wilderness defining our lives, has been particularly strong. During these times, when we've been on the move almost daily, a restless, nomadic urge to move on seems natural to human existence. How quickly man adapts!

And yet, over periods of time, deficiencies in both types of life appear—in either a wholly civilized life or a life spent constantly in the wilderness. It is a life spent contrasting and living alternately in both worlds that seems best to me. Indeed, the very words "civilization" and "wilderness" imply their own dual existence. For without civilization, what meaning would wilderness have, and vice versa?

SATURDAY, AUGUST 19
Today we climbed up to a long, narrow lake just west of the Jago River. An oil exploration crew had been there before us. The beauty of this lake is now diminished by the presence of over 20 scattered oil drums, the tattered ruins of a large, domed shelter and litter of all kinds strewn about the shores. Man's insensitivity in a wilderness landscape can be appalling. This waste will mar the pristine back country for years.

Fate provided some consolation, however, for amidst the litter we found tins of cooking oil and sugar and a pancake turner that allowed us to turn out whole cakes. We also found a new frying pan handle to replace the one we had lost, and a new teapot top to fill in for our badly battered one. We'd also lost two forks and a cup and found replacements for them as well. Finally, I added a fourth plate from the litter heap as a luxury item—intending to use it as a base for our small stove or as an extra surface to cut up food. Thus, all things considered, we had good scavenging. But I would rather forgo these helpful extras for a clean tundra lake.

MONDAY, AUGUST 21

After we had walked on the tundra for about two hours, the fog gradually began lifting. Abruptly the Sun broke through, turning the land to gold. It is almost impossible to describe the apparently endless and special beauty of this open tundra. I was struck by the absolute flatness of the northern horizon—we were able to see the coastal clouds piled up in the sky over 50 miles away. On either side of us the tundra, rich in vegetation of soft brown hues, rolled out in long, gradual waves, gentle slopes linked by broad basins. Behind us, to the south, the broad plains swept up to the very edge of the granite peaks of the Brooks Range, decked with long white glaciers. The subdued shadings and mild contours of the prairielike tundra seem strangely accentuated by the sharp, savage mountain topography and color. And tonight another mood comes over the tundra as the Sun sets and colors the land a deep burgundy.

WEDNESDAY, AUGUST 23

It is midmorning in camp as I write this day's entry in my journal, while the Sun strives to burn through the clouds above the tundra. We will soon resume our hike northward along the Okpilak River toward the Arctic Ocean. How simply and easily we have moved thus far through this wilderness—and what a contrast (I like to think) to the encumbered movements of the oilmen who try to drag all of civilized life with them, and succeed most in blotting the landscape with their oil drums. In the wilderness, man should move like the nomad, with the aim of leaving few traces.

We have done well in abiding by this ethic, I feel; bits of paper are burned, our few small cans flattened and buried, our campfire stones scattered. All that remains is a patch of slightly flattened moss and grass, no more than a few caribou or a family of wolves would leave behind following a pause for rest.

SATURDAY, AUGUST 26

Morning came with a cold, damp fog hanging over our camp. We ate a full breakfast of eggs mixed with what was left of last night's vegetables, and a meat-and-bacon bar. By the time we were through, the fog had lifted and the Sun was shining. At 1 P.M. we broke camp and headed north. About two miles north of camp we topped a rise and saw the Arctic Ocean in the distance before us.

Under the blue layers of sea and sky, the brown tundra ended in low cliffs running along the coast as far as we could see. A creek widened into a small marsh at the beach. We followed the creek downhill. A series of barrier beaches and islands, the largest of which is Barter, extended from a few hundred yards off the coast to several miles out.

Far off on the northern horizon shimmered widely spaced icebergs. Around us the Arctic tundra glowed golden in the sunlight of late afternoon. Along the edges of a marshy area fresh grizzly tracks showed in the mud. And dead ahead the unreal twin radar towers of the Barter Island DEW-line installation (an outpost of the North American Air Defense Command's Distant Early Warning System) drew our attention, and then our steps.

Fifty miles behind us, now, the Brooks Range thrust a solid wall of snow-covered summits into southern skies.

After crossing the channel to Barter Island with the aid of an Eskimo boatman, we finally found the military commander and a civilian aide. "Walked over the Brooks Range? . . . over the high peak region?" we were asked. "For God's sake—why?"

Why? To know the wilderness.

Where land meets sea along the northern coast of Alaska, there is surprisingly little ice during the summer, owing to the moderating effect of the ocean.

141

Chapter Two

MIRACLES OF WATER

*Quiet ponds, roaring cataracts, boiling springs, dark rivers below ground—all are
the guises of fresh water in its unending cycle of evaporation, condensation and flow*

There's Magic in a Pond

By Marston Bates

I have spent many hours sitting quietly by ponds or
looking into the water as my canoe drifted over the
shallows of a lake, trying to understand what was going on in that world.

A pond, especially, has the fascination of the miniature. It is a world clearly limited by the shores and
bottom and surface; a sufficiently self-contained world,
a world small enough so that one should be able to
figure out everything going on there, describe it,
analyze it, perhaps fit the relationships among the living things into neat equations—and, solving the equations, solve all mysteries. But the mystery of the pond
still eludes me.

The water of the pond is but one link in a single,
vast circulating system that embraces all the water on
the surface of our planet. The sea is the main reservoir
and the Sun is the furnace that provides the energy to
keep the system circulating. We can distinguish easily
enough between the atmosphere and the hydrosphere,
as the Earth's water system is called, but the two are
completely interdependent, with gases from the atmosphere continually entering into solution in the liquid
of the hydrosphere, and with water from the hydrosphere constantly evaporating into the atmosphere.

The warmer the air, the more water vapor it can
hold. Thus in the shortest version of the circulating
system, water evaporates from the surface of the sea
and land into the warm air above it. This warm, moisture-laden air is carried upward by atmospheric circulation and cooled until the water is precipitated out in
fine droplets to make the mist that we call clouds, and
with further cooling, to make individual drops of water
large enough to be pulled back to the Earth by gravity
as rain or snow.

Water is thus constantly circulated from the sea to
the air and back again. But the water in the air is often
carried over land, with all sorts of consequences. The
water that falls and collects on land may flow directly
over the surface to form streams in gullies, the streams
converging into rivers until at last they pour into the
sea again. Or, more commonly, the water may sink into
the soil to unite with the vast accumulation of groundwater that underlies all land. The groundwater emerges
here and there as springs or as seepage to augment the
flowing surface systems of streams and rivers. Often
the flow of these streams is blocked by the contour
of the land, to make pools, ponds and lakes, where the
water is dammed until it reaches some level at which
flow can be resumed. But water in contact with air always tends to evaporate, and in some places the evaporation is fast enough so that the water never accumulates sufficiently to allow free flow again, and we have
dead seas, salt lakes with no outlet, little side eddies in
the great system of planetary water circulation.

Much of the groundwater, of course, is picked up
by the roots of the land vegetation and returned to the
atmosphere through the process called transpiration,
the evaporation of water from leaves. Some of the
groundwater may be trapped in sealed-off pools, out
of circulation, for millions of years; some of it returns
to the reservoir of the sea through underground channels, seeping out in offshore springs.

Thus there are many pathways for the interchange
between hydrosphere and atmosphere; all turn eventually on the evaporation-precipitation sequence. When
water evaporates, it becomes pure water—that is, it
leaves all its dissolved materials behind. But since water
is the most universal of solvents, it never stays pure
very long. Even as rain it picks up not only gases from
the atmosphere, but traces of substances present as
dust in the air. In addition, sea salts are carried up into
the clouds from ocean spray.

Rainwater, however, is still reasonably pure. The

Formed about 12,000 years ago when the last Ice Age was drawing to a close, Canada's Mackenzie River has already laid down 4400 square miles of delta where its silt-rich waters enter the Beaufort Sea.

solution process gets fully underway after the water hits the ground. The chemical nature of fresh water collected on land, then, always depends on the composition of the soil and rocks through which it passes; thus fresh water's chemical nature varies greatly in different places, with important consequences for the organisms that depend on the water. The proportions of dissolved salts in the vast and continuous reservoir of the sea are almost constant, but in the fresh waters of the Earth they vary greatly. The salt content of inland water is never as great as that of the sea, except when the cycle is short-circuited, as in lakes without outlets —where water evaporates into the atmosphere, leaving behind its dissolved minerals and salts. Then, as in the Dead Sea and America's Great Salt Lake, the salt content may become much greater than that of the sea, and with quite different proportions among the dissolved salts. Inland waters, then, always differ from the seas simply as a chemical environment.

They also differ in many other important respects. The seas, as we know, form a continuous system over the face of the Earth. Fresh waters, however, lack this continuity of space. They are also discontinuous in time. For example, the pattern of lakes in north-central North America is a consequence of the last glaciation. Some 15,000 years ago, just yesterday in geological time, there were no lakes there; the whole land surface was buried under a vast sheet of ice. A few of the lakes of the world—notably Baikal in Asia, Tanganyika in Africa and Ochrida in the Balkans—are of much greater age. Some of the great river systems of the world are also quite old, but their histories bear no comparison with the uninterrupted story of the seas.

This has several consequences. From the point of view of evolution, one can, in general, look at the sea as a thing complete in itself, reconstructing the history of its inhabitants without reference to either fresh water or land. The striking exceptions here occur among the vertebrates. It is generally thought that the bony fishes (as distinguished from the cartilaginous fishes such as the sharks and rays) developed in the fresh or brackish waters of river estuaries nearly 500 million years ago; but if they evolved in fresh water, they soon reinvaded the sea and have flourished both

143

Filled by melting snows, the world's highest large lake, Titicaca, gleams among the Andean peaks of Peru and Bolivia at an altitude of nearly 12,500 feet.

braided streams and meandering streams, of beheaded streams and captured streams.

Streams in general tend to be tied up closely with the land through which they flow—shaded through forests, open in meadows, gaining chemical elements from the soil of their beds and shores, receiving organic matter from the plants and animals that fall into their currents. But this relationship tends to become less close as the stream grows larger, until, again, the great rivers of the coastal plains are things in themselves, understandable without reference to the physical or biological characteristics of their shores. Indeed, in this final phase the mud of the river bed and of the shores is apt to be a creation of the river itself, deposited over thousands of years.

Up to this point the transitory nature of all freshwater accumulations has been emphasized. This is much less true of river systems than of lakes. The details within the pattern of a river system shift constantly, changing even in the moment of time open to direct human observation. Channels shift, tributaries of one system are captured by another, uplift or landslides or lava flow may block a course. But the great river systems of the world show a considerable continuity, because these shifts for the most part are gradual; and even when river systems change radically, parts of the old system are carried over into the new.

The largest of the world's rivers in terms of volume of water discharged into the sea is the Amazon. This mightiest of rivers forms a network of water channels that permeates nearly half of the continent of South America. Through the Rio Negro and the natural channel of the Casiquiare Canal, the waterways of the Amazon are directly connected with the Orinoco system in the northern part of the continent. The waters of these two river systems, though widely dispersed, can be looked at as a great inland sea that has had continuity both in space and in time; because of their equatorial location, these rivers must long have provided a stable and favorable environment for life.

The result is a fabulous and special fauna. But this life is all too little known to science, since so much of it is native to the least explored parts of the South American rain forest. No one knows exactly how many species of fish there are in this river-sea, but there must be something like 2000. Among these are some of the biggest fish outside the ocean. (The arapaima reaches 15 feet in length and can weigh as much as 400 pounds.) I have vivid memories of huge fish skulls left on sandbars by fishermen who had been using dynamite: the living specimens must have weighed several hundred pounds.

Here, too, is the electric eel (actually a relative of the catfish rather than a true eel) and the notorious piranha—these latter, I would suspect, the most dangerous of all fish for man, making sharks and barracuda seem almost cowardly. And the biggest of all snakes,

in the oceans and in fresh water ever since. The reptiles, though predominantly a land group, developed many marine forms in the ancient times of their glory, and today are still represented in the seas by a few species of sea turtles, sea snakes and crocodiles. Several groups of mammals—including the seals, manatees, porpoises and whales—have successfully adapted to the sea, evolving from land-dwelling ancestors. But these forms derived from the land are exceptional.

The history of life in fresh water, on the other hand, cannot be reconstructed without constant reference to the sea and to dry land. The clams, crayfish, worms and inconspicuous sponges of fresh water are separate offshoots of basically marine groups. Fresh waters teem with insects—many different kinds of insects representing quite different adaptations of land-living, air-breathing forms to the water medium. Among vertebrates, the amphibians started their evolution in fresh water and have stayed with it, while reptiles, birds and mammals shift between water and land in all sorts of ways. The principal vegetation of inland waters is composed of many sorts of seed plants derived from land ancestors.

Streams—and the erosion always associated with them—are major elements in shaping the physical features of our landscapes. The students of landscape form, of geomorphology, have developed an elaborate and picturesque vocabulary for describing the different types of streams and of stream action. They speak of the anaconda, is at home here, more often in the water than out. The salt sea has contributed many inhabitants to this fresh-water sea—including a dolphin, a manatee and stingrays that nestle in the sandbars high up the tributaries where they rush out of the Andes.

Rays, piranhas, anacondas, electric eels—I make it sound a dangerous place, even without mentioning crocodiles. The upper Amazon or upper Orinoco certainly is not tame country—there are no manicured resorts. But its great fascination comes from escaping into a region of river and forest which man has so far not dominated, and the dangers are easily enough avoided with experience and common sense.

Lakes can be looked at as blockages in water systems —only temporary things on a geological time scale. The great exceptions, occupying ancient depressions in the crustal rock of continents, are Lake Baikal in Siberia and Lake Tanganyika in Africa. A few other lakes, in the evolutionary oddities of their fauna, give evidence of geological antiquity: Lake Posso in Celebes, Lake Lanao in the Philippines and Lake Ochrida in the Balkans. The largest of inland bodies of water, the Caspian Sea, occupies a peculiar place as an arm of the ocean cut off in rather recent geological times— only about one million years ago.

Lakes, in comparison with the sea, are shallow. While the mean depth of the oceans is about 12,700 feet, the deepest fresh-water lakes, Baikal and Tanganyika, only reach depths of about 5700 feet and 4700 feet. Lake Superior, covering an area of nearly 32,000 square miles, is the broadest body of fresh water. Superior has an average depth of 475 feet.

Once they have formed, lakes always tend to grow shallower and smaller from the accumulation of sediment brought into them by streams and from the encroachment of vegetation growing on their margins. They tend also to drain away because of erosion at their outlets. These shrinking processes can be reconstructed easily enough in any particular area by comparing a series of lakes, ponds, swamps, marshes and bogs; one can see that many a fertile farm is located in the bed of some extinct lake.

For animals with amphibious lives, the disappearance of bodies of water does not matter much; as one lake dwindles, they can move to another. For animals wholly bound to the water, though, the problem is more serious. Their only escape lies through the streams and rivers feeding or flowing from the lake. The aquatic fauna of geologically transient lakes differ only in number from the fauna in the streams of the same region. It is necessarily part of the same system, and there is little chance for peculiar forms to develop in a particular lake.

The world of inland waters is remote in many ways, even though we are in frequent and direct contact with it. When we fish a trout stream we may become very

expert in choosing flies and in spotting likely pools, but do we really have any understanding of the world in which our trout is living? And that pond of mine again: I love to watch the dartings of the minnows, the huddling tadpoles in the puddles of the swampy edge, the bright-eyed turtles that stick their heads out to check on me at a safe distance. But always I am an outsider watching a play, trying to find its meanings. The surface of the water, like the footlights of the stage, is a barrier that I cannot cross—except in spirit.

Clouds shadow the jungled delta of the Amazon. The Earth's greatest river, it drains nearly three million square miles of equatorial South America.

Waterfalls Have Many Moods

Few sights are more magnificent than a feathery cascade leaping down a mountain, but when a giant cataract roars, delight gives way to awe for the unbridled power of falling water

No poet has spoken so eloquently of a waterfall as the tribe that called Africa's Victoria Falls *Mosi-oa-tunya*, "the smoke that thunders." But the volume of water flowing over Victoria Falls gives it a ranking of seventh among the world's great cataracts, behind several South American falls and North America's Niagara. The most powerful, standing on Brazil's border with Paraguay, is 374-foot Sete Que-das, whose annual flow averages nearly a million tons of water per minute—compared to an average 75,000 tons per minute over Victoria and 400,000 over Niagara. Venezuela's 3212-foot Angel Fall is the world's highest, followed by cascades in South Africa, California, New Zealand and Norway. In their misty leaps, these thin torrents are as thrilling as the big cataracts in all their tumult.

Roiled by the torrential rains of the wet season, the furious cascades of gigantic Iguaçu swell to engulf the jungle hillsides along the Argentina-Brazil border in raging waters, foam and spray.

We usually think of waterfalls as merely streams or rivers that happen to encounter a sudden, sheer drop in the terrain as they flow along. But geologists point out that most falls have far more complicated origins. Many of the world's tallest falls, for example, are "hanging tributaries"— streams cut off along the rims of glacier-gouged fiords and valleys when the ice sheets of the last Ice Age receded. Some waterfalls occur where crustal subsidence along a fault suddenly dropped the bottom out of an established river system. But the most important force in the creation of a major cataract is usually simply erosion of the river channel by the waters of the river itself. Where a river that has been flowing over hard bedrock, such as granite, abruptly encounters soft sedimentary layers, the latter wear away more rapidly; thus, as time passes, the erosion-resistant bedrock is left standing as a precipice above an ever deepening chasm, pounded by a foaming barrage.

Falling temperature stops this cascade in the Canadian Rockies, masking the sheer wall of a glacier-gouged valley in a veil of frozen motion.

The mists of Cascata delle Marmore, a crashing, 525-foot cascade, give a wild region near Terni in central Italy the lushness of a rain forest.

Roaring down Augrabies Falls in the desolate Kalahari region of South Africa, the Orange River carves an ever deepening canyon.

147

In the eternal epic of water's journey from the mountains to the seas, even the mightiest cataract is fated to be but a temporary phenomenon. No more renowned waterfalls roil the rivers than Africa's Victoria Falls and North America's Niagara. The former stands where erosion undercut soft sedimentary layers beneath the Zambezi River's volcanic bedrock, while the latter's cliff was cut by the glaciers of the last Ice Age. Like waterfalls everywhere, both these colossi are relentlessly consuming their own rock foundations.

Surprisingly, a big waterfall's erosion proceeds chiefly from its base upward. This is because the energy of the falls excavates a "plunge pool" at the foot of the escarpment, undermining it and continuously bringing down chunks of rock from above. As a result, cataracts retreat upriver (Niagara's Horseshoe Falls, for instance, is retreating at a rate of about five feet per year) and decrease in height with the years. Where swift, thin cascades plunge down sheer mountainsides or glacier-gouged valley walls, erosion usually decays the falls in a notch from the top down. But however erosion occurs, every waterfall must ultimately surrender to the persistent leveling force of the water, and its life is brief in the enduring saga of the river.

Scotland's glacier-grooved Highland glens ring with the music of myriad cascades; here a tributary of the Dee writhes down a cliff near Braemar.

Following a heavy rain, waterfalls bloom everywhere on the mist-shrouded bluffs of Milford Sound, one of many breathtakingly steep fiords cut by glaciers from the hard rock of New Zealand's South Island.

In the quiet flow below the Grand Canyon's Mooney Falls, minerals precipitating from the water have built up a set of shallow, secondary falls.

Canada's highest falls, Takakkaw, thunders 1200 feet down an eroded chute in a glacier-scraped Rocky Mountain chasm in Yoho National Park.

Erosion is rapidly eating away the soft central rock of Niagara's Horseshoe Falls, creating the U-shape.

At the peak of central Africa's rainy season, floods of runoff pour 300,000 tons of water per minute over Victoria Falls, raising thunderous clouds of spray.

149

Great Salt Lake— America's Pocket Ocean

By Wolfgang Langewiesche

Most places have some one master fact you need to know. Get hold of it and everything else falls into place. At Great Salt Lake, some 4200 feet above sea level in the state of Utah, the master fact is this: the lake today, 75 miles long and 50 miles across at its widest point, is only a remnant of what was here once —a gigantic inland body of water some 20 times as large, 67 times as deep, holding many hundreds of times the present volume of water.

Between 1873 and 1940 the water level dropped about 16 feet and the shoreline receded as much as ten miles in some stretches. Nearly 800 square miles of new land were thus bared, mostly malodorous mud flats or humpy chunks of land interlaced with ponds—a no man's land, barren, treeless, difficult to traverse, though at a few places a car can get through to the shoreline.

The Great Salt Lake, then, is a lonesome, harsh, even hostile setting today. Salt Lake City is 15 miles from the present shoreline. There are no summer cabins by the lake. There is no fishing because there are no fish. Only a few organisms live in the lake's dense brine: some species of algae and protozoans, a species of tiny shrimp and two species of fly.

In the 1890s there was a resort, a great big glorious pavilion with towers, built on a pier that ran a quarter-mile into the lake. A steam-driven train took people to the end of the pier. But this structure now stands high and dry—deserted, partly destroyed by fire, a quarter-mile inland from the present waterline.

The remarkable shrinking of Great Salt Lake—from the middle of the 19th until the middle of the 20th century—resulted from an evaporation rate higher than the rate of flow into the basin from the mountain streams and rivers that feed it. In the 1940s, as the balance shifted once more, the water coming in, the inflow, began to exceed the evaporation loss, and as a result the lake began rising. By 1950 the water level had climbed to within 12 feet of its mid-19th-century level. And then it began declining again. By the mid-

Rough and crumbly after an autumn rain, the thick crust of Utah's Bonneville Salt Flats is part of a salt desert at the western edge of the Rockies that was once covered by a 200,000-square-mile lake.

1960s the water level has dropped nine feet off the 1950 mark.

The salinity—or concentration of salts—changes according to the lake's water level. In 1950, for instance, the salinity was about 25 percent—or some seven times saltier than the ocean. At times when the water level is lower, salinity often exceeds the saturation point, and the salts begin to precipitate.

Salts are brought into the lake by all the rivers and streams that feed it. Their water contains tiny amounts of dissolved salts and minerals washed from the rock and soil of the surrounding terrain. Since the lake has no outlet, the water can escape only by evaporation. And as water evaporates, it leaves behind its dissolved salts and minerals. Here, then, are concentrated the salts drawn from Utah's mountains—in a kind of pocket ocean in a mountain basin.

A swim in Great Salt Lake is an interesting experience. Like the Dead Sea, the water is so salty that you *can't* sink. You float with head, feet and arms buoyed up out of the water. When you come out, the water dries almost instantly. The salt remains, forming crystals on your skin.

A close-up view of the rain-drenched surface of the Flats discloses the salt slowly recrystallizing.

The drying-out of this huge, ancient lake, which once covered the entire basin, has left an enormous desert on the western and southern sides of the present Great Salt Lake. This desert reaches 160 miles from north to south and some 70 miles to the west—a broad, grayish-white expanse strewn with billions of tons of salt, perhaps the most implacable environment in North America. To attempt to traverse the Great Salt Lake Desert on foot would be to flirt with death. Although a highway crosses this barren plain, one drives it in the knowledge that for 50 miles on either side there are no other roads, no houses.

About 100 miles west of Salt Lake City the highway passes beside the Bonneville Salt Flats, an area 14 miles long and seven miles wide of smooth, white crust that resembles a sheet of ice over a lake. Of course, the "ice" is salt, and on this surface world automobile speed records are regularly established and broken. From an airplane you can see the long, straight black line that serves as a guideline for the speed runs. There is no grandstand; there are no buildings; there is only the white plain and the adjacent naked mountains, a landscape of aridity and emptiness.

How did this salt flat form so many miles away from the present shore of the lake? A depression existed at this spot in the ancient lake bed. As the big lake receded, a super-salty solution was left behind in the depression. When this dried, only the salt remained—hard, level and smooth.

A few feet below the surface the Bonneville Salt Flats remain saturated with water. Drawn upward by wick action, this moisture acts to keep the salt firm. The moisture also keeps the salt cool. A cement pavement in that desert sun soon gets hot enough to fry eggs. But the salt usually stays ten degrees cooler than the air.

During droughts, the water under the salt dries up in spots. As a result the surface becomes crumbly, and potholes form. But in winter the flats are normally flooded by runoff from the nearby high ground, and the salt surface softens and smooths itself out.

How do we know that an ancient lake covered the entire region of the Great Salt Lake Desert? The evidence lies in the old high-water marks on the mountainsides. They are easy to discern. The highest lies 1000 feet above the present lake level. What draws the eye is that the marks are strictly horizontal. Experts have identified more than 20 of the former shorelines. Each gives evidence of the lake once having been at that level for a few hundred or thousand years.

This immense lake of the Ice Ages—called Lake Bonneville by geologists—was fed by the meltwater of ice caps that existed on the surrounding mountains. At its maximum, probably some 20,000 years ago, Lake Bonneville spilled over to the northwest in a mighty waterway that followed the course of the present-day Snake-Columbia river system to the Pacific Ocean.

This mountainside next to the Great Salt Lake Desert is planed into distinct levels; they mark old shorelines of now dry Lake Bonneville as it receded from its maximum depth of more than 1000 feet.

Currently the great excitement around the lake concerns not the whys and wherefores of its geological history, but the legacy of that past in the form of minerals that can be mined today. The water of the Great Salt Lake is about seven times as rich in minerals as sea water. Geologists estimate that some six billion tons of salt are held in solution by the lake waters. Besides common salt, gypsum and potash, the lake contains various compounds of boron, lithium, sulfur, magnesium and chlorine.

The most important of the lake's minerals commercially are its magnesium salts. Another major resource present in the brine is lithium. This soft substance, the lightest of all metals, is used in the manufacture of a diversity of products, including lubricants, ceramics and rocket fuels. In addition, medical scientists have recently experimented with lithium compounds in the treatment of mental disorders.

To extract this wealth, developers follow a procedure similar to the natural process of evaporation that causes the salts to accumulate in the lake. First, a series of shallow evaporation ponds are excavated and brine is pumped into them. The Sun and desert air take out the water, and the salt forms in deposits on the bottom. These are scooped out and refined. The process is ancient. What is new is that modern engineers have learned to control the process so that all the various salts contained in solution—including the rarest—drop out in an orderly fashion, more or less separately, each in its own pond.

The great landmark of this process is the string of evaporation ponds—a salt farm, or more accurately, a solar-powered factory. The oldest of these salt refineries is not at the lake but at the salt flats. There a salt farm has been extracting potash, for fertilizer, from the deep strata deposited by Lake Bonneville during the Ice Ages. Other projects are under construction along the shores of Great Salt Lake, on a still bigger scale.

So this scruffy thing that people disliked has suddenly become a pot of gold. Its mud-flat shores are becoming lined with huge salt farms. In the ground view they won't show up much, but for air travelers they will be the biggest man-made feature in the landscape. Because the chemistry of the brine in the man-made evaporation ponds differs from bed to bed, each bed shows its own characteristic color in the sunlight: blue, brown, silver, purple, milky—like a collection of enormous stained-glass panes strewn across the surface of the desert.

Even in an industrialized state, the Great Salt Lake will retain an air of wonder, a blue gem in a jeweled setting at the edge of a salt-encrusted desert.

The Saltiest Sea on Earth

The narrow blue finger of the Dead Sea stretches between Israel and Jordan in an extension of Africa's Great Rift Valley. The shore along the Sea is the lowest place on land—approximately 1300 feet below sea level. Its water is the world's saltiest, seven times more so than the oceans' and almost twice as salty as the Great Salt Lake. Why so salty? Being cut off by the surrounding land, the Dead Sea loses water mainly through evaporation; and when water evaporates, it leaves an increasing residue of its dissolved salts and minerals. Today the layers of salt that have precipitated over the bottom of the Sea are thought to be hundreds of feet thick and still growing.

We Explored an Underground River

By Lamberto Ferri-Ricchi

The subterranean world is a land of mystery, a place of fantastic shapes, bewildering labyrinths, dark lakes and lost rivers. This is the special province of the speleologist and, in a country where little remains to be explored on the surface, the scientists who specialize in caves are among the few who can still discover truly unknown places. In recent decades underground geology has become the subject of intense study and exploration, especially in Europe where more and more followers are being won over by the fascination of the world of darkness. In fact, speleology has passed the stage of scientific exploration and has become more like a sport.

Perhaps the most exciting advances have been made in the exploration of underground rivers. Until recently, subterranean rivers have resisted even the best prepared expeditions. The risks of scuba diving in dark, confined waterways—in addition to the problems of normal cave exploration—have demanded the development of new equipment and a new breed of space-age explorer, the "speleonaut."

Several years ago I learned of some unusual problems at the Cave of Pastena near Frosinone, some 40 miles southeast of Rome. The Cave of Pastena is widely known for its natural beauty and is one of the few in Italy that can be visited by tourists. A stream leads into the caverns; but whenever there was a heavy storm, floods completely submerged the cave's huge entrance, causing considerable damage to the areas open to tourists and to nearby low-lying farmland.

Since the stream re-emerged from the ground on a hillside several miles away, it was obvious that something was blocking the river deep underground during times of flooding. Several expeditions had tried to locate this blockage but had not been able to penetrate beyond an area about 1000 feet from the entrance, where the roof of the cavern dips lower and lower until it sinks beneath the surface of a lake. A few speleonauts tried to find an outlet from this lake and some had even tried to push upstream from the point where the river re-emerged on the surface, but to no avail.

I already had many years of practice in underwater fishing with scuba gear, and on several occasions had gone into underwater tunnels that were many yards long. Thus the idea of exploring narrow, submerged alleyways was something I was psychologically prepared to do, even though it meant penetrating several pitch-black "siphons"—submerged passageways that were completely filled with water up to their roofs.

After several visits to the underground lake at the end of the main cavern at Pastena, and a series of preliminary hydro-geologic surveys, I became convinced that it would be possible to push farther along the river. I contacted speleologists in Rome and found several who were willing to accompany me. They were outstanding rock-climbers and good swimmers, but they had no experience with scuba equipment; so we trained intensely in the cold muddy water of Lake

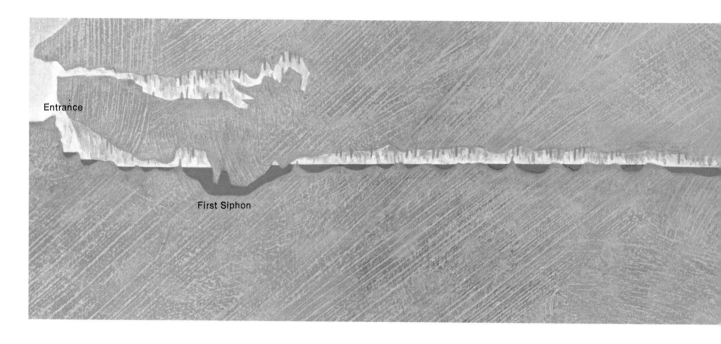

Entrance

First Siphon

The blue gleam of the underground river flashes in the powerful illumination of an explorer's head-lamp as he inches his way down a narrow crevice.

Albano, where conditions were similar to what we could expect in the Pastena caverns. We realistically simulated all the emergencies we could anticipate.

Then came the first dives at Pastena, the first fears and finally the gradual awareness of what we were undertaking. In an early expedition I plunged into the lake at the far end of the entrance cavern. I found a narrow tunnel leading off from the bottom of the lake and followed it for about 65 feet, but I was forced to stop at a tangle of heavy logs that had become wedged at that point during a flood. The logs completely blocked this submerged conduit which was not very wide to begin with.

The next year I made plans to try again after learning that a flood had covered the valley in front of the cave. Perhaps it had opened the blocked siphon. The weather forecasts were good for the period covering our next attempt. This was essential—an unexpected storm taking us by surprise inside the cave could mean the end of all of us.

A group of three expert speleologists preceded me and two other speleonauts, Mario Ranieri and Vittorio Castellani, into Pastena, rigging ropes and ladders in the cave. The equipment needed for diving was already in place at the terminal lake, painstakingly brought there by team members who had to let themselves down into deep chasms on swaying ladders, carrying the heavy air bottles. At the last lake our advance men had set up the base camp and made ready our equipment. The atmosphere, as always, was filled with tension as we prepared for the dive. The usual briefing preceded the dive: I repeated the last warnings, we washed out our mouths with a strong disinfectant to prevent infection from pollution and then went into the water.

Twenty-six feet down, a black opening indicated the course of the river. We were lucky; the passage-

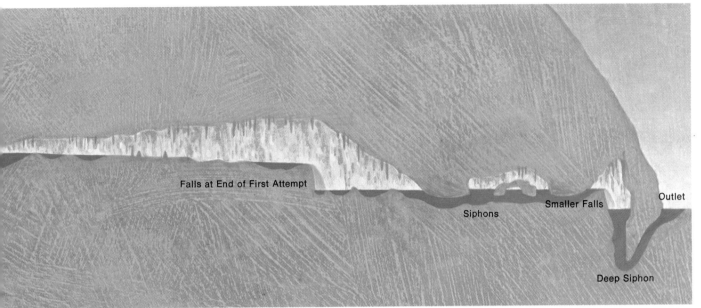

Falls at End of First Attempt

Siphons

Smaller Falls

Outlet

Deep Siphon

Entering Pastena Cave from the left, the Italian team was halted by the large waterfall. They reentered from the right and worked back to the falls.

Wearing wet suits and scuba gear, the "speleonauts" swim across a small lake in search of an outlet that will permit them to continue down the river.

way seemed clear this time. I advanced rapidly. The average diameter of the passage, aside from an occasional narrowing, was about six feet. I moved forward, threading my way through a curtain of rocky spikes, pulling with me the light guideline that was my only contact with the outside world. Farther back were my friends Ranieri and Castellani. By now I had been struggling along for several minutes in the conduit and had lost all sense of the distance I had traveled. Suddenly there was air. I emerged, glanced at the surroundings, and sent a signal on the guideline to Ranieri and Castellani. A few minutes later we were together again. This siphon was 230 feet long.

Now we could set aside our air bottles and explore the new tunnels. The course of the river was wide and changeable and alternated with lakes. We passed through two more small siphons by holding our breath. The length of this section was much greater than we expected. At last the roof of the tunnel rose in a characteristic pointed arch, and we heard the muffled sound of waterfalls. We reached the edge of the falls and found them to be more than 20 feet high. The walls were smooth, and without the help of a rope it was impossible to get down.

We allowed ourselves a brief pause. Thinking about the return trip, we realized that we were very near the limit of our energy and equipment. Our only flashlight was barely able to light up the walls of the large cavern. We went back quickly and arrived exhausted at the first siphon, guided by the reddish glow of the flashlight. One after another we went back through the siphon. This time the visibility was zero. When we surfaced again, our friends, tense and very worried,

Two members of the Italian team emerge from a narrow "siphon"—an especially hazardous underground conduit completely filled with water.

bombarded us with all kinds of questions. A few hours later we were outside, worn out from the long submersion in the icy lake water followed by heavy perspiring in the dry sections of the cavern.

We believed that we had reached a point very near where the river emerges from the ground. We decided to conquer the final segment of the subterranean river by entering downstream at the point where the river surfaces in open country at the base of a steep rocky wall. In three attempts we forced our way through a siphon 190 feet long and 65 feet deep and came out in the middle of a lake into which water fell from a height of a bit less than 20 feet. We scaled the falls and found a second lake where we located the mouth of a second siphon. Our activities in these surroundings turned out to be more tiring than expected, and the cold and fatigue discouraged us from continuing the exploration.

I carefully studied the causes of our lack of success and convinced myself that by using special equipment and by programming our operations with care we could save much time and energy. We began our final attempt on August 15, 1967.

Mario Ranieri swam with me; four other friends followed us and carried out support operations. When we were ready, Mario fastened to his belt a sack containing a hammer, pitons, spring clips and 100 feet of strong nylon tape. He also carried an underwater telephone and a watertight container with instruments and food. I carried a special reel with 400 feet of electric wire, long enough to follow me through the succeeding siphons. Mario left first. The material strapped to us made it hard to maneuver and the guideline paid out with difficulty. During the last 65 feet, which were almost vertical, we began to think we wouldn't make it.

Finally we emerged from the end of the siphon gasping for breath and collapsed on a small rock ledge. A brief rest, then we removed our equipment and scaled the small waterfall located where we had stopped on our last exploration. Time passed quickly and when the first of the double air bottles was hauled to the top of the waterfall, we felt quite tired. We tried to smoke cigarettes but the matches barely lighted and the cigarettes went out almost at once. Perhaps there was an accumulation of carbon dioxide in this section of the cavern which is shut off between two siphons. This would also explain our exhaustion. We had to hurry, so we decided not to bring up the second air bottle.

I put on my underwater outfit and arranged the electric cable reel as best I could. I managed to get through three very tortuous and narrow siphons for a total distance of 250 feet under water. At the end of the third siphon I could finally discard the air bottles and proceed on foot. I was alone. Between me and the outside world were hundreds of yards of stone and long, deep, narrow passages filled with freezing muddy water. The roof of the cave rose swiftly. After 300 feet I heard the sound of a waterfall. I reached it and I had no doubts. It was the same one that blocked the passage when we came the other way the year before. The exploration of the underground river of Pastena could be considered definitely completed.

I got ready for the return trip and after an hour I rejoined Mario. We carried the double air bottles and equipment down from the smaller waterfall. After a last telephone conversation with the base camp outside we submerged. Contact with the icy water, which is always unpleasant, now caused sudden cramps that paralyzed my legs. With Mario ahead of me I got through the last siphon by pulling myself along the guideline. The visibility was reduced to zero. We emerged in the middle of the night to anxious friends and wonderfully hot tea under a starry sky.

Four years of planning and work, failure and success and fear and elation had culminated in success. The surveys and investigations we had carried out enabled us to point out the reasons why floodwater became bottled up in the cave. We were able to suggest ways to eliminate this problem, and within a few years tourists who go to Pastena may once again be able to see the full beauty of the caves.

The 10,000 Geysers of Yellowstone

By Peter Farb

The visible water supply in lakes and rivers represents but a small portion of the water of a continent; the rest lies hidden in natural reservoirs underground. Indeed, most lakes and rivers would dry up if they depended solely upon precipitation. It is the continuing supply from the underground sources that keeps lakes high and rivers running, even during periods of drought.

The water of these subterranean storehouses is constantly replenished by the moisture of rain and melting snow that seeps downward through the soil, through porous rock and through crevices in rock. It may flow for long distances underground or emerge in the springs that flow into streams, rivers and lakes, its final destination being the sea.

The only contact most people ordinarily have with this hidden water comes when they drill a well. But underground waters are capable of highly visible displays. Geysers, consisting of tons of hot water spouting up out of the ground, are the most arresting of these effects. The world's most spectacular geysers are found in Iceland, New Zealand and America's Yellowstone National Park, on a high plateau among the Rocky Mountains in the state of Wyoming. At Yellowstone there are at least 10,000 geysers, hot pools, pots of boiling mud and fumaroles (steam vents).

Several special conditions are necessary for the formation of a geyser. First, seething molten material (volcanic magma) must exist under local layers of surface rock. Leading from the heated subterranean rock to the surface there must be narrow channels with walls strong enough to endure the explosive force of the geyser. Finally, there must be an underground supply of water that reaches the magma in such a way that it is heated and forced upward. The primary difference between a hot spring and a geyser is the degree to which this water has been heated by the molten rock; geysers occur only when the heat and pressure are sufficient to produce steam, and hence cause explosive effects.

While geysers and hot springs appear to be spread in a random fashion throughout Yellowstone's 3472 square miles, they are actually arranged in definite belts. These belts occur along faults or fractures where sections of rock move up, down or sidewise in relation to adjacent sections. These faults provide passageways for the escape of steam and hot water trapped at subterranean levels. As sources of heat, Yellowstone's geysers have created favorable winter living conditions for plants and animals, which cluster around them. Many bears have located their dens near geysers, and

Under layers of overcast, Yellowstone's Old Faithful blows right on schedule, its superheated steam billowing upward to merge with the clouds in the sky.

Faulted Rock

Magma

A geyser's behavior is due to its restricted upper passages. These trap cooler water which acts as a lid until pressure is high enough for an eruption.

157

A solid wedding cake of white rock has built up around Minerva Terrace at Yellowstone's Mammoth Hot Springs. Here the noneruptive flow of mineral-laden water deposits calcium carbonate when it cools.

The delicate colors of Morning Glory Pool result from minerals in the rocks and colonies of algae that flourish in the warmth of this thermal pool.

numerous birds winter in these same warm areas. Ducks, in particular, are often seen in winter in the pools of water kept free of ice by the warm water issuing from the geysers. The temperature of the Yellowstone River rises by an average of six degrees as it flows through the geyser area, and in that stretch of the river, trout are active all winter long.

Old Faithful is the best known but not the most powerful geyser in Yellowstone; others have produced greater eruptions, and Steady Geyser almost never ceases to play. Nor is Old Faithful the most beautiful of the natural founts. But no other geyser of its height erupts on so regular a schedule and with such consistent beauty.

Old Faithful spurts up from the center of a mound measuring about 12 feet in height and 55 feet across at its widest point. This mound is built up of minerals that have been deposited there by the geyser itself. Geyser water is extremely rich in dissolved minerals—it is the kind of water commonly called "hard water." Anyone familiar with hard water knows that a coating of minerals soon forms on the inside walls of kettles and pans used to boil this liquid. This residue is left by the water when it evaporates.

Precisely the same sort of thing occurs at the base of the jet of a geyser. The mound of minerals deposited by Old Faithful is surpassed in intricacy by the deposits of other geysers at Yellowstone. Formations of rock have been created in shapes that suggest cauliflowers, sponges, chairs, flowers and necklaces. Sometimes terraced formations are created, as at Yellowstone's Mammoth Hot Springs, by water flowing down hillsides. The brilliant streaks of color in these deposits are due largely to the great variety of minerals carried to the surface by the geysers. Some of the coloration, though, is caused by algae that have adjusted to living at high temperatures.

Old Faithful's eruptions do not occur like clockwork, as many believe. For years past, eruptions came every 60 to 65 minutes. Today its interval is far less regular, with eruptions sometimes coming as little as 30 minutes apart, at others as much as 90 minutes apart. The geyser almost always gives advance warning of an eruption: a short spurt, then a graceful column begins to rise. At first it seems to grow with great effort, but soon a jet shoots upward, playing at heights between 115 and 150 feet. At its crest the jet is broken by the breeze. The water unfurls into a dazzling shower of droplets that, capturing the sunlight, shimmer with the colors of the rainbow. The full display lasts between two and five minutes, expelling as much as 12,000 gallons of water during a single eruption.

What I have just described is only the visible part of the spectacle of Old Faithful. Concealed beneath it is an intricate plumbing system that connects with hot magma or heated rock deep beneath the surface. As groundwater seeps into this system, it is heated at the bottom, just like water in a teakettle. Whereas in a teakettle the hot water can rise by convection to change places with the cold water above it, the narrow tubes of the underground geyser system hinder the free circulation of water, so that the upper part of the column of water remains much cooler than the rapidly heating water at the bottom of the column. This overlying column of water exerts pressure on the water below, thus raising its boiling point as though it were being heated in a pressure cooker.

The water trapped under pressure at the bottom of the tube grows hotter and hotter. Eventually its temperature rises far above the normal boiling point of water. The vapor produced is prevented from escaping as steam by the overlying mass of cooler water. Finally the water in the upper part of the column warms and expands, some of it welling up out of the mouth of the geyser. The escape of this water abruptly decreases the pressure on the superheated waters at bottom, which change to steam. The sudden expansion of the superheated water into steam occurs in a powerful explosion that forces all the water and vapor up out of the geyser.

A Geyser's Tricky "Clock"

An intriguing new study reveals that the "clock" governing a geyser's eruptions can be reset by the Moon's gravitational pull and by earthquakes. American geologist John Rinehart found that when the Moon was aligned with the Sun—at New Moon and Full Moon—the resulting strong gravitational stresses in the Earth's crust altered the timetable of one geyser by more than an hour. Old Faithful, furthermore, has proved to be sensitive to changing crustal stresses before and after earthquakes. The graph shows that eruptions in the months leading up to a quake in nearby Montana in August 1959 occurred at intervals shorter by several minutes than its normal every-65-minute tempo; then, shortly after the quake, its interval leaped abruptly to 68 minutes. Geologists surmise that crustal stresses must act to warp a geyser's fault system—the subterranean "plumbing" that channels water and steam to the surface. Although the mechanism isn't precisely understood, clearly geysers are far more sensitive to outside events than originally believed.

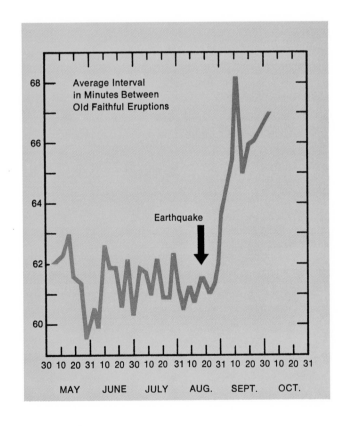

159

It is remarkable that Old Faithful is even as regular as it is, for only a slight variation in the amount of groundwater seepage or in the temperature of the subterranean rock could alter its schedule enormously, perhaps putting an end to its displays entirely. Most other geysers are irregular by comparison, with intervals varying in some instances from minutes to years.

Yellowstone also has thousands of springs that do not erupt as geysers; they range from quietly steaming pools where there is only the slightest turbulence to hot springs that constantly bubble like caldrons. They are all potential geysers which lack one or more of the conditions that produce spectacular eruptions. Most often, the reason they do not erupt lies in the shape of their underground plumbing, which permits the heated water to escape before the buildup of the required pressure. Some of the most attractive of the noneruptive pools are the sulfur springs; many discharge only small amounts of water, but thick crusts of sulfur rim the pools with brilliant yellow.

Iceland's Hot Water Wonderland

By Julian Kane

An Icelandic saga tells of a 10th-century hero who stood on a high promontory on the south coast of this rugged North Atlantic isle, keeping a steadfast watch for the sails of Viking marauders. Today this same peak, located near Hveragerdi, stands almost 20 miles inland from the coast. Intense volcanic forces have raised the rock shelf that in the 10th century lay submerged off the southern coast, adding new area to the island.

Such renovations are not unusual occurrences on this northern island, for Iceland is the scene of more active geologic forces than any other comparably sized segment of the Earth's surface.

With an area today of nearly 40,000 square miles, Iceland is the largest exposed portion of the Mid-Atlantic Ridge, a huge, largely submarine mountain chain that stretches in a sinuous pattern for about 12,000 miles along the Atlantic Ocean Basin's center line, from the Arctic to the Antarctic. It is part of a 40,000-mile-long, worldwide undersea range, called the Mid-Ocean Ridge system. Bending to the east between the tip of South Africa and the coast of Antarctica, the ridge winds its way into the Indian Ocean and the Pacific as a globe-girdling seam in the Earth's crust. Other visible segments of the Mid-Atlantic Ridge are the Azores, St. Paul Rocks, Ascension Island, St. Helena and Tristan da Cunha.

What forces have created Iceland and the rest of the Mid-Atlantic Ridge? Many geologists now think that the floor of the Atlantic Ocean Basin is spreading outward on either side of the ridge. As the sea floor continues to split farther apart at its center line, new volcanic material wells up through faults along the ridge. This process also forces the continents on either side of the ocean to move farther apart. The continuing geologic activity along the ridge may well be part of a huge convection cycle of molten rock rising from the

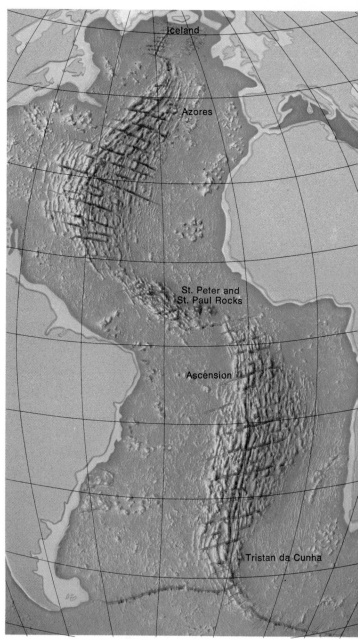

Snaking down the floor of the Atlantic basin, the undersea Mid-Atlantic Ridge surfaces in the far north to create Iceland, in mid-ocean to form the Azores, Peter and Paul Rocks, Ascension and the world's most remote settlement, Tristan da Cunha.

Just below the Arctic Circle, a scalding hot river pours from the vents of one of Iceland's many geo-thermal springs. A pumping station located on the river supplies hot water to homes in a nearby town.

hot mantle that extends beneath the crust for hundreds of miles into the earth. In Iceland large tension cracks form rift valleys that bisect the country, causing earthquakes and volcanic eruptions along the rift zone. All are believed to be manifestations of the changes occurring far below in the Earth's hot interior.

The bedrock of Iceland is composed mostly of basaltic rock and ejected volcanic debris. Thus Iceland's geology resembles that of the ocean floor, which also consists of basalt; on the continents there is an overlying layer of granite, which is largely absent on Iceland. The bulk of Iceland's rock solidified into its present form 40 to 60 million years ago. Most of the recent subsurface activity, however, has been manifested in a zone—approximately 50 by 250 miles in size—along a northwest-to-southeast axis through the center of the island. Because of Iceland's history of volcanic activity, fossils are rare and geological dating is mostly limited to that which can be done on the basis of the radioactive isotopes in its rocks.

In addition to containing a great number and variety of volcanic cones, Iceland is endowed with active fissure flows that form lava plains and lava plateaus. These are made up of successive layers, the result of outpourings of extremely fluid lavas, capable of covering many square miles before congealing in essentially horizontal layers. When magma pressures build up from below, the molten material readily finds its way up to the surface through the cracks located in the

rock near the island's center, and fissure lava flows result.

A sinuous ridgeline five miles long juts out above a snowy plain some 30 miles from Iceland's capital, Reykjavik. Down the center of this ridge runs a broad fissure. The walls of hardened lava on either side of the fissure were believed to have been creeping farther apart. Geologists suspected this fissure, called the Almannagja, to be a visible section of a rift in the Mid-Atlantic Ridge. Recent measurements have indeed confirmed that the Almannagja fissure is slowly widening, with the land on either side spreading outward like the sea floors themselves. The steep, eroded edges of hardened lava flows in the fissure walls frequently exhibit faceted rock columns, formed when shrinkage cracks developed in the cooling lava.

The Reykjanes Peninsula is a stubby arm of land extending off the southwestern coast. It is on the protected northern shore of this peninsula that the capital city of Reykjavik is located. The peninsula experiences frequent earthquakes. The quakes' persistence, coupled with the birth of new *gjas* (tension cracks) and hot springs in the area, indicates that additional volcanic eruptions may be expected in the near future. Scientific investigations have been designed to predict the start of volcanic activity in order to avoid loss of life and property.

Infrared detectors are capable of locating areas where the ground heat is increasing significantly. These

161

detectors are put to work in surface arrays, and they are also used to monitor Iceland from airplanes and space satellites. In recent years, five regions have been identified where rising temperatures indicate potential eruptions. One of them is the volcanic isle, Surtsey, off Iceland's southern coast, which appeared unexpectedly above the surface of the sea in a burst of fiery lava and ash in 1963. According to written records, Mount Hekla, Iceland's most celebrated volcano, has experienced major eruptions approximately twice per century since the 12th century. Seen from the treeless lava plains at its base, this elongated composite cone, 4920 feet high, stands stark against the Arctic sky.

Hekla's most recent powerful eruption, in 1947, began as a tremendous discharge of rocks and ash that reached to the stratosphere, darkening the entire countryside. The winds of the upper atmosphere carried some of the cinders and ash as far as Scandinavia, 1000 miles east of Iceland. The lavas, which for more than a year poured out intermittently from vents along the summit, had temperatures of approximately 1900° F. When these lava outpourings finally ceased, they had—along with the new pile of ejected debris—added 450 feet to the height of Hekla's cone. After the eruptions ceased in the spring of 1948, dense volcanic gases continued to seep down the flanks of the volcano, collecting in nearby valleys where grazing animals were occasionally suffocated by the fumes.

Periodically, parts of Iceland are devastated by a turbulent flooding known as *jökulhlaup*. These floods are sudden rushes of billions of cubic feet of melted glacial water, resulting from the eruptions of periodically active volcanoes that melt out large pockets in the undersides of ice sheets. Immense volumes of ice are melted and retained under a glacier until a weakness develops in the confining ice, when the pent-up waters burst through the rupture. The waters surge across the countryside, carrying huge ice fragments and boulders at velocities as great as 60 miles per hour. Farmhouses, barns, people and livestock in the flood's path are swept up in the torrent.

Two active volcanoes, Katla and Grímsvötn, have their flanks and summits almost completely covered by broad ice sheets. They have produced most of the country's recent large floods. The more fearful is Katla, lying beneath the Mýrdalsjökull glacier in southern Iceland. When it erupts, about twice per century, it unleashes a *jökulhlaup* that rushes pell-mell across the vast Mýrdalssandur plains toward the south coast like a major river at flood level.

Iceland's coastal areas south of the big Mýrdalsjökull and Vatnajökull ice sheets contain wide, flat *sandurs*—broad plains made up of roughly sorted, irregularly layered sand, gravel and erratic boulder deposits left by glacial meltwaters and successive *jökulhlaup* deluges. Many glacier-fed streams meander southward across

Floods called jökulhlaups *occur in Iceland where glaciers cover volcanoes. Eruptions melt the ice from the bottom up, releasing a deluge when the melting and eruption break through the surface.*

Boiling through the mud, an Icelandic hot spring vents a plume of steam. Hundreds of such springs provide heat for homes and hothouses, where large varieties of flowers, fruits and vegetables are grown.

these plains toward the sea, while erosion and deposition along their banks cause the streams to shift their beds constantly, to the discomfort of cartographers and local farmers.

Iceland is the home of the geyser (the term comes from the Icelandic word *geysir*, meaning "gusher"). One of the world's three major geyser areas is located about 35 miles northwest of Mount Hekla. North America's Yellowstone Park and the region of Lake Taupo and Lake Rotorua on New Zealand's North Island are the other two great geyser locations on the Earth. Iceland's Grand Geyser has largely subsided during this century after having spouted frequently for hundreds of years. Its huge cone of mineral deposits—left behind by the evaporation of the hot waters after each of its eruptions—is surrounded by about two dozen active geysers of various sizes.

About 700 hot springs are scattered across Iceland. The hottest (more than 160° F) are located in the central rift area of volcanism and geologic activity at fissures in the crust. In some cases, boiling temperatures cause hissing vapors to escape from vents under pressure, after which the steam condenses in the cold air as turbulent, misty jets. Wherever soil or clay covers these vents, noisy, bubbling mud pools form.

Many an Icelander's house is heated by the warmth of these geothermal waters. Almost every building in the city of Reykjavik is heated in this manner. The

hot water is tapped by means of boreholes drilled 1000 to 2000 feet into the Earth. With a temperature of about 220° F at its underground sources about ten miles from town, the water reaches Reykjavik's homes at about 212° F.

Hydrothermal energy is of vital importance to Iceland's economy. Few trees grow on the island at present, and no significant coal or petroleum is found in the volcanic bedrock. The birch forests that blanketed parts of the country centuries ago never recovered from extensive cuttings by early settlers.

Hveragerdi is a small inland village located approximately 25 miles east of Reykjavik. Although less than 200 miles below the Arctic Circle, the village is well known for its fruits, vegetables and flowers—even tropical varieties—which are made possible by its hydrothermally heated hothouses. Many homes and commercial buildings are built right over the hot springs. Several years ago one family got more hydrothermal heat than it wanted when a crack opened in the kitchen floor and scalding water gushed into the house.

Such incidents might vex the rest of us, but Icelanders are willing to deal with relatively small inconveniences in return for their inexpensive, pollution-free and readily available source of heat. With the benefit of centuries of experience, they have learned to cope with, use and appreciate their country's uncommon, if sometimes troublesome, geologic resources.

Chapter Three

THE LAND'S SURPRISES

From ancient times to the present, man's imagination has been excited by strange rocks, tantalizing ores and curious sands tucked away in odd corners of the Earth

Ayers Rock— Home of the Primeval Spirits

By Victor Carell

It was raining a little that day in September many years ago as we left Alice Springs in our truck. Our destination—some 275 miles to the southwest across some of the most striking terrain of central Australia—was Ayers Rock. At that time the region was a sacred reserve of Aborigines; there was no road, only barely distinguishable tracks. It was a time when special permits were required to enter such reservations, set apart by the government for the Aborigines.

Today the Rock is well known to tourists. Air-conditioned buses and small planes take people there, to the conveniences of modern hotels and guided sightseeing excursions. Only the spirits of the Aborigines remain. No longer do the tribes of the Pitjandjara congregate at the sacred monolith to carry on their communal ceremonies in this once-great center of their tribal beliefs and home of the spirits of their heroes.

By the time we were past Heavitree Gap heading south, the sky had cleared and it looked as though the saying "It never really rains during the Dry" would prove true. My wife and I had left Alice Springs mindful of the advice of our old friend Ted Strehlow, anthropologist and expert in Aboriginal languages. "Find Maggie Springs water hole right away. If you don't find water, return immediately. That country can be very dangerous if you run out of water."

South of Alice Springs the land had a look of desolation. But the farther we traveled, the greener grew the country. Although it was the dry season, there had been unusual rains. The plains looked like a beautiful parkland, except for scattered areas of red sandhills and occasional dry lake beds encrusted with salt. Our way led through a series of small hills that were like roller coasters of yielding sands. We approached each hill at high speed, fearing that otherwise the truck would bog down.

By late afternoon we could see the purple crown of Mount Conner in the distance. A soft wind rustled the plains of yellow grasses stretching before us to the base of the mountain, which stood, misty and mysterious, some miles from our track. We decided to camp for the night, and drive the remaining short distance to Ayers Rock come morning.

Ayers Rock is one of the world's wonders. The enormous monolith's sand-blasted and water-eroded walls rise almost vertically for 1143 feet above the surrounding flat plain. There are no foothills—just the stark silhouette of this enormous loaf of red sandstone that is some five and one-half miles in circumference. To geologists, Ayers Rock is a "monadnock"—named after the American Mount Monadnock in the state of New Hampshire. A monadnock is an erosion-resistant remnant of mountain—a single stump of rock representing the last remains of an old, eroded mountain range.

Early the following day we caught our first glimpse of the red monolith. We had driven up a sandhill, and there, in distant solitude, we could see Ayers Rock rising from a broad, flat plain of grass, trees and undulating hills of red sand. We pressed forward in excitement. But after our first glimpse from the top of the sandhill we seemed to be a long time reaching the Rock. Its very size makes distances deceptive. Each time we glimpsed it anew it appeared still larger, until we were overwhelmed by its bulk as we drove into its shadow, with its ramparts blocking the sky above us.

We drove around the base of Ayers Rock until we came to the depression that is the site of Maggie Springs. Water trickled down through a ribbed pattern of channels cut into the rock.

A contrast of sunlight and shadow emphasizes the dramatic vertical ridges and crevices that millions of years of erosion have sculpted on the face of central Australia's stark and mysterious Ayers Rock.

A few dark clouds raced across the sky. The wind was gusty. As we sought a camping site, sparsely scattered large drops of rain began falling. So, to be safe, we moved toward higher ground. Following a path among the boulders, we found, under an overhang, an oblong cave—a wide, grinning mouth in the face of the mountain. The floor of the cave was of fine white sand. Its back wall, away from the weather, was covered with many Aboriginal paintings. They were built up layer upon layer—here, red on yellow, black circles on strange figures and ghostly white footprints of birds and animals. These totemic symbols were obviously of different dates. Many of the brighter patterns were superimposed upon earlier, faded drawings.

As the rain began falling heavily, we carried our supplies up into the cave and brought in some wood. Heavy storm clouds moved low in the atmosphere, trailing their rain across the face of the rock. Looking up into the giant semicircle of sandstone about us, we could see the curtain of moisture like a silver veil. Collecting into streams on the flanks of the summit, the rain formed countless small waterfalls. Spilling over

ledges, these cascaded down into the basin of earth about us.

Later the rain eased and the Moon came out. We had already cooked and eaten our evening meal, so we went for a short walk. It was an unforgettable night. We moved among the immense shadows cast by the ghostly Moon. Reappearing between scurrying clouds, the Moon cast its light over the wet Rock. A moment later the Moon disappeared, pitching all into darkness. As we moved along the northern edge of the Rock, the Moon shone forth again, illuminating in the north face the deep erosion pit whose shape suggests a gigantic human skull. As we stared in awe at this formation, rain started to fall once more, forcing us to return to the cave for the night.

Next morning it looked like a long stay, for the rain continued without respite. With torrents of water pouring down the crevices, Maggie Springs had overflowed. A river was tumbling around the base of the Rock and racing out across the plains, and the saying "It never really rains during the Dry" seemed open to question, to say the least. Under the dark, sullen sky

Just before a rainstorm, gum trees near the base of Ayers Rock seem to huddle for shelter within the shadow of the moody, time-battered monolith.

With the coming of a rare torrential rain, countless waterfalls cascade down the grooved face of Ayers Rock, then vanish when the storm ends.

the red rock now had a wet, gunmetal-gray appearance.

It was awesome—that tremendous fall of water. Cascades leaped down the Rock literally by the thousands. The air reverberated with a thunderous tumult. But there was something fanciful in the sight too. It all looked rather as though a giant had emptied a bottomless jug of water over the bald head of a wrinkle-faced old gentleman. Sometimes the water, caught by a sudden gust of wind, would fly back up into the air above the rock, like a stray silver lock blown into rebellious disarray.

Sitting in the cave on bare stones and looking at the towering rampart opposite, my wife and I meditated on the forces that had left this heavy lump of stone alone in the midst of a great plain—ageless forces of wind and water erosion, sudden heavy rains such as the one that had momentarily trapped us. The thousand-foot dash of water which began as a slowly creeping film at the crest of the Rock, then as it gained momen-

tum shredded into separate torrents that ribboned down to the earth, made us aware of our own, Lilliputian size in our cave. The cave was wet and cold, yet a haven from the power of the weather.

Our most outstanding memory of the cave is the stark symbolism of the Aboriginal paintings; their yellows, dull reds and blacks, seen in the flickering light from our fire; the smoke of the fire circling past them, before finally, reluctantly mixing with the dense, wet night outside. It was most fantastic—seeing these vivid pictures, these expressions of Aboriginal artistry, as though we were seeing the art of our own prehistoric ancestors still being carried on today in a form unchanged over the intervening thousands of years.

Often we found ourselves wondering what the cave drawings represented. Why were they drawn? We know that much of the painting resulted from the Aboriginal's need to depict the sacred epic stories of his culture's heroes; it gave form to the totemic ancestors, allowing the men of the tribe to commune more closely with their own spirit origins. Anthropologist C. P. Mountford in his book *Ayers Rock* says that the Rock in the Aborigine's conception has "remained unchanged from when, at the close of the creation period, it rose miraculously out of a large flat sandhill." The Aborigines credit ten totemic "Dreaming Time" spirits with

the creation of their art symbols, the caves and the strange formations of the Rock. Some of the eroded shapes were believed caused by battles between totemic ancestral snakes, or were sites where ancestral spirit wallabies camped or where spirit bird women made love with spirit dingo men.

However, other drawings must have been made merely to tell some story of personal importance to the artist himself; probably the animal footprints that abounded were a record of a noteworthy or unusual hunt. We saw footprints of emus, wallabies and dingoes, several sketches of animals and snakes. Obviously some of the drawings in this rock gallery were of ancestral beings. The stylized designs, the concentric circles, the rectangles filled with lines, the whole series of circles within circles are known to represent long journeys made by ancestral figures and the features of the country passed over. But only each artist himself, I feel, could explain his own drawing.

In spite of our fire and all our clothes and blankets, we felt the cold badly. Our sympathies lay strongly with the Aborigines who formerly took shelter under the Rock to avoid the wind and the rain, probably huddled, as we were, shivering about a smoky bonfire.

At dawn on the fourth day the rain finally began to let up. The next morning there was a faint circle of pink touching the edges of cloud. Then the Sun made its long-awaited appearance above the horizon. The Sun's light lit up the drawings of our cave, touching each with a splash of gold. We ate a hurried breakfast and then, carrying our cameras, we set off to walk around the Rock.

Ayers Rock is magnificent from every angle: high, steep ridges running knifelike up, up to the sky; crevices cutting deep into the heart of the rock; walls of round, smooth stone going straight up for a thousand feet, dwarfing the large gum trees that ring the Rock's base.

The wind and the weather have brought about the most curious formations. Besides the huge skull-shaped cavity gouged out by erosion, there is another hollow cavity formed in the classic shape of a bell. Strangely shaped caves abound, and many have totemic drawings. But most impressive of all is the hugeness of the Rock itself. Looking to right or left along its base, one can see only a very small part of the whole; and as one looks up along the sheer wall toward the sky, Ayers Rock seems to continue endlessly. To the northeast

Large, dim caves along the base of the sandstone monolith were used until recently as sanctuaries for religious rituals by Aboriginal tribesmen who considered Ayers Rock the center of their universe.

we saw the plain jeweled with lakes from the heavy rain. On one of the tall trees at the base of the Rock we found an eagle's nest containing young birds. What better place for eagles than this inland promontory?

As we rounded the northwest corner the Sun's rays were falling upon the curious pile of rocky turret formations known as the Olgas. Across the green sweep of plain, the summits of the Olga mountain range some 20 miles away glowed claret red and gold above a purple foundation, looking like a gigantic heap of opals held in a hand gloved in russet velvet.

We returned to our cave just before dark. We were exhausted by our day's explorations. That night was the last we would spend in the cave. Next morning we packed, filled our water containers from Maggie Springs and started off.

As we drove alongside the Rock, it was glowing red in the morning sunshine. And it still shone when we took a final look from a sandhill 20 miles away. We headed south across country toward the Musgrave Mountains, carefully avoiding the soft wet patches in the desert sands and the waters remaining in hundreds of pools. Already a miracle was happening: tiny new green shoots were springing up everywhere as grasses and wildflowers responded to the life-giving waters of the rain.

Vaguely reminiscent of the recesses in a human skull, these odd, shadowy pits in Australia's Ayers Rock were hollowed out by deep erosion.

The Riddle of Mexico's Strange Stone Spheres

By Robert S. Strother

Strewn about an eroded mountainside in the state of Jalisco in western Mexico are hundreds of ancient, huge stone spheres that look like gigantic bowling balls abandoned by the gods. They range from four to 11 feet in diameter, and their discovery in 1967 was the unexpected beginning of a scientific detective story.

Until then the existence of only one of the mysterious stone balls was known, except to a handful of Mexican farmers living near the rugged slopes of Sierra de Ameca, who knew that others existed but apparently saw nothing remarkable in their presence. The previously known specimen, six feet in diameter, stood on a natural rock pedestal at the entrance to a long-abandoned silver mine, the "Piedra Bola" (Stone Ball), located high in the Sierra de Ameca, some 50 miles west of Guadalajara. Because of its almost perfectly spherical shape, this ball was long assumed to be man-made—perhaps a religious symbol chiseled from rock by a tribe of pre-Columbian Indians.

But in 1967 this assumption was shattered. Ernest Gordon, an American mining engineer who had been superintendent of the Piedra Bola Mine years before, returned to Mexico to make a prospecting trip through the remote, overgrown area. Within a mile of the old mine he was amazed to discover four more giant stones as symmetrical as the one he remembered, but in a somewhat more weather-beaten condition. A fifth stone in the new group was deformed, yet recognizably spheroid in outline.

Gordon knew of reports by archeologist Matthew Stirling that described smoothly finished stone spheres hewn from granite by pre-Columbian Indians in Costa Rica. So Gordon sent photographs of his own discovery to the archeologist-author and offered to guide him to the site of the Jalisco spheres.

In December 1967 Stirling flew to Mexico and soon set to work in the Sierra de Ameca digging the earth away from some half-buried balls. In doing so, he exposed 17 more stone spheres whose existence was unsuspected. Indeed, the mountain seemed to be full of buried giants. Stirling's party might have gone on digging there indefinitely had not a local workman thrown down his spade in mutiny. What was the sense of all this digging, he wanted to know, when by climbing over the next ridge Dr. Stirling could find enough *piedras bolas* to satisfy anybody, all standing out in plain view?

A stiff one-hour climb confirmed what the Sierra de Ameca villager had told Dr. Stirling. Almost at the summit of the mountain, the party came upon a stone

Originally believed to be the work of pre-Columbian Indian sculptors, these precariously perched stone spheres in Mexico's Jalisco state proved to be products of unusual volcanic eruptions.

ball larger than any they had seen until then. There it stood, 11 feet in diameter, silhouetted against the sky at the very top of the ridge. Once Dr. Stirling's party reached the summit of the ridge, they looked down the far slope and saw scores of big stone spheres among the trees descending to the valley floor. A few of these big stones were pear-shaped, and two were welded together in a curious dumbbell formation. But most, except for weathering scars, were almost perfectly spherical and remarkably uniform in size, with a common diameter of roughly six feet. Apparently some had tumbled from their original positions into the ravine below, where they stood amid the debris of several that had rolled down and smashed into pieces. One still in place had been split in half, evidently by the heat of a forest fire. On the crown of another grew a little plumelike tree.

Dr. Stirling had already begun to doubt that the stones were man-made sculptures, and with this new discovery he became convinced that they must represent some unique creation of nature. There was no sign that anybody had ever lived on the mountain—no pottery shards or artifacts of any sort—and now the large number of big spheres showed conclusively that no tribe of Indians, no matter how industrious, could have carved all of them. Surely they had to be of natural origin. But the mystery remained. How had the spheres been formed?

Hearing about the mysterious spheres, I recently made a journey into the Sierra de Ameca to inspect them firsthand. In the old colonial town of Ahualulco de Mercado 40 miles west of Guadalajara, I hired an ancient taxi equipped with special tires for traveling through mud. It was piloted by Miguel Hernández, a local driver skilled in negotiating barely existent trails obstructed by boulders and cavernous potholes. Hernández, who is one of the half-dozen people in Ahualulco de Mercado who know about the stone spheres, explained that the badly rutted track we were forced to follow had once been a well-cobbled road—prior to its abandonment by a mining company 40 years ago.

After a rough three-quarters of an hour, we arrived at the Tiro Patria Ranch, where the old mining road ended. There Don Antonio Martínez, the amiable 70-year-old patriarch of the region, set some of his nine children to work saddling horses and mules for the five-hour trip up and down the mountain.

Leading the way on a white mule, Don Toño remarked rather cheerily that the upward climb might seem difficult, but that it would prove simple compared with the descent. The trail, visible only to him, snaked across rocky ground made more treacherous by loose, fist-sized stones disguised in many places by a cover of fallen leaves. As a novice rider busy dodging overhanging branches and thorn bushes, I was gratified that I needed to worry only about retaining my balance in

Buried by time, hidden by foliage, some of these mysterious Mexican spheroids are hard to find, even though they may be as large as 20 feet across.

the wooden saddle, for my horse, walking at the heels of the lead mule, knew its business, and rarely made a misstep among the camouflaged clutter of stones.

A half-mile out three young dogs from the ranch squeezed into our file, trotting along almost under the hoofs of the horses. At intervals they broke away from the trail on long, silent forays into the surrounding forest on the scent of deer or mountain lion. Much of the time the trail took a sharply twisting course. On especially sharp turns our scrambling mounts sent showers of rocks bounding down the mountainside while Don Antonio calmly pointed with his short braided whip to points of interest in the valley far below. The reason for the region's isolation became more obvious as the trail grew steeper.

The first of the spheres lay in a ravine and came into view after one hour of riding uphill from the Tiro Patria Ranch. It proved to be six feet in diameter. Slightly farther down in the ravine were the fragments of a second, shattered spheroid. In another ten minutes the path had taken us into an eroded mountain valley with nearly vertical walls where there were numerous free-standing 20-foot columns of rock. On top of one of these columns, just as some natural force had left it, stood one of the great stone spheres. This fantastic clearing was bounded on one side by a boulder as big as a ten-room house.

Higher on the trail the spheres occurred with greater frequency. Finally we reached the site of Dr. Stirling's main discovery. Here I saw a stone ball 11 feet in diameter, balancing rather precariously on the main ridge—an astonishing sight against the azure of the sky. Huge as this sphere was, Don Antonio predicted that it would soon yield its pre-eminence. He pointed to a domed mound between the trees a few yards away. Closer inspection suggested that it had to be a buried spheroid which, judging by the arc of the protruding dome, would be at least 20 feet in diameter when fully uncovered.

"*Es el jefe de todas*," Don Toño told me. "The chief of them all." But the chief of what? That remained the question.

In March 1968 a scientific expedition visited the Sierra de Ameca to try to solve the riddle of the spheroids. The expedition was organized through the combined efforts of the National Geographic Society, the Smithsonian Institution and the U.S. Geological Survey. It was under the leadership of U.S. geologist Robert L. Smith.

The main part of the puzzle quickly fell into place. Dr. Smith had once examined some natural two-foot spheres in New Mexico, and had identified them as being composed of the volcanic glass called obsidian. He saw at once that the giant Mexican spheres were of the same material, and concluded that they must have been formed, like those in New Mexico, within immensely deep deposits of volcanic ash.

The geology of the region led Dr. Smith to decide that a volcanic avalanche occurred some 40 million years ago.

"The ash flow deposits formerly must have covered much of the Sierra de Ameca, but erosion has removed all but a few remnants," the geologist states.

"The spheres were formed by crystallization at high temperatures," he suggests. "The ash was 75 to 85 percent hot glass, at a temperature somewhere between 1000° F and 1400° F. At these temperatures, with slow cooling, volcanic glass can crystallize. Crystallization began at numerous nuclei, progressing outward from each on spherical fronts until inhibited by declining temperature or coalescence with neighboring spheres.

"Only in one case did I observe a sphere still enclosed in its original context of ash. Since the ash is comparatively soft, it easily erodes away. Erosion has left the spheres as bare, visible bodies upon the terrain."

Dr. Smith admits that it may be impossible to prove that the crystallization process proceeded precisely the way he outlines it, even when samples of the volcanic rock and ash are thoroughly analyzed in the laboratory. But the broader question of the origin of the Sierra de Ameca spheroids has been answered. They had a natural, fiery beginning in the ancient volcanoes of the region millions of years before man walked the Earth.

New Zealand Has Stone Spheres, Too

Rock forms along Moeraki Beach, about 40 miles north of Dunedin on New Zealand's South Island, look very much as if they were related to the volcano-born stone spheres of Jalisco, Mexico. But the spheres at right are concretions—balls of mineral matter that collected around nuclei within layers of mud on the sea floor, shrinking and cracking as they hardened. After the sea bed was raised as beach, the waves wore away the mud, baring the spheres. Thus dissimilar forces, volcanism and sedimentation, produced a set of misleading twins.

Another World Beneath Our Feet

Dazzling and mysterious, vast networks of caverns twist and turn wherever subsurface water has tunneled into soluble rock, creating a wonderland of strange shapes and colors

For centuries water worked underground—sometimes in roaring currents, sometimes in slow, steady drips—to produce fantastic landscapes in the dark. It was only left for man, with his curiosity, to find and explore them. In virtually every country of the world, we can visit the magnificent grottos that nature has created. Some are stupendous in scale. For example, America's Carlsbad Caverns in New Mexico extend some 23 miles along an underground route. The largest of the many chambers is a dreamlike study in weird shapes and colors that covers 14 acres and rises more than 200 feet in some places. Smaller and less famous caves often are no less exquisite in detail. Stalagmites rise from the floor and stalactites hang from the ceiling, suggesting some alien kind of life. Where light strikes, odd pale colors are revealed that have no natural counterparts on the surface. Water is the leading agent of this underground artistry. It flows and seeps through cracks in the rocks, dissolving minerals as it moves, depositing them later when it evaporates and dries out. Thus, delicate formations build up over thousands of years, undisturbed by wind or weather. The overall effect is that of a fantasy land, totally unlike anything to be found in sunlight and the open air.

The King's Palace in New Mexico's Carlsbad Caverns is decorated with intricate draperies of water-made forms. This chamber is open to the public as part of a three-mile tour that descends 829 feet.

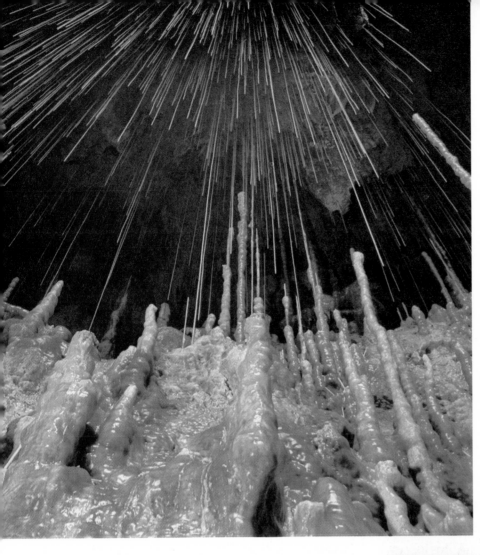

Resembling trunks of palm trees, the odd stalagmites (above) are in the "Green Grotto" of Sardinia.

Caught in a camera's flash at left, fine stalactites in the Monte Soratte caves near Rome look like a shower of shooting stars.

Enormous, icicle-like stalactites in New Mexico's Carlsbad Caverns (right) formed from the slow trickle and evaporation of mineral-bearing water.

Water dripping to the floor from a cave ceiling leaves minerals that form stalagmites, much as a child at the beach makes castles of dripping sand.

Down to the Deeps for Gold

By J. D. Ratcliff

It is a glittering real-life counterpart of the fabled El Dorado. So far it has produced over 30,000 tons of gold with the end not yet in sight. Known as the Golden Crescent, this extraordinary piece of real estate stretches in a broad arc east and southwest from Johannesburg, South Africa, for some 300 miles. Here, three fourths of the free world's annual gold production is mined. At officially set 1971 prices, the gold it has produced is worth $33 billion—and billions are yet to come. But the existence of this fabulous treasure trove was not even dreamed of until 1886.

The countryside around Johannesburg is veld—a grassy, rolling plain. The geological explanation is that about 2.6 billion years ago it was an inland sea surrounded by mountains. This huge lake was fed by tumbling rivers which eroded gold-containing gravels from the adjacent mountains and transported the material to lake shores where it was deposited. The sorting action of the water caused the gold-bearing gravels to be deposited first followed by the lighter silts. Over millions of years this sediment slowly compressed into

rock. During periods of volcanic activity, hundreds of feet of lava were dumped on top of the sedimentary rock. And in subsequent wet periods the swollen rivers brought down more gold-bearing sediments into the ancient basin, depositing new strata on top of the lava. Thus a geologic layer cake was formed—with narrow gold-bearing "reefs," ranging in thickness from an inch to 20 feet, sandwiched between layers of lava and sedimentary rock.

At some later stage in the history of the Earth, there was violent earthquake activity in the region. This heaved, twisted and broke the gold-bearing rocks—hoisting some portions up 300 feet, dropping others as much as 500 feet. That's why today, as a mine is progressing nicely, a reef may suddenly disappear.

Until late in the 19th century, most of the world's gold production came from sites in California and New South Wales where great gold discoveries had occurred in 1849 and 1851, respectively. Then in March 1886 Petronella Francina Oosthuizen, a widow who owned a farm five miles outside the settlement that would become modern Johannesburg, hired two itinerant odd-jobs men, George Harrison and George Walker, to enlarge her home. Although the exact chain of events isn't known, it seems probable that they were digging foundation stone out of a rock outcrop when something aroused their curiosity. They decided to pound one piece of rock to powder and pan it. Gold! The two ragged nondescripts had found a reef to dwarf anything ever seen.

The rush was on. Harrison sold his claim for $50, wandered off into the hinterland to do some more prospecting. According to one story, he was eaten by a lion. Walker got $1500 for his claim and died in poverty in 1924.

The men who work the area today hardly fit the picture of dusty prospectors equipped with pick, shovel and mule. They are highly trained geologists, geophysicists and mining engineers backed by entrepreneurs with millions of dollars to spend. If an area looks promising, core drilling follows—an expensive business costing as much as $200,000 for a single drill hole. If the rock samples brought up indicate a reef that is economically feasible to mine, a shaft is sunk.

Recently I visited one of these new mines—Western Deep Levels, 43 miles west of Johannesburg. It is one of the biggest and most costly gold-mining ventures ever undertaken. With an estimated life of 60 years it is expected to yield $1.7 billion worth of gold—enough to stock the treasury of a major nation. Shaft sinking for this huge enterprise began in January 1958 and employed 15,000 men.

Only superlatives can describe this operation. Eventually, it will go 12,500 feet below the surface, deeper than man has ever before gone for gold. It is pumping about 260 million gallons of water a day from water-

In a shaft more than two miles below the surface, two workers in South Africa's Western Deep Levels gold mines prepare to blast a solid wall of rock.

A mountain of gray tailings—the rock debris remaining after gold is extracted from its ore—rises from the complex around the No. 2 Shaft of the Western Deep Levels mine near Johannesburg.

bearing strata. For every ton of rock brought to the surface the mine pumps tons of air back into the ground.

Heat has always been a great barrier in deep mines. As one descends into the Earth temperature increases about 1° Fahrenheit per 180 feet of depth—evidence of the great pressure of overlying strata and the heating effects of radioactive isotopes deep within the Earth's interior rock. At 12,500 feet rock temperatures rise to 135° Fahrenheit, and since water must be sprayed on rock surfaces to hold down deadly, lung-choking dust, the humidity is all but unbearable. To take care of this problem Western Deep Levels has an air-conditioning system which will eventually have five times the capacity of that used to cool the United Nations complex of buildings in New York. It will hold maximum mine temperatures to a barely tolerable 85° Fahrenheit.

A visit to the mine begins with a 3000-feet-per-minute descent in one of the triple-decked "cages" that can carry 120 workers per trip. Ears pop and your heart seems to be crowding into your mouth in a drop that appears to have no end. Finally the cage slows and halts. You are at the 6600-foot level. Change here for another cage that goes on down to 10,000 feet. Here an inclined shaft begins that goes to 11,400 feet. And from this level a vertical shaft begins, reaching eventually to 12,500 feet.

All of this time, effort and treasure has been spent to get to the Carbon Leader Reef, an ore seam that is often no more than two inches thick. But the Carbon Leader is one of the richest gold seams ever found. There is no visible gold in most gold-bearing ores—they look like grayish rock. In the Carbon Leader, however, flecks of gold are often actually visible. To

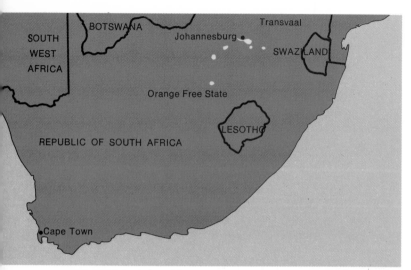

The major goldfields of South Africa, indicated by light areas, spread in a "Golden Crescent" around Johannesburg, the gold capital of the modern world.

mine this thin seam, 36 inches of valueless rock must be drilled, dynamited and shoveled out of the way. Some two tons of rock must be hoisted to the surface for each ounce of gold recovered.

How are these minuscule amounts of gold sorted from such huge volumes of rock? First, the ore is ground to talcum-powder consistency, then agitated with a cyanide solution. The cyanide dissolves the gold and carries it away in solution. Then intricate chemical and filtering steps extract the gold from the cyanide solution. Many mines work ore seams that are rich in uranium as well as gold. In such cases, after the gold is extracted, the residues go to a uranium extraction plant. South Africa has produced well over a billion dollars' worth of uranium oxide since the first uranium plant was opened in 1952.

After the minerals are extracted, the waste—powdered rock in solution—is pumped out into an area that may cover 100 acres or more. After a layer of waste rock dries and hardens, a stepped-back retaining wall is installed; then another layer is poured. Thus a great mound grows, much like a pyramid. Lately, mine companies have begun to plant grass and trees on waste piles, making a series of green hills on the previously flat Transvaal plains.

As gold comes from mine refineries, it is about 89 percent pure. The final refining step takes place in the Rand Refinery, a group of buildings surrounded by green fields on the outskirts of Germiston. Day and night a glittering stream of molten gold, 99.6 percent pure, is poured from crucibles into 400-ounce ingots—at a rate, in recent years, of more than a billion dollars' worth a year.

Waste cannot be tolerated when dealing with a product of such great value. Electrostatic precipitators extract gold from flue gases. Old crucibles are ground

up, flooring from old refineries, even clothing of workers is processed to extract what gold they contain.

The final product of the refinery is the gold ingot. At the Rand Refinery, bars of gold weighing 25 pounds—worth $14,000 apiece at official 1971 market prices—are packed in a small wooden crate about the size of a shoe box. Each box holds two of the ingots. Regular shipments of these boxes go to the South African Reserve Bank, then by ship to the London Gold Market. Finally the ingots are purchased either by the central banks of the various nations of the world or by commercial dealers for use in jewelry making and dentistry or in the electrical, electronics, ceramic, aircraft and aerospace industries.

It may seem ironic that government banks around the world purchase gold that has been brought up from deep underground at great cost and labor only to be put back underground in vaults. Traditionally, the solidity of any nation's currency has rested on the gold that backed it. As yet, international monetary experts have found nothing else as convenient upon which all men can agree as a common standard of monetary value. As long as this is so, men will continue to descend ever deeper into the Earth in its quest.

In an air-conditioned tunnel 7500 feet below ground, workers pursue an elusive vein of gold in the fabulously rich Carbon Leader Reef.

The Mystery of the Singing Sands

By Paul Brock

Scientists are trying to solve a puzzling mystery—the phenomenon of "singing sands." At Britain's University of Newcastle-upon-Tyne, extensive field and laboratory experiments into sands and beaches that "sing," "whisper," "squeak," "roar," "ring," "hum" and "shriek" have been carried out during the past decade. These investigations have attempted to explain the cause of singing sands, which have mystified mankind for centuries.

Musical sands occur in many widely scattered places on the Earth's surface. Perhaps the best known exist on the island of Eigg in the Inner Hebrides off Scotland's west coast. In his book, *The Cruise of the Betsy*, the 19th-century British geologist Hugh Miller (1802–1856) gave an eloquent description of the sand at Eigg: "I struck it obliquely with my foot . . . and the sound elicited was a shrill sonorous note, somewhat resembling that produced by a waxed thread, when tightened . . . and tripped by the nail of the forefinger . . . at each step, and with every blow the shrill note was repeated. My companions joined me; and we performed a concert. . . . As we moved . . . an incessant *woo, woo, woo* rose from the surface, that might be heard in the calm some 20 or 30 yards away."

Sonorous sands have been found at many other beaches and deserts around the world, including Long Island and Massachusetts Bay in the continental United States; the Hawaiian Islands; the west coast of Wales; the Northumberland coast of Britain; the island of Bornholm, Denmark; Kolberg, Poland; a few spots in Australia including the coast of New South Wales; Brazil and Chile; and several deserts in Asia, Africa and the Middle East.

When walking over musical sand, the foot sinks deep, as the grains are easily displaced. The highly polished surfaces of millions of grains set up a continuous vibration and produce a prolonged sound like a note of music.

Charles Darwin was one among many 19th-century scientists to be intrigued by the growing mystery of the singing sands. In his book, *A Naturalist's Voyage Round the World*, a journal entry dated April 19, 1832, reads: "Leaving Socêgo [in the vicinity of Rio de Janeiro, Brazil] . . . we retraced our steps. It was very wearisome work, as the road generally ran across a glaring, hot, sandy plain, not far from the coast. I noticed that each time the horse put its foot on the fine siliceous sand a gentle chirping noise was produced."

Three years later Darwin reported singing sand in

The irregular shapes of "singing sands," as shown in this enlarged Hawaiian sample, rule out the theory that all grains must be round to make noise.

the valley of Copiapó, Chile: "Whilst staying in the town I heard an account from several of the inhabitants of a hill in the neighborhood of what they call 'El Bramador'—the roarer or bellower. . . . So far as I understand, the hill was covered by sand, and the noise was produced only when people, by ascending it, put the sand in motion."

Other allusions to sonorous sands are scattered through the writings of a thousand years. The *Arabian Nights* mentions them, and old Chinese chronicles tell of singing sands occurring in the Gobi Desert of central Asia. Marco Polo related hearing the notes of supernatural musical instruments and drums drifting above the Asian desert sands.

The phenomenon of singing sands probably accounts for an old legend of a buried monastery somewhere in the Sinai Peninsula. The legend refers to a monastery that had been covered by a mountainous sand dune. Its bells continued to give off drawn-out ringing notes that were sometimes heard by nomads and other travelers passing in the desert. Tales were told of travelers' camels being frightened by underground music as they came within earshot of this mysterious mountain.

Some 200 years ago European pilgrims to Mount Sinai brought back stories that seemed to confirm the existence of the buried monastery. They too had heard a prolonged, steady ringing in a certain region of the

The "barking sand" along the beach at Kauai in the Hawaiian Islands is composed of pulverized lava, shell and coral. The strange sound produced at times when people walk on it is like the barking of a dog.

desert. It sounded as though the *nakous*, or suspended metal bar that serves as a bell to Arabian priests, was being struck rapidly and continuously. But the place where they heard it was deserted. Neither priests nor any other human beings were in sight.

So the Gebel Nakous, or Mountain of the Bell, became a legend. Travelers who ventured into this region of the Sinai in the 19th century were not content with the mythical explanation of the sounds. They traced them to the peculiar character of the sand which covered one side of a particular mountain. This sand had been deposited by the strong westerly wind that blows almost constantly across the peninsula. And when the wind blew strongest the mountain was said to emit a prolonged ringing note.

Hearing of the puzzling ringing mountain, the Scottish naturalist Sir David Brewster (1781–1868) visited Sinai and conducted an investigation. "The Mountain of the Bell," he reported in his *Letters on Natural Magic*, "is situated about three miles from the Gulf of Suez in that land . . . in which the granite peaks of Sinai and Horeb overlook an arid wilderness. . . ."

Brewster instructed one of his Bedouin guides to climb up the "musical" slope of the mountain. It was not until the guide had reached some distance, Brewster relates, that he perceived the sand in motion, rolling down the hill. At first he thought the sounds might be compared to those of a harp when its strings first catch the breeze. As the sand became more violently agitated by the increased velocity of the descent, however, the noise more nearly resembled that produced by drawing a moistened finger over glass. As the avalanche of sand reached the base, the reverberations attained the loudness of distant thunder, causing the rock on which Brewster sat to vibrate.

The first truly scientific study of singing sands did not come until the 1940s, when British physicist R. A. Bagnold investigated the phenomenon. Speaking of the "song" or "booming" of desert sands, Bagnold wrote: "I have heard it in southwestern Egypt 300 miles from the nearest habitation. On two occasions it happened on a still night, suddenly—a vibrant booming so loud that I had to shout to be heard by my companion. Soon other sources, set off by this disturbance, joined their music to the first, with so close a note that a slow beat was clearly recognized. This weird chorus went on for more than five minutes continuously before silence returned and the ground ceased to tremble."

Bagnold found that singing sands often occur in two general localities—on the seashore and on the slip-faces (or leeward slopes) of desert dunes and drifts. He applied the word "whistling" to the sands of Eigg and to beach sands in general, while using the term "booming" to describe desert sands. Tests showed that beach sand emitted a squeak or whistle at a frequency of between 800 and 1200 cycles per second (in the range of high C on a piano). The tone could be produced, he wrote, "by any rapid disturbance of the dry top layer—walking over it, sweeping it with the palm of the hand, plunging a stick vertically into it."

The sound emitted by desert sand, he found, is much lower in frequency when disturbed in the above ways—132 cycles per second. But when desert sand flows downslope in an avalanche, he discovered, it may attain surface velocities that make it hum quite audibly at roughly 260 cycles per second (about middle C) or at even higher pitches, depending on the speed of the avalanche. When sand from the Kalahari Desert in South Africa was taken from its desert atmosphere to Pretoria, it lost its "vocal qualities" unless kept in airtight containers prior to testing. Or the vocal quality could be restored by heating the sand to 200° C. These facts suggest that humidity may destroy the sonorous voices of at least some desert sands.

Recently the Newcastle-upon-Tyne scientists A. E. Brown, W. A. Campbell, J. M. Jones and E. R. Thomas have followed up on Bagnold's testing techniques. First, they place samples of singing sand in an evaporating dish, then they strike the samples with a blunt-ended round wooden rod, causing them to "sing." A breakfast cup and the handle of a wooden hammer serve the same purpose, they report.

They discovered that roundness of grain is not an essential characteristic of singing sands; rather, uniformity in the size of the grains is most significant in making a volume of sand exhibit musical properties. Moreover, the presence of fine particles impairs the singing of the sand and sometimes stops it altogether. When the grains are polished, unpolluted by other material and nearly all of the same size, the sand sings.

The ability of such sand to "sing" is destroyed by constant pounding, but is restored after the fine fragments produced by such pounding are removed by sieving, washing or boiling.

But what actually gives singing sands their musical properties? A general explanation of singing sands is advanced by the Newcastle-upon-Tyne scientists. It is clear, they say, that a shearing motion must occur between two or more layers of sand if there is to be any hint of music. When the layer of sand is thin and not confined (as in a thin dry surface layer on a damp beach) only an oblique blow will produce the sound.

Striking from above proves a convenient way of producing shearing motions in laboratory experiments, but this is effective only when the sand is supported by the sides of a container.

Under certain conditions, say the British researchers, a shearing motion can make a restricted volume of beach or desert sand vibrate almost like a volume of air within an organ pipe. But as to exactly how this happens, Bagnold still has the final word: "There is as yet no real explanation," he says.

Quicksand— Nature's Terrifying Death Trap

By Max Gunther

The low, swampy land south of Florida's Lake Okeechobee teems with exotic subtropical wildlife—a naturalist's paradise. One summer morning two American college students, Jack Pickett and Fred Stahl, shouldered heavy packs and headed into the dense growth in search of parasitic plants. As they were walking along the sandy bank of a small, nearly dry stream, Pickett, who was in the lead, suddenly cried out, "It's soft up here! Stay back!"

He had stepped onto what looked like dry, sun-baked sand. But the caked surface crumbled oddly beneath his boots, and he sank up to his ankles. Trying to reach firm ground, he floundered forward a few more steps. But with each step he sank deeper, until the strange marshmallowy sand had engulfed him up to his knees.

"It's quicksand!" he shouted. "Help me!"

Stahl knew it would do no good to plunge into the quicksand and try to rescue his friend. Both would then be trapped, and there was nobody for miles around to help them. He ran into the brush where he saw a long-fallen tree branch.

Pickett continued to struggle. With a tremendous effort he managed to free one foot from the ghastly trap. His other leg, however, sank up to the thigh. The sand around him was quaking now, like a monstrous bowl of jelly. He lost his balance and fell forward on his chest.

Stahl, who had raced back to the stream with the fallen branch, extended it to his friend. But Pickett couldn't quite reach it. Stahl yelled, "Take your pack off!"

The heavy pack was forcing Pickett down into the quaking sand. But the pack's catch was at his chest, submerged, and he couldn't move his hands through the mire to unhook it. He strained to hold his head up; still the sand rose swiftly to his chin. He gave a last terrified cry as the sand rose to cover his mouth and nose. Only his panic-stricken eyes showed.

"Try to grab the branch!" Stahl shouted.

Again Pickett struggled to lift his hands from the mire, but this only forced his head deeper. Frantically Stahl, using a rock as a fulcrum, pushed the branch into the sand and under Pickett's chest; desperately he tried to pry him upright. But the branch snapped.

All that showed of the victim now was the sole of one boot and his pack, both rapidly sinking. Fred Stahl sat down on the rock and buried his face in his hands. When he looked up again there was nothing left to see except a stretch of level, dry-looking sand.

This cross-section shows how a quicksand bed is formed. Water under pressure, in this case from a river, flows through underlying beds, then percolates upward through the sand, forcing grains apart.

Elsewhere in North America, near Bearden, Arkansas, a hunting party hiking along the Ouachita River broke out of some dense brush and stopped abruptly, some members of the party gasping in horror. On a level stretch of sand lay a man's head, apparently disembodied, the eyes staring up at the sky. Members of the party started forward again, but halted after another few steps, suddenly realizing what the situation was. They were looking at a bed of quicksand. The man in it had sunk to his head and starved to death.

Quicksand is one of man's oldest nightmares. To be trapped in it seems a peculiarly hideous way to die, and fiction writers and films have exploited its gruesome fascination with relish. In fact, quicksand has played so big a part in so many tall tales that it is hard to know where the line stands between fact and fiction.

What are the facts?

Though quicksand exists in many parts of the world, little was known about its composition until recent years. The most popular theory was that quicksand was made up of rounded sand grains. Unlike the jagged grains of ordinary sand, the theory went, the rounded grains acted like miniature ball bearings, rolling with so little friction that any weight, such as a man's, would sink rapidly. Another theory held the quicksand grains were lubricated with slime or some other slippery substance that made them slide out from under a weight. But no one knew.

Our present understanding of the phenomenon of quicksand stems from certain experiences in World War II. Then, during the 1944–45 invasion of Germany, Allied armies discovered they needed a lot more knowledge of how to move troops over unstable soils.

An adventure in April 1945 illustrates the problem that quicksand can pose to military forces. When an American supply convoy near Wismar was attacked by Nazi bombers, the lead driver, Cpl. Roger Jonas, immediately turned off the road into what looked like a sandy meadow. He felt his truck lurch and tried to open his door, but found it jammed shut. Then he stuck his head out the window and saw to his dismay that the truck was slowly settling into the meadow like a sinking ship. The sand was already halfway up the door.

Jonas climbed out the window and up onto the cab roof. In a few minutes the sand was climbing up the windshield. Between bomb bursts the corporal could hear a weird sucking noise, as of a man slurping soup. The sand reached the cab roof and Jonas climbed onto the tarp-covered cargo. In quiet terror he watched the cab disappear below him and the sand climb upward.

Finally he leaped for the roadway. Sinking to his knees, he pitched forward, frantically grabbing a tuft of grass on the road embankment. The roots of the grass held, and he pulled himself from the trap. By the time the air attack ended, his truck had completely vanished.

The American Army's interest sparked a number of scientific studies of quicksand. One was conducted by Dr. Ernest Rice Smith, geology professor at Indiana's DePauw University. He spent days studying a quicksand bed in a pasture not far from the university. The bed was near a small stream, and its surface was a mottled yellow-green from the pond slime growing on it. When he threw a rock into it, the sand quivered unpleasantly and seemed almost alive.

What made this sand different from commonplace sands? Smith scooped up a bucketful and later examined samples of the sand under a microscope. The grains turned out to be an ordinary mixture of shapes —some were round but most were jagged. So the round-grain theory was out. The lubricated-grain theory didn't seem to hold up either. Though Smith kept his sample of sand moist and the green algae continued to thrive, the sand wasn't quick any longer. It was just as firm as beach sand.

Back in the pasture, Smith talked to the owner of the farm. "Funny thing about that stuff," the farmer said. "Sometimes it's quick, sometimes it's firm. Come back in August and you'll be able to dance on it."

August—the dry month. Could water be the answer then? But ordinarily, Dr. Smith knew, moist sand supports weight as well as dry sand. Maybe the answer, Smith reasoned, lay not in the amount of water but in its flow. Where water is just sitting in sand, the sand is not quick. But suppose the water is flowing through it in some peculiar way. . . .

Smith canvassed other geologists and found that many had been speculating along similar lines. To settle the question, Smith and several other scientists built experimental devices in which water could be made to flow through sand in various ways.

One of the most sophisticated models was built by Prof. Jorj Osterberg of Northwestern University. His device was a large tank filled with ordinary sand and equipped with hose connections so that water could be made to flow in at the top and out the bottom, or vice versa. To complete the experiment there was a plastic dummy filled with lead shot so that its specific gravity was roughly the same as a man's—that is, it floated in water, with the top of its head above the surface.

When the sand in Professor Osterberg's tank was dry, the dummy could be placed in standing or lying positions on the surface, barely making a dent in it. When water was poured in from the top, the dummy still would not sink. But when water was forced into the tank from the bottom, welling up through the sand, the dummy sank to its neck. Upwelling water, as from a spring, forces the grains apart slightly, the researchers discovered, and makes the sand mass swell. Each grain then rests partly on the cushion of water instead of solely on other grains.

Researchers discovered that some kinds of quicksand are quicker than others. They found that the finer the sand, the slower the upwelling of water that is needed to make it quick. With fine sand and a fast upward flow of water, the result is what soil engineers call a "super-quick" condition. A human being sinks in it immediately, though it may look as firm as concrete. Where the water flow is slow, or the sand grains are coarse, the result is slow quicksand. Someone can take a few steps into it, and usually is still able to turn around and get out.

If a person does not sink immediately in quicksand, and keeps his head, he may be able to float in quicksand as in water. Since quicksand obeys the laws governing the displacement of liquids, a body will sink in the stuff only until it displaces its own weight in liquid; then it begins to float. And since quicksand is heavier than water, a person floats more readily in it than in water.

But encounters with quicksand are sought after by no one, including the experts—who recommend possible survival techniques. For all that scientists have learned about the workings of quicksand, its chief fascination still lies in its potential deadliness.

How to Escape From Quicksand

An encounter with quicksand is seldom as horrible as it is depicted in stories and films. But quicksand pits are found in virtually every part of the world; thus it is possible to stumble on one unknowingly. If you ever get into this situation, try to follow these rules: First, don't panic. Quicksand seldom kills its victim. Second, warn off companions. You might need their help. Third, get rid of packs and items that can weigh you down. Fourth, lean backward in a spread-eagle position. This should allow you to float. Finally, gently squirm or roll your way toward firm ground. It is a good idea to carry a pole when walking in suspected quicksand areas for use as a rescue aid.

Part Four

MARVELS OF THE SEA

Cresting on a Hawaiian beach, a giant breaker typifies the power and beauty of our planet's seas.

THE RESTLESS OCEANS

The rushing change of tides, the sweep of sea currents and the wild roar of breaking surf are vivid reminders that the Earth's waters are in a state of unceasing motion

The Eternal Force of the Tides

By Peter Freuchen

The greatest force that acts upon the water of the oceans is the tide. Each small particle in the sea, even in its deepest abysses, responds to the gravitational pull of the Sun and the Moon. Of course, so does each particle of rock and soil on dry land. But these can respond by moving only slightly whereas the water may rise and fall many feet. The planets of the Solar System and the distant stars also exert a gravitational attraction on the Earth, but so faintly as to have little measurable effect with respect to the tides.

Both the Sun and Moon raise tidal bulges on the Earth's surface, but the Moon is so much nearer to the Earth that it is the more powerful factor in the ebb and flow of the tides. Everyone who has lived by the sea has noticed that the tides usually change every six hours and that the time of high tide is about 50 minutes later each day, which corresponds with the daily time changes in the rising of the Moon.

The level of the tides also varies with the Moon's revolution around the Earth. Twice each month—at new moon and full moon—much higher tides occur. These highest high tides are called "spring tides." Spring tides come when the Sun, Moon and Earth are lined up so that the pull of the Sun coincides with that of the Moon. When the Moon is in its first and third quarters, it and the Sun and the Earth form the three points of a triangle. Then the pull of the Sun is working at cross-purposes to that of the Moon and the flow of the tide is relatively lower. These tides are called the "neaps."

The shapes of coastlines strongly affect the nature of local tides. For example, at Cape Columbia, in the Ca-

nadian north, where Admiral Peary started on his trip to the North Pole, the tide averages only four inches. But in Minas Basin at the head of the Bay of Fundy, between Nova Scotia and New Brunswick, the rise is as much as 53½ feet above low water, the largest tidal range in the world. The funnel shape of the Minas Basin causes this buildup of tidal waters in the narrow end at the head of the bay. About 100 billion tons of water are carried in and out of the bay by each change of the tides.

One thing to remember with respect to the complexities of the tides is that the oceans are divided up into a great many basins whose contours are determined by the surrounding land and the uneven bottom. The shape of these basins can act either to reinforce the normal ebb and flow of the tides or to diminish them.

For years the behavior of the tides at the island of Tahiti was a riddle to sailors and scientists. High tide in Tahiti is not in phase with the Moon. Instead it occurs every day at noon and midnight. Low tide is at 6 o'clock in the morning and evening. One would think, therefore, that the Moon did not exist at Tahiti, with the sea obedient only to the Sun.

The explanation is that the island lies at the pivotal point in an ocean basin. Its water is set in oscillation by the gravitational pull of the Moon and swings to and fro like water in a pan. But if you tilt a pan of water gently up and down, you will notice that the water in the middle remains at nearly the same level, while on the edges it sloshes up and down much more energetically. The water behaves like a seesaw pivoting on its axis. Tahiti is near the fulcrum of such an axis with regard to the lunar tide. Only the Sun works on the water here, and the tides correspond to its rising and setting.

Tides are of the greatest importance in navigation. In the open sea, of course, the ebb and flow of the tides are not felt, but it is a different story when a ship nears the coast. The vast amount of water put in motion by

One of the world's great tidal bores surges up the Amazon. Created by high tide at the river mouth, the bore takes the form of a low, breaking crest that travels 300 miles upstream at a speed of 15 knots.

tides creates powerful currents in many harbors. Where the tides are strong, even the biggest ships must wait for an auspicious time to move in and out of harbor. Some of the largest vessels afloat, for instance, sail to and from New York City. Although the entrance to New York Harbor is protected by the Narrows, and most of the piers are built far up the Hudson River from the harbor's mouth, the biggest liners wait for slack water so that the tidal currents will not slam them into their piers.

Certain big inlets with narrow entrances have tidal currents so violent that navigation is entirely out of the question except at slack water. Soendre Stroemfjord, in Greenland, for example, has the appearance of a smooth and friendly fiord. But jutting up at its narrow entrance is a small island called Simiutak, which in the Eskimo's language means approximately the same thing as plug or stopper. The inlet really is like a big bottle emptying and filling itself, and the island is like a cork that does not quite stop up the neck. Standing on this island, I have seen the tide rushing in and out like a furious river. The big body of water inside the fiord is drawn out and returned twice a day. In the past the Eskimos in their skin boats had only about two half-hour periods a day when they could cross the fiord's strong currents. Even now, in powerful modern motorboats, one must exercise caution or risk being swept out of control in currents of eight to ten knots.

The same thing happens in other parts of the world, and not always in a fiord. In the waters north of Australia, the growth of coral reefs has created channels through which the tides surge at speeds up to ten knots. Where tide and wind work together, almost any nar-

High tide in Canada's Bay of Fundy (above) almost covers a large weir used to trap salmon. The scene changes dramatically at low tide (below), and a truck can be driven out to retrieve the catch.

Fundy Tides—Why So High?

The world's most spectacular tides are a matter of daily routine in the Bay of Fundy, an inlet of sea that cuts about 170 miles into Canada's Atlantic coast between the provinces of Nova Scotia and New Brunswick. Main reason for the gargantuan tides—which rise to 53 feet at some points—is the Bay's unusual funnel shape. Not only do its shorelines converge sharply from mouth to head, but its bottom also slopes upward. There is no outlet to the sea at the head of the Bay; thus when the tide sweeps in, its waters converge and build up as the Bay gets narrower. At low tide, broad reddish mud flats lie bare beneath vertical cliffs that stand high and dry. Then twice each day the sea comes rushing in to fill this immense natural funnel, and the landscape is transformed into a seascape.

row passage can be hazardous to navigate. A well-known example is Pentland Firth between Scotland and the Orkney Islands. Here the current running in and out of the North Sea is so dangerous that captains generally make a long detour to the north. The conditions are worst when strong northwest winds combine with an ebb tide, resulting in the heavy, breaking seas the Scots call the Swelkie. Many sailors have lost their lives here. Since the days of the Vikings, in fact, Pentland Firth has been feared by seamen. It is said to be haunted by the ghosts of the drowned who howl and call out to sailors passing by on dark winter nights. Even on today's modern ships—when the strait is almost as well lighted as an avenue in a major city and the engines throb powerfully and steadily—everybody feels easier when Pentland Firth is left behind.

Sailing directions, the invaluable navigation publications used by sea captains the world over, are full of warnings about tidal conditions, compiled from the experiences of countless seamen. For instance, *The Alaska Pilot* tells captains that tides create the greatest dangers to be found in the waters of the Aleutian Islands. In Alaska, too, can be found a prime example of the phenomenon called the "tidal bore." The Alaskan bore builds up in Turnagain Arm in Cook Inlet along the southern coast of the peninsula. Cook Inlet has exceptionally high tides, often more than 30 feet above low water. In fact the British explorer Captain James Cook (1728-79), for whom the inlet is named, sailed into Turnagain Arm at high tide. Afterward he found his ship standing high and dry, embedded in the mud. He was not able to get it off until six hours later.

Tidal bores are curious phenomena created by unusual circumstances but found in several parts of the world. The inlet into which the tide flows must be a shallow river mouth or fiord with obstructions at the entrance sufficient to delay the flow. Combined with a strong wind from the right quarter, the tide will rise faster than it can pass over the obstructions. When the tide reaches a height sufficient to surmount the obstacles, it breaks over them and comes rolling in with a steep, high front. Tidal bores occur at several places around the world besides Alaska, including the Severn and Trent rivers in Britain, the Bay of Fundy, the Gulf of California, the Amazon River and the Tsientang River in China. The Tsientang and Amazon bores are the world's largest—walls of water 15 to 25 feet in height that roar upstream at speeds between 10 and 15 knots at the time of the highest spring tides.

I personally witnessed the advance of a bore at Turnagain Arm in Alaska. It could be heard approaching like the thunderous pounding of surf some 20 minutes before its arrival. It did not alarm the local people as they were well prepared for it. They had drawn all small craft up on the beach a safe distance before the Turnagain bore put in its appearance.

The tremendous power of the tides as a possible source of energy has long fascinated men of an inventive turn of mind. In past centuries, tide mills were operated in England, Wales and Holland—and in New Amsterdam (New York) where they were built by Dutch settlers. Today on France's River Rance, which empties into the sea on the coast of Brittany, a large tidal power station generates about 500 million kilowatt hours of electric power annually. Here the tide reaches a height of some 40 feet. Two-way turbines tap the force of the currents both as the tide enters and leaves the river channel.

There are tide motors in use along many national coasts. I have seen some in Britain—but not on a scale great enough to light a city or run a big factory. The most common tide motor employs floats that slide up and down on posts, and turn gears inside the factory. The incoming tide forces the floats up. At ebb tide they go down, transmitting power while moving in either direction. Another device entails a reservoir with an opening into which the tide pours, turning a wheel one way as it flows in and the other way as it flows out.

With an eye to the future, the Soviet Union now operates an experimental tidal power station at Kislaya Bay in the Arctic Ocean near its border with Norway. As the energy needs of industrial nations continue to increase, the day is fast approaching when in many parts of the world man may turn to the tides to help light his cities and operate his industries. In harnessing the tides, man will be drawing upon the gravitational energy of the whirling spheres.

The incoming tide, funneling into Britain's relatively shallow Severn River, generates a tidal bore that churns the estuary with onrushing waves.

Hidden Rivers in the Deep

By J. D. Ratcliff

No one stands on their banks to admire their beauty, no one writes songs or poems about them. Yet they are the Earth's greatest rivers—the dark, silent, mysterious currents that flow through the seas.

The ocean currents are the Earth's circulation system —and are quite as vital as man's bloodstream. Their impact on life is enormous. They give San Francisco its fogs, Norway its ice-free ports, Peru and West Africa mercifully cooling breezes. But for them, Britain might be an ice-covered wasteland.

Currents churn the seas like gargantuan plows, bringing up rich mineral nourishment from below. A tremendous amount of the Earth's vegetation is in the seas, and this vegetable matter is the food base for the fish— including the more than 50 million tons caught commercially every year and sorely needed by a hungry world.

Although man had sailed the seas for centuries, no one even suspected the existence of ocean currents until 1513. Sailing along the coast of Florida, which he had discovered and claimed for the King of Spain, Ponce de León observed something totally baffling. Although he was sailing south with a respectable following breeze, his little ship was moving north! He had found the mighty Gulf Stream—more important, as it

In this Apollo 9 space photograph, puffy cumulus clouds at right outline the Gulf Stream's path as it nears the North American coast at Cape Hatteras. The clouds form in warm air rising above the current.

turned out later, than the "fountain of youth" he was seeking.

Not for two and a half centuries did anyone attach importance to the discovery. Then one of history's most inquiring minds, Benjamin Franklin, began asking questions. At the time, he was postmaster general of the American Colonies. Why, he asked, did mail packets traveling directly from Britain take two weeks longer to cross the North Atlantic than did merchant ships? He put the question to a New England whaling-ship captain, Timothy Folger.

There was, Folger said, a great current flowing across the North Atlantic toward the British Isles. Sailing west to the colonies the mail packets bucked it all the way. Whalers avoided it. Why, Franklin asked, hadn't mail-packet captains been informed? Folger replied dryly: "They are too wise to be counseled by simple American fishermen."

Since the Gulf Stream was fed by warm water from tropical seas, the simplest way to map it was by taking temperature readings. Franklin distributed thermometers to whalers, and the first crude outlines of the Gulf Stream system emerged. To this day astute yacht skippers in Newport-Bermuda races still use thermometers to locate the current.

A century later another remarkable man—who did his sea studies from an office in Washington—began to trace global current patterns. Lt. Matthew Fontaine Maury of the U.S. Navy, housebound as the result of a crippling accident, handed out special logs to captains of naval, merchant and fishing vessels. Would they, he pleaded, make daily notations of winds and currents wherever they were on Earth?

Years of patient work went into plotting the results. The broad picture of ocean currents began to emerge. Great whirls, thousands of miles wide, rotated in a clockwise direction in the North Atlantic and North Pacific. In the Southern Hemisphere the flow was in the opposite direction.

In the North Atlantic a huge ocean river sweeps along north of the Equator, soaking up heat. Then, as the Gulf Stream, it heads up the east coast of the United States and out to sea. South of Iceland a submerged mountain chain divides the flow. Part of it goes north of Scotland and finally peters out on the shores of Scandinavia. The other branch moves down the coast of Europe and Africa as the Canaries Current.

The North Pacific has a similar whirl. The Kuroshio Current runs up the coast of Japan and cuts eastward to form the south-bound California Current. The whirl then circles westward as the Equatorial Current.

Some of these massive circular rivers move as fast as five knots, and some of them have flows of 50 million tons of water a *second*. What pushes them? Temperature and density differences have some effect, but prevailing winds are the main propellants. Why do they

twist? This deflection is caused by the rotation of the Earth.

There are subsidiary currents as well. In sunny seas, water cools at night, becomes heavier and sinks, giving rise to vertical currents. And there are subsurface currents. Cold water is heavy and sinks. It settles to the bottom and flows out from polar latitudes, spreading over ocean floors like syrup on a plate. Frigid water from the Antarctic has been found 8000 miles from home—on the bottom off Virginia and off Spain. Conversely, frigid Arctic water from the Labrador Current has been found off Argentina.

When cold polar waters clash with warm surface currents, the great fishing grounds are formed because of mineral nourishment brought up from below by the

Subtropical vegetation thrives at Logan Gardens in a corner of Scotland warmed by the Gulf Stream.

churning. When surface life dies, it sinks, decomposes and forms a carpet of fertilizer on the ocean bottom. When currents plow this nourishment to the surface, a population explosion among microscopic marine plants takes place. Minute marine animals feed on the tiny plants. Next, filter feeders, from small herring and sardines to the largest sharks and whales, whose open mouths strain nourishment from the seas, consume the floating life. Finally, such oceanic predators as tuna and cod eat the smaller filter feeders.

In the North Atlantic, Newfoundland's Grand Banks fishing grounds are located where the cold Labrador Current sweeps down and crashes into the warm Gulf Stream. In the Southern Hemisphere, the cold Benguela Current flows up the west coast of Africa, and the Humboldt or Peru Current moves up the west coast of South America creating vast fisheries in both areas. The wealth of food on these fishing grounds is difficult to comprehend. African fisheries are bountiful enough to attract Japanese fishing ships from halfway around the world. Off Chile, fish-eating birds feed in the Humboldt Current. The amount of manure the birds drop gives some idea of the number of fish they consume. Chile marketed over a million tons of this natural guano fertilizer each year before the advent of chemical fertilizers.

Normally the great sea currents flow more or less steadily. But occasionally winds upset the pattern—sometimes with disastrous consequences. About once a decade a current of warm water pushes south from the Gulf of Panama and displaces the cold Humboldt Current. Fish that normally live on the surface dive to colder levels, but the plankton dies and rots. Birds perish by the million. Winds soak up the warm water and drop it ashore on normally dry coastal Peru, causing disastrous floods.

While currents make some areas of the seas incredibly fertile, others are more barren than the Sahara. Thus the Sargasso Sea to the southeast of Bermuda is one of the Earth's most lifeless areas. It lies in the stagnant center of a huge eddy and lacks the currents needed to bring up nourishment from the deep.

Currents act much like a conveyor belt in distributing life over 70 percent of the globe. Sea beans from the West Indies are often found thousands of miles from home, washed up on the beaches of Europe. The original home of the coconut palm was Malaysia, and since many grew along the shore, the nuts dropped into the sea. Currents picked them up and distributed them throughout the South Pacific. But the big role played by the currents is distributing fish life. Estimates indicate that larvae of 80 percent of all food fish are distributed by currents.

The most noted and dramatic of these migrations is the journey of the European eel. For five to eight years eels live a fresh-water existence in the rivers of Europe. But when the time comes to breed they start an epic

Inside the Gulf Stream

In a historic voyage of exploration made in 1969, the research submarine *Ben Franklin* carried six scientists on a 30-day underwater journey along 1444 miles of the Atlantic Ocean's Gulf Stream. The mission's leader was the vessel's designer, Swiss oceanographer Jacques Piccard. In addition to cameras and sonar, the submarine's complex of data-gathering equipment included special devices for analyzing sea water and collecting marine-life samples. The *Ben Franklin* remained submerged at depths ranging between 600 and 1800 feet, while the Gulf Stream's force carried it on a northward drift. The journey began off the Florida coast. Scheduled to last 30 days, it ended successfully when the crew was picked up 300 miles south of Nova Scotia. The voyage's big surprise was finding that Gulf Stream marine life is far more sparse than had been believed. Just as interesting was the discovery that the Stream is not a single current, but a network of many interweaving currents running on a parallel course. However, Piccard summed up the voyage aptly, admitting: "We unlocked many more questions about the Gulf Stream than we answered."

Divers check out hull of the submerged Ben Franklin. *The 49-foot craft can dive to a depth of 2000 feet.*

The Ben Franklin, *which drifted for 30 days in the Gulf Stream, can support six men for six weeks of underwater research. It gathered over one million items of oceanographic data on the Gulf Stream trip.*

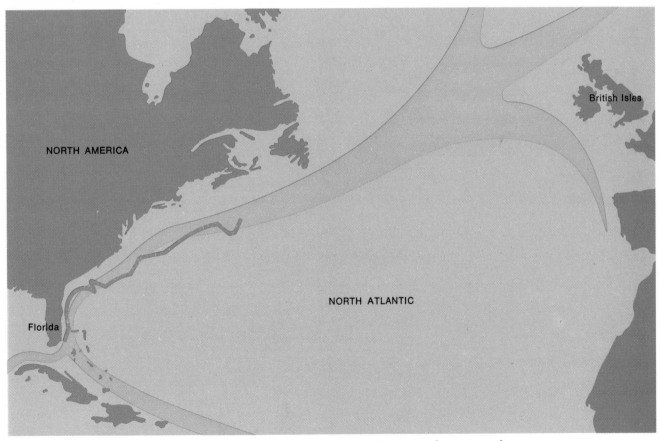

NORTH AMERICA

British Isles

NORTH ATLANTIC

Florida

The Gulf Stream is a system of currents that cross the Atlantic, then diverge as they approach Britain. The Ben Franklin, *its route shown in red, drifted along 1444 miles of the Gulf Stream's path.*

migration—to the Sargasso Sea. For unknown reasons eels find this watery desert attractive for their nuptials.

Once baby eels hatch they face a staggering challenge: a journey of perhaps 3000 miles back to the rivers from which their mothers came. Except for the assisting currents, they would never make it. Even so, the journey takes three years.

It is much the same with Pacific salmon. After two to four years at sea the mating urge awakens in these handsome fish. They begin a sad migration—back to the rivers where they were born. Once the female has laid her eggs in a fresh-water stream, physical deterioration sets in. Shortly after egg laying she dies. For a few months her young tarry in the relatively safe fresh water. Then they swim downstream to the sea where they remain until maturity.

Although surface currents are reasonably well understood, what goes on at greater depths is less well known. Great currents that profoundly influence life on Earth are certainly there, waiting to be discovered.

For example, Townsend Cromwell of the U.S. Fish and Wildlife Service was in the Equatorial Pacific in 1952 studying tuna migrations. Surface currents were pushing his drifting ship westward, but long fishing lines over the side were being pulled in the opposite direction. Clearly there was a great subsurface river flowing deep beneath the ship. It was like a latter-day Hernando de Soto discovering another Mississippi—

only the Cromwell Current is now known to be much bigger. In 1963, oceanographers from the University of Miami found a similar wrong-way, subsurface current along the Equator in the Atlantic.

Sea studies have a number of immediate practical objectives. Just as airlines are demanding better information about the jet streams that give high-flying planes an enormous push, shipping companies want more facts about the precise locations and speeds of ocean currents. The reasons are obvious. A tanker bound from the Gulf of Mexico to the east coast of the United States can save as much as $10,000 in fuel bills by using the push of the Gulf Stream. Savings for Japanese bulk carriers hauling iron ore and coal from Australia also appear to be substantial when their courses are plotted to make the most of Pacific currents.

As the world food shortage becomes more acute, another idea may have practical application. Because of lack of circulation many vast tracts of the seas are virtually lifeless. A small, properly shielded atomic reactor dropped to the bottom in such an area would warm up rich bottom water and start it rising toward the surface. With an artificial circulation established, surface waters would be fertilized, plant life would sprout and fish life would follow.

Thus there is a lot of unfinished business in the ever fascinating seas.

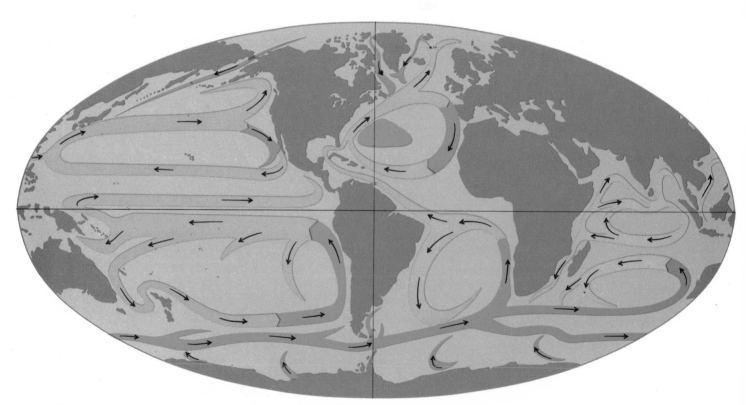

Currents are actually rivers flowing through the world's oceans. The warm currents are shown here in orange and the cold currents are in blue. The Sargasso Sea, near Bermuda, is shaded in green.

*For more than a thousand miles a bleak desert extends to the water's edge along the coast of Peru.
Offshore, the cold Humboldt Current is rich with life. But the Current inhibits rain over the land.*

When Pacific Currents Shift,
Life Comes to a Dead Land

Nowhere is the effect of ocean currents more dramatic than along the coast of Peru. Land of the storied Incas, Peru, whose boundary touches the Equator, might be expected to be a lush, tropical country like its northern neighbors, Ecuador and Colombia. But in reality the coastal plain of Peru is a 1200-mile desert that is almost entirely barren. Paradoxically, the seas alongside the desert teem with life, forming one of the richest fishing grounds in the world. The key to this paradox is the Humboldt Current. Moving north from frigid Antarctic waters, the Humboldt

Current churns up enormous quantities of nutrients for sea life. But where the Current hugs the coast, it traps the land between its chilly waters and the Andes Mountains, which form a moisture barrier to the east. Off northern Peru, the Humboldt Current abruptly turns west. The sea and land above the turn are truly equatorial in climate, warmed by a weak southerly current, *El Niño*. For reasons that scientists do not fully understand, the Humboldt Current sometimes starts its turn farther down the coast, permitting *El Niño* to flow farther south than usual. And suddenly the desert comes to life.

One plant that can survive in the Peruvian desert is Tillandsia purpurea, *a rootless air plant that traps and conserves moisture from the coastal mist.*

As El Niño *moves south, rain-bearing clouds billow over the warm current and drift inland toward the high peaks along the western edge of the Andes.*

For months on end the Peruvian coastal plain receives no measurable rainfall. Some moisture is provided by a heavy mist—the *garúa*—that rolls in from the sea. When the mist is thickest—from May to September—it revives patches of cactus, moss and lichens here and there in the desert.

But it is *El Niño*, a name derived from the happy coming of the Christ Child, that changes the desert most dramatically. Around Christmas *El Niño* moves farther down the coast as the Humboldt Current swings westward. With *El Niño's* warm water comes rain. Seeds that have lain dormant throughout the hot, dry months spring to life in a miracle of growth and color. Some desert farmers can even grow crops of fine cotton when *El Niño* favors them.

Two or three times a century *El Niño* pushes several hundred miles down the coast beyond its normal limit. Marine life, accustomed to the Humboldt's cold water, dies in enormous quantities, bringing starvation to millions of birds. Ashore, floods race over the ground before the vegetation catches hold. In ancient times, the Incas tamed these current-borne vagaries of climate with an extensive irrigation system, but today the land lies at the mercy of the sea.

The sandy soil of coastal Peru, ordinarily incapable of supporting plant life, was moistened in 1970 by one of the wettest seasons of the century.

During the unusual wet season, normally barren hillsides were carpeted in green, and succulent, flowering plants advanced rapidly over the desert sand.

In 1970 the shifting offshore currents altered weather as far south as Callao, and vegetation filled in areas that had been bare for many years.

Face to Face
with a Tidal Wave

By Francis P. Shepard

The term "tidal wave" has had an ominous sound in Hawaii since April 1, 1946. At that time my wife and I were living in a cottage at Kawela Bay on northern Oahu Island. On the previous day, a Sunday, the beaches and reefs had been swarming with people, and the cottages were alive with activity. Fortunately, almost everybody went back to Honolulu that night.

Early the next morning we were awakened by a loud hissing noise which sounded as if dozens of locomotives were blowing off steam directly outside our house. Puzzled, we jumped up and rushed to the front window. Where previously there had been a beach we saw nothing but boiling water sweeping over the top of a ten-foot ridge behind the beach. The water was coming directly at the house. I quickly grabbed my camera, forgetting such incidentals as clothes, glasses, watch and wallet. As I opened the door I saw that the water was not advancing any further, but was retreating rapidly down the slope.

By that time I realized that we might be experiencing a tsunami, or seismic sea wave. My suspicions were confirmed as the water moved swiftly seaward, and the sea level dropped several feet, leaving the coral reefs exposed in front of the house. Fish were flapping about and jumping up and down where they had been stranded by the retreating waves. Trying to show my erudition, I said to my wife, "There will be another wave, but it won't be as exciting as the one that awakened us. Too bad I couldn't get a photograph of the first one."

Was I mistaken! In a few minutes, as I stood at the edge of the beach ridge in front of the house, I could see the water starting to rise and swell up around the outer edges of the exposed reef. It rose higher and higher and then came racing forward with amazing velocity. "Now," I said, "here is a good chance for a picture." I took one, but my hands were unsteady that time. As the water continued to advance, I shot another one, fortunately a little better. As it built up in front of me, I began to wonder whether this wave really was going to be smaller than the preceding one.

I called to my wife to run to the back of the house for protection, but she was already on her way. I followed her just in time. As I looked back I saw the water surge over the spot where I had been standing a moment before. Suddenly there was the terrible sound of glass smashing at the front of the house. On the right a wall of water came sweeping toward us down the road that was our escape route from the area. We were also startled to see that there was nothing but kindling

left of what had been the house next to ours. Finally the water stopped coming on and we were left on a small island, protected by the undamaged portion of our house, which withstood the blows, thanks to its good construction and the protecting trees.

My confidence that the waves were getting smaller was vanishing rapidly. Having noted that there was a fair interval before the second invasion (actually 15 minutes as we found out later), we started to run along the beach ridge toward the slightly elevated main road. As we ran, we met some wet and very frightened Hawaiian women wringing their hands and wondering what to do. With difficulty we persuaded them to come with us. As we hurried along, another huge wave came rolling in over the reef and broke with shuddering force against the ridge at the top of the beach. Then, rising as a monstrous wall of water, it surged after us. We reached the safety of the elevated road just ahead of the wave.

When Hawaii's 1946 tsunami struck, author Francis P. Shepard awoke to see a wall of water surging toward his beach cottage. While he fled with his wife, successive waves almost destroyed their home.

There, wearing a motley array of costumes, other refugees were gathered. One couple had been cooking breakfast when the first wave came in. It lifted their house off its foundation, carried it several hundred feet and set it down so gently that their breakfast kept right on cooking. Needless to say, they did not stay to enjoy the meal.

We walked along the road until we could see nearby Kawela Bay. From there we watched several more waves roar onto the shore. They came with a steep front like the tidal bore that I had seen move up the Bay of Fundy and up the channels on the tide flat at Mont-Saint-Michel in Normandy. We could see many ruined houses, some of them completely demolished. One house had been thrown into a pond right on top of another. A house was floating out in the bay.

Finally after about six waves had moved in, each one apparently weaker than the one preceding it, I decided I would go back and see what I could rescue

from what was left of our house. I had just reached the door when I became conscious of a powerful mass of water bearing down on the place. I rushed to a tree and climbed it as fast as I could and hung on, swaying back and forth under the impact of the wave. The water soon subsided, and the series of waves that followed were all minor in comparison.

After the excitement was over, we found half of our house still standing and started to pick up our belongings. I chased all over, trying to find the books and papers that had been strewn about by the angry waves. We finally discovered our eyeglasses undamaged, buried deep in the sand and debris covering the floor. My waterproof wristwatch was found under the house a week later.

"Well," I thought, "you're a pretty poor oceanographer not to know that tsunamis increase in size with each new wave." As soon as I could I began to look over the literature on tidal waves and felt a little better

197

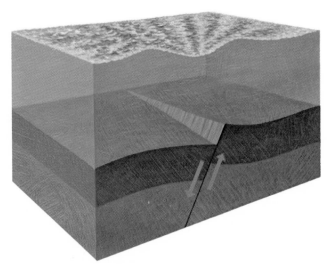

Seismic faulting on the sea floor causes a part of the floor to fall. Water surface above also falls, forming a huge, moving wall of water—a tsunami.

when I could not find any information to the effect that successive waves increase in size. Yet what could be a more important point to remember?

Today there are tidal-wave alerts when reports of earthquakes indicate dangerous possibilities, or when early waves arrive at other islands along the general route, or when the tide begins to fluctuate in an abnormal fashion. These warnings are important. Most of the 159 people who were lost during the 1946 tsunami could have been saved if they had run to higher ground when the waves first began. The Hawaiians are early risers and are usually attuned to the varying moods of the ocean. Almost everyone was conscious of a sudden diminution of the noise of the breakers when the sea withdrew. Most people ran to see the strange sight of the reefs being laid bare, and many went out on the reefs to pick up stranded fish. The 1957 and 1960 tsunamis were almost as destructive to property in Hawaii as that of 1946, but thanks to the warning system no lives were lost in either instance.

The meaning of the Japanese word *tsunami* is "large waves in harbors," a good name, as it takes a tremendous disturbance to produce large waves in sheltered bays. Tsunamis do not have anything to do with tides, although the approach of the waves on an open coast where there are no reefs looks like a rapid rise of the tide, hence the common term, tidal wave.

Most tsunamis have their origin in the great faults in the Earth's crust that surround the Pacific Ocean. The waves are caused by a sudden dropping or lifting of a segment of the ocean bottom, resulting in a displacement of large amounts of water. When the ocean bottom drops suddenly during an earthquake, surface water is sucked into the depression. When waters flowing from both sides of this depression come together, the surface of the water rises, then subsides, and waves

move out in all directions under the force of gravity. Alternatively, if the bottom rises, the water is lifted and moves outward. These displacements of the ocean bottom are caused most often by earthquakes, although volcanoes and submarine landslides can also produce displacements massive enough to generate tsunamis.

The waves are frequently most violent in their effect in a direction at right angles to the fault line at their point of origin. Since the Aleutian Trench, south of the islands of that name, runs east and west, movement along the faults that bound the trench produce waves that are most significant to the north and south, as were those of 1946 and 1957. Almost no one lives along the exposed south side of the Aleutian Islands, so that little property damage has resulted in that area, although in 1946 the water at Scotch Cap on Unimak Island rose more than 100 feet, destroying a lighthouse and flowing over a 100-foot terrace. The Hawaiian Island group, more than 2000 miles to the south, had waves that washed up to a maximum height of 57 feet.

Tsunamis move at an enormous speed in the open ocean, averaging about 450 miles an hour. They have very long periods (time intervals between crests), commonly 15 minutes, and have distances of as much as 100 miles between crests. The waves took about four hours to reach Hawaiian shores after the Aleutian earthquake of 1946. As the waves reached shallow water they slowed down, so that they advanced at a rate of only about 15 miles an hour as they approached the coast. As their energy became confined to shallow water, they grew in height. The exposed north coasts of the Hawaiian Islands had large waves, whereas lesser heights were observed on the protected south side of the islands.

Investigations that followed the 1946 tsunami resulted in some conclusions that should prove helpful in lessening the damage done by future calamities of this sort. The increasing height of the successive waves was perhaps the most important lesson. We found that in some places the second or third waves were the largest. But elsewhere the seventh or eighth waves reached the greatest height. On the western coast of Hawaii (the big island) some waves actually came in during the following night after an interval of 18 hours and reached heights greater than those experienced the preceding morning. These surprising reports were confirmed by a considerable number of sources, but are not readily understood. These waves may have been reflected off submarine cliffs near Japan so that they finally arrived at their destination after making what is comparable to a three-cushion shot in billiards.

In the tropics, coral reefs have helped protect man from tsunamis. The widest reef in the Hawaiian Islands, at Kaneohe Bay, is found on the north side of the island of Oahu and therefore on the side from which the waves approached. Yet this wide reef seems to have stopped the waves. Most people living in its lee were not even aware that a tsunami had occurred. Heights of not more than one or two feet were all that could be found by careful investigations along this shore. Other areas where reefs had smaller widths were less fortunate, as at Kawela Bay. Here we were living behind a small reef, and the water rose 10 to 19 feet. However, the height of the raised water level behind these reefs was in almost every case less than in adjacent areas, where the water came in unimpeded.

Similarly the existence of a submarine valley or canyon off a coast definitely has an important effect. Tsunamis are greatly reduced when they pass over a submerged valley. Conversely, waves traveling over a submarine ridge are particularly large. For example, there are three ridges extending down the slope on the north side of Kauai Island, and over these the waves attained their greatest heights. So if you would like to live by a beach in Hawaii, build your house behind a coral reef with a submarine valley out in front.

The Wild Whirlpools of Norway

By Olga Osing

One of the world's strongest "maelstroms," or giant whirlpools, is located on the coast of Norway, just north of the Arctic Circle. This is the Saltstraum. Long the subject of Norse legends and stories, this formidable current was once believed to be caused by a sulphurous underground fire that alternately sucked water deep into the earth at great speed and then violently ejected it.

We know now that the Saltstraum is generated by powerful currents that rip through channels connecting two large fiords. There are three channels. The northern and southern ones are rarely used for ship traffic. The third or center channel, the Storstraum, is

The whirlpool's appearance of being able to suck men and ships to the sea bottom has spawned many exaggerated sea stories. Nevertheless, even large modern ships do well to stay clear of whirlpools.

Norway's Saltstraum occurs four times each day—at tide changes—on the Storstraum channel. Its whirlpools are heralded by a tremendous roar of rushing water, which can be heard for several miles.

the most important connection between the two fiords, both for traffic and for water masses. And this is where the maelstrom occurs.

The Storstraum channel has a minimum depth of 300 feet and is about 2000 yards long and 150 yards wide at its narrowest point. During spring tide periods of peak velocity, the current reaches more than 10 miles per hour. During the six-hour cycle of an ordinary spring tide, it has been calculated that 100 million cubic yards of water flow through this channel.

The surging currents that create the Saltstraum vary with the phases of the Moon. During new and full moon, the current is strongest; it is weakest during the first and last quarters. Westerly and southwesterly winds also play a part in the Saltstraum's performance. When they blow strongest, larger masses of water are forced through the channel.

Four times a day, when the tides change, a tremendous amount of water races through the restricted waterway with the roar of a huge cataract. The noise can be heard for several miles. The sea takes on an appearance of fullness, like a water-filled cup on the verge of overflowing. Hundreds of small whirlpools begin to form. They grow larger and larger, gathering speed. Some of them reach 30 feet in diameter and are often 25 or 30 feet deep. Legends have it that some of the whirlpools are "bottomless," and their black depths do give that impression. As the water whirls, the air above it also begins to move, producing an eerie, moaning sound.

A signal station is located at each end of the channel —and with good reason. When the Saltstraum flows at its strongest, it can be dangerous. Boats and ships make no headway against it.

In 1905 the Swedish iron-ore ship *Heros* tried to navigate the Saltstraum despite the warning signal from the station at the end of the channel. The captain tried to turn back, but his ship was swept against a small island. The crew managed to scramble onto the island, but in a short time the ship broke up and the wreckage was swept away by the current.

A red ball by day and a red light by night mean that the channel is not navigable. Two red balls or lights means that the channel is safe for ship traffic. The ebb and flow periods of the Saltstraum are published in the newspapers at Bodö, the nearest city, so that local shipping lines can adjust their schedules to take advantage of the channel when it is safe for navigation.

Many kinds of sea life are carried by this current— herring, mackerel, crabs. Lying in wait for the smaller fishes is the coalfish, a member of the cod family. With all this ready food at hand, the coalfish's flesh becomes very tasty. Then, too, the constant battle with the current makes the flesh firm and well knit. The coalfish, a popular food fish in this part of the world, has been known to grow to a length of four feet.

Fishing has been popular here since ancient days, and most of the inhabitants regard it as their secondary occupation. Tourists look upon it as a great sporting challenge. Often, as the whirlpools swirl by, fishermen stand along the shore trying their luck. When the coal-fish are numerous and when the water appears to be dying down, dozens of boats can be seen on the channel. But the water may not be as safe as it looks—treacherous eddies may be building up beneath the boats. The eddies may be small at first and soon disappear. A boat can get away from a whirlpool that is not too deep. If it is caught by one, a boat may rotate with the water. Some old-time fishermen claim that slapping a whirlpool with their oars will divert the current to such an extent that their boat can escape—at least that is the fishermen's story. Another old tale is of a fisherman who was pulled into a whirlpool. A few seconds later he reappeared from the depths and exclaimed, "You should have seen the coalfish down there!"

Not everyone has been that lucky. Parish registers record that in the last 150 years at least 28 fishermen have been drowned in the Saltstraum's treacherous waters.

Sometimes when the Saltstraum is in action, flocks of eider ducks frolic among the whizzing whirlpools. The male ducks are handsome black-and-white birds; the females are a more somber brown color. These little ducks ride on the waters to the edge of the whirlpools searching for food. At the right moment they dart up into the air and away. Sometimes a duck does not make it fast enough. Then he disappears head first, and after a minute—if he is lucky—he will come bobbing to shore.

There are many farmhouses along the shore. Many of the residents are direct descendants of the Viking chief who called himself "King of the Saltstraum." Legend has it that he had a wooden castle nearby, owned most of the land in the vicinity, had many wives, much wealth and power.

Maelstroms occur in many other parts of the world. Among the best known are the whirlpools in the Naruto Strait of Japan's Inland Sea and those in the Strait of Messina between Sicily and the southwestern tip of Italy. The Messina Strait maelstrom is the fearsome "Charybdis" of Homeric legend. Ancient Greek sailors believed Charybdis to be a female sea monster who devoured men and ships as they attempted to pass the rock of Scylla on what is now the Italian coast.

The legendary power of maelstroms to suck boats and ships to their doom at the bottom of the sea has been embelished by sailors' tales and fiction writers for centuries. Actually the downward suction of the maelstrom is relatively weak, and most maelstrom accidents are caused by powerful tidal currents that sweep small craft against rocks and shoals.

Nevertheless, man's fascination with this awesome phenomenon continues as strong as ever. During Norway's tourist season, people arrive at the Saltstraum in cars, on bicycles and on foot, laden with cameras and fishing rods. A café is located on a strategic point and wide windows face the maelstrom. On cold days people sit at the tables, drink coffee, listen to the jukebox and watch the water. The Saltstraum is a spectacular sight at any time, but in stormy weather, when the clouds are darkening with rain and wind, the sight of the wild whirlpools is one that can never be forgotten.

A Whirlpool Ride for Tourists

The doughty little boat skirting whirlpools is a Japanese tourist craft venturing into the treacherous Naruto Strait, near Osaka on Japan's Inland Sea. Tide changes in the constricted waters hurl currents through the Strait at speeds up to 10 knots, and resulting whirlpools often measure 50 feet across. The Japanese boat does not actually enter the dangerous whirlpools for fear of being swamped. But even skirting them, its passengers get a wild ride and a memorable close-up look at a very unusual sea phenomenon.

The Ways of the Waves

By James Nathan Miller

Most of the summers of my life have been spent at the edge of the sea watching the surf and, like everyone else who watches it, wondering about it. The surf: frolicsome sometimes, vicious at others—but always and above all, a mystery.

In windless spells, when day after day the air is motionless yet the breakers grow bigger and bigger, from how far off must these waves have come, rolling silently across the sea, hump after hump like a solemn procession of elephants? And why, long before they hit the beach, do the elephants' heads begin to turn white, then their bodies, until finally they are all froth and turbulence? How big can these monsters get—and how dangerous?

For centuries such questions have puzzled mankind. In fact, only in recent decades have oceanographers been able to tell us how waves are born at sea and travel many miles to perish on the shores of islands and continents.

One may begin by asking, what is a wave? It is distinctly not, as some may think, a mass of water moving across the face of the sea. Observe a piece of driftwood as a wave comes along. You'll see the wood move toward the wave a bit, up, then forward a bit, and down again. When the wave has passed, the wood is still right where it started. Hence, fact No. 1 about a wave: unlike a current, which actually transports water, a wave merely moves *through* water. It is nothing more than a pulse of energy propagated through the water by the oscillations of water molecules.

The most familiar ocean waves are generated by wind. (The rare but intensely dangerous seismic sea wave called a tsunami is generated by an earthquake.)

A wave is energy pulsing through the water. As a wave rolls toward shore, its shape is distorted by collision with the shallow sea bottom. The resulting rise and break of its crest makes a whitecap.

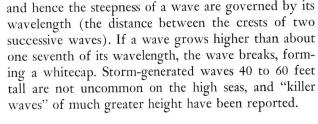

Biting into the surface of the water, the wind stirs up ripples; these mound up; the wind pushes them higher and higher and, at the same time, the oscillations reach deeper and deeper.

Note that "deeper and deeper." Ocean swell carries half of its wind-given energy *under* the surface. Divers on shallow coastal bottoms can actually see this energy expressed; it sways seaweed back and forth as the swell passes on the surface above. English long swells have washed one-pound stones from the bottom into lobster pots set at depths of 100 feet. Along the bottom off Ireland's west coast, rocks weighing several hundred pounds are often rolled to and fro like bowling balls by long surface swells. But it is the top half of the wave, the energy carried in its crest, that puts dread in a sailor's heart.

The size of a wind wave is influenced by the wind velocity, the duration of the wind and the expanse of open water over which the wind blows. The stability

and hence the steepness of a wave are governed by its wavelength (the distance between the crests of two successive waves). If a wave grows higher than about one seventh of its wavelength, the wave breaks, forming a whitecap. Storm-generated waves 40 to 60 feet tall are not uncommon on the high seas, and "killer waves" of much greater height have been reported.

In 1966 the liner *Michelangelo* was steaming through 30-foot waves in a fierce Atlantic gale when suddenly out of the night loomed a monstrous wave rising, the captain estimated, 60 feet. It smashed into the ship with such enormous force that it flattened the three-inch steel bow plates, tore a 30-by-60-foot hole in the bridge, and crumpled steel cabin walls in the ship's interior, killing three people.

Scientists have an explanation for great waves like this. They point out that the surface of the sea is the parade ground for the "wave trains" of many different storms. At any given spot on the sea, some of the waves observed are the swell from old storms in faraway regions while others are of more recent and nearer origin. As these waves from different directions combine, they can either diminish or reinforce one another. Often the crest of one will engulf another's trough and the two waves will cancel one another out, presenting the appearance of a calm expanse in the midst of turbulence. But at other times two—or three or four—crests may join, and for a few fleeting seconds climb up on one another's shoulders to build a huge wave.

The largest wave ever reliably reported was during a gigantic Pacific typhoon that hit the U.S. Navy tanker *Ramapo* in 1933. For seven days the winds of the vast storm system had been blowing in a constant direction over thousands of miles of unobstructed ocean, creating awesomely high seas. According to scrupulously documented affidavits and trigonometric calculations of the ship's officers, a procession of waves was observed that rose taller and taller, from 80 feet to 90 feet to 100 to 107—until one was seen that towered 112 feet from trough to crest. Eyewitness accounts such as this must inspire wonder, for we may suppose many greater storms with even bigger waves have gone unseen and unrecorded.

The waves we see breaking on the beaches at the seashore are nowhere near as high, generally, as those observed far out at sea. Wave recorders on North American coasts show that about 80 percent of the breakers are less than four feet, and that generally they get as high as ten feet only in winter storms. Even on the leeward coast of Oahu in Hawaii, noted for its high surf, a swell that produces 20-foot waves is remarkable enough to make the record books. Along the northern Pacific coast of North America, winter breakers often reach heights of 35 to 40 feet, as high as anywhere else in the world.

Waves may travel for great distances over the

In Sea-state One (above) a faint breeze ripples the mirrored sunlight on the Caribbean. Sea-state Three (below) shows wavelets no higher than two feet, and occasional whitecaps made by 7-to-10-knot breezes.

oceans. As soon as they stop wasting their energy in the collisions and turbulence of the wind that spawns them, they are transformed into fully developed "seas" —regular waves of the maximum size for wind of a given velocity. Thereafter, they become gently undulating "swell," moving across the ocean toward a remote shore. As swell, wave action can efficiently transport a storm's energy halfway around the world.

This fact was borne out by research conducted after World War II. For instance, in 1949, during a period when seas were running high in the North Atlantic, British researchers analyzed a series of wave trains breaking on Britain's Cornish coast and concluded that they must all have been generated four days earlier by a storm at a point off the North American coastline, 3000 miles away. Checking weather maps, they discovered that there had indeed been a hurricane off Florida at that very time and place. Today wave recorders in the Pacific frequently show that California surfers are being propelled by energy packed into the sea only a few days before by high winds off Australia.

By observing waves as they move through deep water, we can estimate the speeds at which they are traveling. To gauge a wave's speed, in miles per hour, count the number of seconds between two arriving crests, then multiply by 3.5. For example, two waves arriving ten seconds apart are traveling at 35 miles per hour. Even a small wave collapsing on a beach releases a large amount of energy; dynamometers on piers have shown that often waves smack into obstacles with a pressure of 50 tons per square yard (nearly 100 pounds per square inch). And sometimes the waves demonstrate their power in wild ways. In a December gale on the Oregon coast, waves picked up a 135-pound rock and heaved it through the air, knocking a 20-foot hole in the roof of the Tillamook Lighthouse keeper's house 100 feet above sea level. At Wick, Scotland, waves swept from the breakwater a solid block of concrete weighing 2600 tons.

Why do waves sometimes break when they are still far offshore? The answer is simple. As soon as the underside of a wave's energy pulse scrapes bottom, the behavior of the entire wave is radically changed. In a swell moving over deep water, the underside extends half as deep as the swell's wavelength, or the distance between successive crests; thus a 300-foot-long swell touches bottom at a depth of 150 feet.

An observer standing on a promontory can actually see a wave change its shape and speed as it approaches the shoreline. First, the wave slows down as it begins to drag on the bottom. Then, as it continues rolling landward against the sloping beach, the entire wave rears high out of the water. Finally, unable to hold itself up, the rising crest flops over into foam. This is why big waves—those with long wavelengths and consequently deep undersides—will foam briefly as they cross a submerged reef, while shorter (more shallow-rooted) waves will cross the same underwater barrier without foaming at all.

Heavy surfs are not sure harbingers of approaching storms. Since the wave trains manufactured by a storm usually move faster than the storm center, if both are going in the same direction the waves will indeed reach the coast first, giving warning to the inhabitants. But sometimes the storm itself moves in another direction—like Atlantic hurricanes, which often pile up waves on North American shores while moving out to sea.

There is no validity to the commonly held belief that every third wave—or ninth, or twelfth—is bound to be a big one. Wave size can vary according to a pattern, however, where very regular wave trains are arriving at the same beach at different speeds—their crests combining to produce a big wave, then gradually falling out of synchronization until the crest of one cancels the trough of the other, and then gradually combining again. Often an observer on the beach will be able to discern the cycle in the surf's rise and fall.

The chances of strong undertow developing at the seashore are usually not very great. Though the foaming breakers look like a river of white water rushing at the beach, very little water actually does flow up the sandy slope. And when this water flows back into the surf, it gets no farther than the next incoming breaker, which drives it upslope again. Undertow, in other words, can *not* sweep one out to sea; the worst a heavy backrush can do is knock a person off his feet and send him sprawling into the surf's edge.

Then how can we account for well-authenticated reports of drownings in which swimmers, trapped in the "undertow," were unable to battle their way back to shore? Sometimes waves shape a sandbar off the beach that traps the undertow backflow in a long pool. This water then flows sideways until it finds a break in the sandbar that permits it to rush back to sea. At this spot, the outward-flowing mass actually does form a strong current. Once beyond the breaker line, though, this so-called rip current balloons out and vanishes. The key difference between this and "undertow" is that rip currents are narrow (perhaps 10 or 20 feet wide); they can be escaped by swimming parallel to the beach for a short way, and then coming in through the breakers.

So, as you lie on the beach in the summertime listening to the hiss and churn of the surf, remember to respect the power of the waves. True, they can be wonderful companions when they're in a good mood. But should we venture among them when they are angry, we may find ourselves suddenly being banged by a 50-ton battering ram.

The Atlantic seascape at left shows Sea-state Five, characterized by whitecaps and the presence of very light spray blown by 22-to-27-knot winds. At right, Atlantic gale winds of 48-to-55 knots create Sea-state Seven. The heavily rolling sea is covered with foam, while high spray reduces visibility.

OCEAN LEVEL

UNITED STATES

Martha's Vineyard

Bermuda Rise

Continental Shelf

Kelvin Seamount

Strange Landscapes Beneath the Sea

By E. P. Lay

Ever since man took his first hesitant voyage in a boat, he has tried to understand the ocean. His superstitious mind attempted to explain the disasters that occurred at sea, so he reasoned that his gods sent the storms that tormented him, sea monsters snatched ships into watery depths and unsuspecting fleets sailed off the edge of the Earth. Although man made these explanations quite fanciful, he mistakenly (and unimaginatively) pictured the unseen ocean floor as a flat, sediment-covered plain.

Until the 20th century, man's limited knowledge of the ocean floor came from the ancient method of casting a weighted rope overboard to measure the depth of the sea. From the globs of sediment that clung to the weights, man guessed that an oozy mud covered the whole of the ocean floor, obscuring sea bottom, sunken ships, treasure and even lost civilizations. Beginning in the 1920s, with the aid of the newly developed echo sounder, startled oceanographers began to realize that the ocean floor was extremely irregular and not the flat smooth plane it was thought to be.

To chart the ocean depths, the echo sounder measures the time it takes sound pulses to travel from a ship on the surface to the ocean floor, and return as echoes. Echoes that bounce quickly back to the surface outline huge mountain ranges thrusting up jagged peaks. It takes much longer for echoes to return from the deep trenches in the sea floor—the greatest of which, the Mindanao and the Mariana trenches of the western Pacific, form giant gashes reaching down some seven miles below sea level. In addition to giving oceanographers a picture of the varied shape of the sea floor, echo sounding acquainted them with the deep ocean's true range of depths—which generally lie from 12,000 to 18,000 feet—and outlined the rims of the ocean basins. Within the unprobed sea floor existed geological wonders never suspected until strangely behaving echoes aroused man's curiosity.

Oceanographers' soundings have established that the ocean floor is divided into three distinct areas: the continental shelf, the continental slope and the deep-ocean basin. The continental shelf borders on continental land areas, and some nations have agreed that, in a legal sense, it is a part of the land out to a depth of 200 meters (about 656 feet). Actually the shelves vary a great deal from this idealized definition and sometimes extend out from the continents for hundreds of miles. They vary from flat, terrace-like plains to irregular,

Atlantic Seamount

Azores

Mid-Atlantic Rift
Mid-Atlantic Ridge

OCEAN LEVEL

SPAIN

Fissure
—17,500 feet

Josephine Seamount

Gettysburg Seamount

Gibraltar

Continental Shelf

This profile of the Atlantic Ocean shows bottom features along a 3200-mile stretch from Massachusetts to Gibraltar. It is based on echo-sounding data collected during the voyages of several research ships.

rough terrain. A combination of sediments—rocks, sand, mud, silt, clay and gravel—blankets the shelves. Sand forms the most common sedimentary material. It consists mainly of coarse particles eroded directly from the land, transported by rivers, currents, ice, wind and volcanic eruptions.

Biologically the continental shelves are shallow, sunlit seas supporting immense numbers and varieties of animal and plant life. The energy of the sunlight permeating these waters is used in the process of photosynthesis by a variety of algae and other marine plants, including the minute diatoms. Such tiny free-floating plants form the phytoplankton, the basic "producers" of the sea. These drifting organisms provide photosynthesized proteins, starches and sugars to marine animals.

The drifting organisms are not all tiny plants. The larval forms of many marine animal species, such as starfish, urchins and corals, begin life in the plankton stage. These tiny drifting larvae are part of the zooplankton. The zooplankton also includes the holoplankton, animals that spend their entire life cycles as minute, free-floating organisms. The dinoflagellates are among these.

The continental crust actually ends near the place where the continental shelf drops rapidly to the ocean floor. This sharp descent is called the continental slope,

and here the deep sea truly begins. Geologists know that the slopes generally drop at from 100 to 500 feet per mile. They are generally cloaked by sediments composed mainly of mud, a little sand and small amounts of gravel. In some areas the steepness of the slope is quite dramatic. For example, the drop-off along the western coast of South America from the top of the Andes Mountains to the bottom of the Peru-Chile Trench measures some 42,000 feet. Here there is no shelf to break the slope's sharp grade from the coastline to the edge of the trench—the near eight-mile descent occurs over a horizontal distance of less than 100 miles. The steepness of this slope dwarfs any other on Earth; most are much more gradual. Many descend like hillside terraces in a series of basins and plateaus.

The ocean-floor trenches generally run parallel to the continental slopes. Often the trenches lie next to rows of active volcanoes, and many earthquakes are generated in their vicinity—evidence that the deep-ocean trenches are the sites of powerful geological activity. It is not surprising that the deepest ocean-floor trenches are located along the perimeter of the "Ring of Fire"—the active volcano belt that encircles the Pacific Ocean basin. Recent geological research suggests that the Pacific Ocean basin is shrinking in area as a result of the movement of continental plates along the edges of the basin. Many geologists now believe that

207

the trenches are chasms where the crustal plates of the ocean floor descend into the Earth's interior as they are overridden by continental plates.

Spectacular canyons are known to exist in the continental slopes. Many scientists believe that undersea "turbidity currents" may help carve such canyons. Turbidity currents are rapidly moving streams of water loaded with sediments. They probably begin as underwater mudslides. Water-saturated material begins to move down the incline of a continental slope. Gathering up rocks and gravel, it pours down the slope with increasing momentum, cutting deeply into whatever lies in its path. When it reaches a level area, the current slows, depositing its load of debris. Geologists believe that the heads of canyons were once above the sea, and have since sunk or "drowned" by the rise of sea level. Turbidity currents may have kept these canyons cleansed of sediments since their submergence.

The open sea beyond the shelf margin is called the oceanic region. The top layer of water is penetrated by sunlight, permitting plants to carry on photosynthesis. The oceanic region, though, supports less life than the shelf area. Most oceanic life is "pelagic" (free-swimming). In contrast to the shelves, few animals are able to live on the ocean bottom. Larvae are less common in the deep ocean, and the plankton is primarily holoplankton—creatures whose entire life cycles are spent as free-floating organisms. (The European eels that breed in the Sargasso Sea and whose larval stages develop as marine plankton are one obvious exception to this rule.)

As the continental slopes continue to descend, they reach the deep ocean basins where the depth averages 15,000 feet. The deep-ocean basins comprise half of the Earth's surface. Oceanographers estimate that 90 percent of the Pacific deep-ocean basin is rough terrain, as opposed to the smooth "abyssal plains" that are more common in the Atlantic basin. Abyssal plains are believed to result primarily from the undisturbed piling up of sediments by turbidity currents.

Traversing every deep ocean floor is an impressive ridge. The first to be discovered was the awesome Mid-Atlantic Ridge. This huge mountain range, soaring more than 30,000 feet above the adjacent sea floor in some places, extends from north of Iceland to below the tip of South Africa. Peaks rising above the surface create islands such as Ascension Island and the Azores. Between Antarctica and South Africa, the Mid-Atlantic Ridge curves eastward, extending around the world in a 40,000-mile-long mountain chain called the Mid-Ocean Ridge. Many underwater earthquakes occur in a rift running down the ridges' centerlines.

The ocean's "abyssal zone" begins at a depth of 6500 feet, and extends downward to the ocean bottom. There are no "producers"—photosynthesizing plants—in water this deep. Bits of organic matter filtering down from the sunlit regions of the ocean, as well as dead marine animals falling from above, provide food

The Ocean's Mineral Gardens

The biggest of them look like cannonballs, while the smaller nodules sometimes suggest potatoes, or clusters of grapes. Actually, the mysterious objects in the picture at left were photographed on the floor of an ocean basin, where they have been growing for millions of years. Composed mainly of manganese, a common element in seawater, the nodules were formed by a chemical process called ionization. Over an incredibly long period, individual atoms of manganese (as well as other elements—notably iron) are drawn out of seawater to form in thin layers around any hard object lying on the sea floor. Objects serving as a nucleus for the process are legion, and include minute glass beads from volcanic action, sharks' teeth, whalebone and ordinary rock particles. Although many details remain to be learned about how and why manganese clusters "grow," there is no question that these mineral gardens are a potentially rich harvest ground for imaginative industrial enterprise.

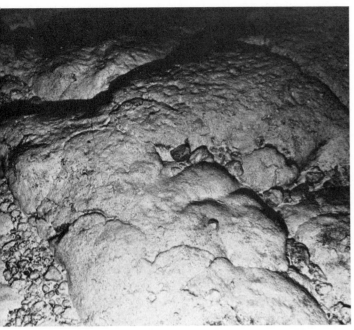

As continents drift, fresh lava (above) wells up at a depth of 10,384 feet along a ridge in the South Pacific. Below, a rare photograph shows a submarine landslide scouring a canyon off Baja California.

for the life of the abyssal depths. Life prospers in these dim regions according to the amount of food that rains down from the lighted waters far above. Bacteria and scavengers transform much of this organic debris into inorganic matter after it reaches the ocean floor. Adapting to their darkened environment, many sea creatures at these depths have modified eye structures. Like the familiar firefly, some have bioluminescent organs that create light in darkness.

Besides the sediments that are deposited by turbidity currents on the abyssal plains, there are three other general types of sediments on the deep-sea floor. "Calcareous ooze," found in warm, comparatively shallow waters, is composed primarily of marine organisms' shells and skeletons rich in calcium carbonate. In deeper and colder waters are found "red clay" sediment, a material that is largely inorganic, and "siliceous ooze," consisting mainly of diatom skeletons (which consist of opal-like silica).

Among the strange and picturesque features of the ocean floor are scattered individual mountains that rise from the sea bottom, but lie submerged under several thousand feet of water today. These isolated peaks, rising a few thousand feet from their bases, are called "seamounts"; seamounts with flattened tops are known as "guyots." Most guyots are found in three general areas of the Pacific: in a line along the Mid-Pacific Ridge, in another group between the Marianas and Marshall Islands and in a third grouping southeast of the Kamchatka coast of northern Asia. The stacking of lava from repeated volcanic eruptions is believed to have created the seamounts and guyots. The guyots' smooth, flat tops indicate that these mountains once stood above the surface, where the action of waves leveled off their peaks.

Something must have subsequently caused the guyots to sink. Geologists think the drowning of the guyots may have involved two processes: the great weight of the volcanic mountains may have depressed the sea floor over the long periods since their creation, while from time to time the level of the oceans was also rising, helping to submerge them.

Undersea canyons, mountains and ridges influence the circulation of sea water. These formations block and channel the movement of deep water and also aid in the stirring and overturning of the seas, greatly affecting world climate. But geologists are not yet completely sure how the canyons and trenches are created, nor of the exact part turbidity currents play in undersea geological phenomena, nor of precisely how the guyots were leveled, not to mention the role of the mid-ocean rifts in the triggering of marine earthquakes. Every year, as new equipment is perfected for exploring the deep-sea floor, man learns a little more about these mysteries. Thus he can pierce the secrets of the enigmatic sea and discover in this immense frontier one more clue to the mystery of our evolving planet.

Chapter Two

THE SEA AND MODERN MAN

The sea has been explored and put to use ingeniously in recent years, but along with the benefits it has bestowed come warnings of its limitations as a natural resource

We Made the World's Deepest Dive

By Commander Don Walsh, U.S. Navy

I had been fighting all night long just to stay in my bunk. The U.S.S. *Lewis* was rolling through heavy Pacific swells in her best destroyer escort manner. From midnight on, explosions from the stern made everyone aware that we were making depth soundings. I may have slept a little, but when someone shook me fully awake at 6 A.M., it was a relief. I could get up now. All I had to do that day was to get into a small steel ball and dive to what was then believed to be the bottom of the deepest part of the world's deepest ocean, the "Challenger Deep." We estimated this spot, which is part of the Mariana Trench, to be about six miles in depth at the point we had chosen. I had the feeling, for a moment, that it might be wiser to stay in bed.

I dressed and went on deck. Sunrise and the start of the dive were nearly two hours away. The swells looked high and mean, and the wind seemed to have freshened during the night.

A mile astern I could see the lights of the Navy tug *Wandank*. Behind her, riding on 600 feet of cable, was the *Trieste*. Towing the *Trieste* nearly 200 miles from Guam had not been easy. She is about as well suited to the high seas as a house is to travel on a superhighway. Jacques Piccard, who would make the dive with me, was aboard the *Wandank*. He is the son of Auguste Piccard, the Swiss scientist who designed the bathyscaph. Jacques helped his father build the *Trieste*. No man has made more dives than Jacques, although squeezing his 6-foot-7 frame into the bathyscaph's cramped sphere is quite a trick.

I climbed up to the bridge and joined Dr. Andreas B. Rechnitzer, the scientific director of the project. At

the moment, Andy was trying to find the deepest possible place for us to make the big dive. "Throw one in," he said to a young sailor sitting beside him. The sailor spoke to the fantail, using his headset. There was a blast that could be heard all over the ship as three pounds of TNT exploded just beneath the surface. Andy punched a stopwatch. Fourteen seconds later his Fathometer headphones crackled as the echo from the bottom reached us.

A quick calculation told us the depth here was 33,600 feet, figuring the speed of sound through water at 4800 feet per second. "Son," said Andy, "we have really found you a hole. Now please see *one* animal down there. That's all it takes. Just one of anything."

When I boarded the *Lewis'* whaleboat, it was 7:30 and beginning to get light. Rain was falling around us in patches. The whaleboat was pitching violently against the steel hull of the *Lewis*, rising and falling as much as 10 or 15 feet. I think that trying to get aboard that whaleboat was the most difficult thing I did all day.

On the *Trieste* two men were waiting for us: Lieut. Lawrence Shumaker, assistant officer-in-charge of the bathyscaph, and Giuseppe Buono, the master mechanic from Naples who, like Jacques, had been with the *Trieste* since she was built. Larry and Giuseppe were going to handle the tricky topside work that had to be done before we could make the dive.

I clambered into the conning tower and started down the long ladder to the sphere. At the bottom, I raised the hatch into the sphere itself. The familiar odor of rubber and solvent reached me as I slid through the hatch and started preparing the sphere for the trip. I looked over the instruments and batteries, checked the bilges, saw that the oxygen and air regenerators were in good condition. I switched on the tape recorder into which I would dictate notes during the trip, and said, "This is dive number 70, U.S. Navy Electronics Laboratory, Walsh and Piccard."

The bathyscaph is an unusual craft. It is basically a

Surfaced on the rolling blue Pacific swell, the bathyscaph Trieste *undergoes preliminary trials before moving into position for its record-setting descent into the abyssal darkness of the Mariana Trench.*

great, buoyant, compartmented tank with a small passenger gondola attached to its bottom. It operates like a balloon. But in place of the balloon's lighter-than-air gas, the bathyscaph uses lighter-than-water gasoline. To descend, a diver releases gasoline, just as an aeronaut releases gas. To rise, he drops iron-shot ballast from two huge ballast tubs protruding from the bottom of the tank. Sea water enters the tank freely during a dive to equalize inside and outside pressure so the tank will not be crushed.

I heard the hatch at the top of the shaft open and then the squish, squish of Jacques' wet shoes as he came down the ladder. His feet poked in through the hatch and shortly he was inside the sphere with me, all 6-foot-7 of him.

Behind him came Buono. There is a certain formal leave-taking ceremony we go through with Buono before each dive. "*Mille grazie. Arrivederci,*" he says to Jacques, and Jacques says it back to him. Then, in his Neapolitan-accented English, he thanks me and says

good-by and I do the same to him. Then all three of us —Buono outside, Jacques and I inside—lower the big steel hatch into place and tighten the bolts.

Now, if everything went well, we would be on our way within ten minutes. Through the porthole in the middle of the hatch, Jacques made an all's-well signal with a flashlight to Buono, crouching in the antechamber. We watched as Buono disappeared up the ladder. In a minute he had opened a valve, and in three minutes the passageway through the tank was flooded—and we were trapped in the sphere for the duration of the voyage.

Then Larry and Buono carefully opened the ballast tanks, letting an extra two tons of sea water run into the float. If we had figured everything exactly right, this would add enough weight to start us down. It did. In fact, we started down so rapidly that Larry and Buono had a lively time getting off the *Trieste* and back into their rubber boat.

In the sphere, of course, we knew nothing of this,

Settling to the bottom at a spot in the rugged Mariana Trench (dot) the Trieste *registered man's deepest descent into the oceans—35,800 feet.*

but three things always happen on the *Trieste* when she's on her way, and they all happened at once: the needle on our sensitive depth gauge began to quiver downward, the rocking motion of the sphere became perceptibly less violent and—why this happens we have never been able to figure out—the stern settled by a degree or so.

After we had been under way for four minutes I called the *Lewis* on the underwater telephone and reported that we were all right and passing 250 feet.

At 300 feet we encountered the thermocline, a layer where the water temperature drops sharply. Since the cold water was denser than the water we had been passing through, we became relatively more buoyant and stopped. We had expected this. Part of our standard diving procedure was to use this brief time to make a final instrument check. Then, by releasing a little gasoline from our maneuvering tank, we got rid of some of our excess buoyancy and started down again.

At 600 feet we entered a zone of deepening twilight where colors faded off into gray. By 1000 feet the light had gone completely. We turned out the lights in the sphere to watch for the luminescent creatures that are sometimes visible at this level. We saw very few. Eventually we turned the cabin lights back on and briefly tested the forward lights that throw a beam in front of the observation window. Formless plankton streamed past, giving us a sensation of great speed.

We were now dropping fast, about four feet per second. It was getting colder in the sphere, and we decided to put on dry clothing. It was quite an operation: two grown men changing clothes in a space 38 inches square and only five feet, eight inches high.

During this time there had been little conversation. Friends often ask what we do along the way, how we keep from becoming bored, what we talk about. The fact is that most of the time we are both too busy either to be bored or to talk very much. There are too many instruments to watch, too many adjustments to make.

Then there are minor incidents such as a small leak that always developed in one of the hull connectors—a place where wires from lights and instruments on the outside of the sphere pass through the hull to the recording apparatus inside. The leak started at about 10,000 feet. It is an old friend, a tiny drip, drip, drip. I timed the drips and found no change from before, which meant that it had not become more serious. We expected it to disappear at about 15,000 feet, when the water pressure packed the plastic sealer in more tightly —and it did.

Up to this point we had managed to maintain voice contact with the people on the *Wandank* on the underwater telephone. But now, at 15,000 feet, we lost them. We were truly on our own except for a crude system of tone signals Larry and I had worked out. By means of a special key, the underwater telephone can send out a tone that sounds something like a radio time signal. These tones carry farther than voice transmissions. In our code, all even-numbered signals are for good news: two means all is well, four means we are on the bottom, six means we are on the way up. The bad news comes in odd numbers: three means we are having mechanical difficulty and are coming up but not in distress, five means something has gone wrong and we are coming up in an emergency. So far, we had never used the odd numbers.

At 18,600 feet and again at 24,000 I called Jacques' attention to the depth gauge, noting that we were surpassing previous record dives. He grinned and waved.

At 27,000 feet we checked our rate of descent to two feet per second by dumping some shot ballast. We were not too sure of the underwater currents here and we did not want to go crashing into a wall of the trench by mistake. As we neared 30,000 feet I started thinking about the changes we planned to make when we got within 1000 feet or so of the bottom. I was running through a mental checklist when we heard and felt a powerful, muffled crack. The sphere rocked as though we were on land and going through a mild earthquake.

We waited anxiously for what might happen next. Nothing did. We flipped off the instruments and the underwater telephone so that we could hear better. Still nothing happened. We switched the instruments back on and studied the dials that would tell us if something critical had occurred. No, we were descending exactly as before.

We dumped more ballast, checking our speed to one foot per second. At 33,000 feet, only about 600 feet off the expected bottom, we turned on our sensitive Fathometer, which always before had quickly and accurately picked up the floor for us. It showed nothing. We continued to slide on down. It also showed nothing 100 feet later—or 100 feet below that. At 36,000 feet Jacques asked me wryly whether I thought we could have missed the floor somehow.

We checked our speed to half a foot a second and continued. At that rate, time and distance pass very slowly, and I think for the first time in the dive we both had the feeling of awe that comes from exploring the totally unknown.

I did not take my eyes off the Fathometer and Jacques never stopped watching out of the tiny porthole with its weak probe of light. No bottom was in sight at 36,600 feet, or at 37,200. But at last at 37,500 feet the Fathometer traced the beginnings of the bottom. Soon Jacques could see a difference in the effect of our light in the water, as the rays reflected off the bottom. As we approached the floor I called the Fathometer readings to Jacques in fathoms: "Thirty . . . twenty . . . ten . . ." At eight, he called that he could see the gray-white floor.

As we sank through the clear water near the bottom, we had a tremendous piece of luck. Peering through the tiny porthole, Jacques spotted a fish. It appeared to be browsing, searching for food along the ocean floor. It looked like a sole or flounder, flat with eyes on the side of its head. It was about a foot long. Our sudden appearance in his domain, with our great light casting illumination such as he had never seen before, did not seem to bother him at all. After we watched the fish for a minute, he swam slowly off into the darkness again, beyond the range of our light. This was an exciting event. The fish was obviously a bottom-feeder, which means that it must spend all of its life at these tremendous depths, under enormous pressure.

At 1:10 P.M. we sank gently onto the soft floor. A great cloud of silt rose around us. We had found the bottom at 37,800 feet by our gauges. Later, when the gauges were precisely calibrated at the Office of Naval Research, the true figure was found to be 35,800 feet.

The silt cloud was still around us, so for the moment we could make no more visual observations. I keyed the underwater telephone four times, the signal that we had reached the bottom. Then, with no expectation that I would be heard, I called on the voice circuit: "*Wandank, Wandank*, this is *Trieste*. We are on the bottom of the Challenger Deep at 63 hundred fathoms. Over." To our complete astonishment Larry's voice came back: "*Trieste, Trieste*, this is *Wandank*. I hear

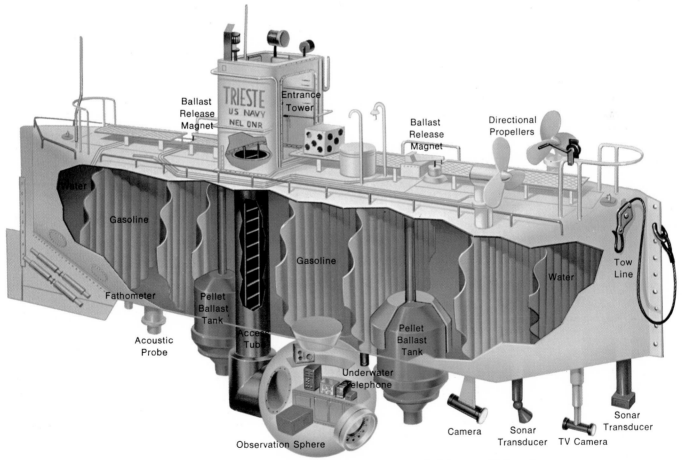

The Trieste's *observation sphere, with 3.5-inch-thick steel walls, is suspended from a 60-foot-long submarinelike hull. The gasoline tanks and pellet ballast are used to control the craft's buoyancy.*

you faint but clear. Will you repeat your present depth? Over." We could sense the excitement in his voice. "*Trieste*, this is *Wandank*. Understand six three zero zero fathoms. Roger. Out."

Solemnly Jacques and I shook hands. Then he unrolled a Swiss flag he had brought along, and I unrolled a U.S. flag.

While we waited for the silt cloud to settle, Jacques had a quick view of the second and last piece of animal life we were to encounter. What seemed to be a small shrimp, bright red and perhaps an inch long, floated by in the middle of the mud cloud. We were elated. To have seen not just one but two live creatures at these depths was staggeringly good luck.

After about ten minutes the water cleared and the bottom became visible again. I switched on the light and looked out. I saw the ocean floor, which looked flat, but I also saw what had jarred us at 30,000 feet. Across the outer window ran a series of cracks, stretching from one side to the other.

This was worrisome. The crack presented no threat to our safety, but if the window should shatter we would not be able to blow the water out of the long passageway up to the conning tower. We decided we had better get to the surface as quickly as we could—we had been on the bottom for 20 of the planned 30 minutes anyhow. So we dumped two tons of ballast, signaled that we were coming, and started up.

The trip to the surface took three hours and 27 minutes—an hour and 11 minutes less than the trip down. As we rose, we saw a curious thing. Mud which had adhered to the bottom of the sphere when we pulled up now flowed upward past our window, giving us the illusion that we were going down again. Mixed with the mud were flecks of paint from the sphere itself. At the depth we reached, the pressure is almost nine tons per square inch. This is enough to compress the sphere by two millimeters, which loosens small flecks of paint.

We reached the surface at 4:57 P.M. Usually at the end of a dive we blow the passageway quickly and violently. This time we tried it slowly and gently so as not to jar the cracked window. I fed first one, then a second, then a third bottle of compressed air into the system as Jacques watched for results through the hatch window. Finally, ever so slowly, the water level moved down past the window. A stream of bubbles broke from the air line, meaning that the water was gone from the chamber.

We wasted very little time opening the big hatch and climbing out. We closed it carefully behind us again and hurried up the long ladder, opened the conning tower hatch and came out into the sunshine and that wonderful fresh air. Off to the west, the *Lewis* was bearing down on us and right behind her came the *Wandank*, breaking all speed records for tugs. She looked great. So did everything else.

Trapped in an Undersea Avalanche

By Jacques-Yves Cousteau

The sea muffled the shouts of "*Bon voyage*" coming from the tender. In our bathyscaph we sank into green silence toward an undersea canyon off the Mediterranean port of Toulon. It was almost night at a thousand feet. In our lights I saw snowlike flakes apparently falling upward. Otherwise there was no sensation of motion. I could have been in a calm room at night in the Alps. The concentration of organisms increased in density at the 2000- to 3000-foot level; the red squids appeared almost subliminally and left their ghosts of phosphorescent ink. I touched off the camera often, hoping to capture one of them at the focal point, although aware of the long odds against it. The real photographic mission would come on the bottom.

My colleague Houot stooped over me and reached around, handling the controls. I looked up at the pressure gauge—4500 feet—and asked, "Could you slow her down?" He released a shower of iron pellets from our ballast chambers and reduced our falling speed to a few inches a second.

He read the sonar graph: "The bottom is about two hundred feet below." That was odd. The down-directed sonar beam was recording bottom far short of our most carefully corrected hydrographic survey of the Toulon canyon. At the edge of the lights I saw, just under our bow, an amorphous yellow shape. "Mud," said I. "A cloud of mud directly ahead. We're down already."

Houot replied, "That's absurd. The echogram is still showing two hundred feet of clear water under us. It can't be wrong."

I was an equally devout believer in vertical echo sounding. I said, "If it isn't the bottom, what am I looking at—a squid giving off a yellow cloud as big as a house? Or . . . or did we touch the side of the canyon with our bow?"

The lights were still stabbing down through clear black water, with the dull yellowish presence as a backdrop. The lights picked up a vague reflection below. "It's getting brighter," I announced. "I can see our two forward lights overlapping on the ground maybe eighty feet below."

Houot retorted, "The gauges show four thousand, eight hundred feet. Is it really the bottom? We're short, aren't we?"

Still going down, I saw five sharks and a big rayfish that shook its wings and flew away. The guide chain touched. Its clanking dispersed the sharks, and unquestionably the *F.N.R.S.-3* had grounded in 4920 feet of water.

Descending into an undersea canyon off Toulon, Cousteau's bathyscaph struck the ledge at left and set off a mudslide. Trapped by billowing clouds of ooze, the vessel hit the opposite wall before escaping.

But we were 380 feet higher than the place we should have hit. Had we drifted during the descent? I looked out and reported, "We are on a wavy mud shelf at the edge of a vertical cliff." Houot could not believe it. "Look for yourself," I said, letting him lean over me. He gazed for some time and got off my back with a perplexed expression.

"It's a shelf all right," he declared.

I said, "And believe it or not, the yellow cloud I saw came from a mud wall we hit on the way down."

The water outside was clear. We sat on our tiny deck and talked things over. The echo sounder used to chart the canyon was at the root of the anomaly. Its beam expanded with depth and was unable to detect the kind of steps we had landed on. Instead, it averaged them, giving the false impression of a smooth decline. We decided to turn the ship 90 degrees to port and take off for the canyon floor.

While we were reaching this decision, the bathyscaph had settled, coiling her long heavy guide chain into the mud until the gondola itself touched earth. Houot dumped some ballast and lifted the vessel. To swing our bow around, he ran the starboard motor ahead and the port astern. But owing to the embedded chain, the *F.N.R.S.-3* would lift only five feet. Houot ran both motors full ahead to pull her out of the mud. The boat tugged hard but made no progress. Here was a dismal sort of anchorage.

Then things began to happen. The bathyscaph suddenly took off. Through my window I saw an enormous chunk of hard mud tumbling off the ledge below.

It dislodged more mud, which sank in a slow-motion explosion. The lights rebounded from a blooming, spreading, climbing yellow boil.

"Houot, we've started an avalanche!" We laughed nervously. An uncomfortable thought came to mind. Suppose we had triggered a turbidity current? Some oceanographers are convinced that high-speed mud currents scour the sea floor following landslides at the head of a submarine canyon. The avalanche is thought to pick up velocity when it converges in the bottom of a canyon. It then bursts out of the canyon at great speed and rolls hundreds of miles across the deep sea floor before it slows down and settles. If the *F.N.R.S.-3* was caught in anything like that, we were in for a pretty bad ride.

"I think we'd better keep both motors going and get away from here," I said—a sentiment readily approved by Houot. For 20 minutes we cruised slowly over immense clouds that churned higher toward the bathyscaph. We thought we might find an undisturbed area by crossing the canyon on a compass bearing, even though it meant sailing through the cloud tops.

It was a mad crossing. We passed through alternating black space and ocher billows that blanked out my window as though cardboard had been pasted over it. In the open stretches I could see yellow mud cloud peaks ahead to the limit of sight. "How could one chunk of mud fill a canyon?" I wondered as I watched particles streaming into the glass. Suddenly they stopped. I looked again. The particles were stationary. Yet our motors were whirring smoothly. "Stop the motors," I said to Houot. "We are not moving." Had we collided with the other side of the Toulon canyon?

In the solemn silence we heard only the gentle sigh of the oxygen system. Houot looked out of the after port. "Nothing but cardboard," said he.

I said off-handedly, "Let's wait for the current to wash the stuff away." Houot had nothing to say. In the absolute quiet within and without, we looked out out of our respective windows with our backs to each other, both unwilling to voice the thought that possessed us: we had unloosed a mud cascade from the other canyon wall, and the bathyscaph was entombed. I looked at the depth gauge: the needle seemed cemented to the dial.

"Let's just relax and wait for the mud to settle," Houot said.

We had 22 hours of oxygen left in our rebreathing system. We sat down, carefully arranged our cramped legs on the tiny deck and talked in low tones, two perplexed men a mile under the sea. A foot outside my window we had fastened an iron bracket with a baited fishhook. I kept looking for the hook as a sign that the mud was thinning out. The window remained entirely opaque. We had the sea bottom to ourselves.

An hour passed. I cleared my throat to overcome the possibility that my voice might crack and said, "Well, the dive is lost for photography. What say we surface, *mon vieux?*" Houot got up and pressed his fingers into both shot buttons like a skyscraper tenant impatient for an elevator. He held the magnets open until hundreds of pounds of ballast had fallen from the chambers.

I watched the pressure gauge and vertical-speed indicator. They did not budge.

Through my porthole not a mud particle moved.

We remained outwardly cool and professional while hot, non-technical thoughts raced through our minds.

The vertical-speed indicator was the most sensitive recorder of motion aboard. We both watched it closely. Just one slight flicker of the needle was all we wanted to see. The dial remained as still as a photograph.

We had dropped enough weight to be soaring upward at a good speed, but the *F.N.R.S.-3* remained as dead as a fly in amber.

"We must have forgotten something," I said. We reviewed everything we knew about bathyscaphs—a syllabus of physics and a litany of operation. The answer was almost simple. During the hour we had been waiting in the deep, the gasoline in the envelope had cooled enough to offset the weight of shot pellets Houot had just dropped. He squirted pellets again. "We're climbing!" I announced. Mud specks skidded down my window. We were enclosed by them for the first 800 feet of ascent, but there was definitely no turbidity current at work. We felt no pull of current on the deep-boat. But we had definitely triggered a

submarine landslide and nearly buried ourselves. We broke into clear black water, and I watched the yellow clouds until they faded from the light and the Sun commenced faintly to tinge the sea.

The next time I met the oceanographer who surveyed the Toulon canyon for us, I said, "Remember that canyon off Toulon that you charted so carefully? You'll have to do it again. Houot and I have just wrecked it."

Safe Havens for Sea Life

By James H. Winchester

Peering through the glass plate of my mask, I float on the clear, sun-speckled sea, looking down into a captivating corner of an underwater world. Seven or eight feet below, rooted to the white sand bottom, is an enchanted coral garden—a maze of imaginative shapes softly glowing with color, tans and pinks, oranges and greens. Everywhere the brittle coral structures branch delicately like the antlers of stags—creating the aura of a petrified forest. Sunken flowers and ferns, yellow and purple, sway gently in the current.

I kick my flippers and glide on. Curious fish swim up close to look at me. With colorful spots like a second pair of eyes near its tail, one fish appears to be swimming backward. Others, searching for marine worms, nuzzle up clouds of sand with their snouts. As I dribble bread crumbs into the water, my hand is mobbed by a school of small lavender and orange gammas about the size of minnows. At the approach of a long, slender trumpet fish, the little fish scatter. A 300-pound green turtle slides across the bottom. My head underwater, breathing through a snorkel, I hear strange crunching noises. Then I see that several rainbow-striped parrot fish are biting bits of coral from the reef. A brown octopus no bigger than my hand darts to cover in the reef, its color changing first to yellow and then to crimson.

Seventy percent of the globe is covered by water, and with my flippers and snorkel I was exploring only a tiny fraction of the beckoning sea floor—Virgin Islands National Park, an underwater preserve established by American conservationists in the warm waters of the West Indies. The underwater park, on St. John Island, is but one of a growing number of marine recreational areas around the world that reflect man's increasing interest in exploring the frontiers of the oceans. Perhaps the most ambitious of the new marine observatories is one being planned by Japan. Elevators will take sightseers to the bottom of a government park in Japan's southerly coastal waters. Once on the sea bottom, the visitors will be conveyed in comfortable seats through pipelike steel and concrete corridors filled with plate-glass windows to view the wonders of the deep.

Presently most of the world's underwater parks are located in the Western Hemisphere. These marine preserves are meant to be much more than playgrounds. Besides providing for recreation, they should permit protective control of offshore waters by marine ecologists. At the first World Conference on National Parks, held in the United States in 1962, the establishment of underwater parks to provide offshore marine sanctuaries was widely discussed—and the saving of coral reefs was given top priority.

In recent years many coral reefs in the Caribbean have been despoiled by amateur divers and entrepreneurs catering to the islands' tourist trade. Indeed, massive abuse in collecting coral has long been common. Coral reefs are created by tiny, jellylike animals called polyps. A coral reef begins when individual polyps attach themselves to rocks, shells or other objects on the bottom in tropical waters. The polyps acquire oxygen from the sea—oxygen liberated by photosynthesizing algae. Food in the form of drifting plankton and dissolved organic matter is brought by the currents. Because the coral polyps' existence is tied

to that of the algae, coral reefs can grow at depths no greater than around 180 feet—below this not enough light penetrates to permit algae to carry on the process of photosynthesis. The brittle material we call coral is the polyps' protective external skeleton. The tiny animals absorb calcium salts from the ocean, allowing them to build these calcium carbonate structures around their bodies. New generations of coral polyps attach themselves to the skeletons of dead polyps. In this way the coral reef grows larger—layer upon layer, generation upon generation. Expanding at the rate of only a few centimeters a year, some present-day reefs have been developing for 100,000 years and more.

Not only corals were becoming endangered. Many varieties of shells, among them the giant queen conch —once common among the reefs of the Florida Keys and West Indies—had already been hunted to near extinction. Spear fishermen, too, had been allowed "open season" from the Caribbean to the Mediterranean Sea. Meanwhile, in California, bathers thoughtlessly snatched starfish from their shallow tidal pools until the stocks were almost depleted. Unplanned for and unprotected, marine life was threatened in many places.

One of the primary goals of the new parks, therefore, is to prevent the widespread destruction of off-

Flippering along an underwater trail off Buck Island Reef National Monument in the U.S. Virgin Islands, tourists nose down for a closer look at a living coral community on the shallow bottom.

Inside the sloping seawall of an encircling barrier reef, a West Indian yacht brings new visitors through a shallow, brightly sunlit lagoon that narrowly rings the overgrown shores of Buck Island.

Faces down in the clear water, snorkelers trace the underwater trail whose starting point is marked by a signpost projecting from the lagoon.

shore environments. "Our off-shore regions were in danger of becoming aquatic deserts," I was told by Charles Mehlert, an official of the California Department of Parks and Recreation.

At Trunk Bay, St. John, the site of Virgin Islands National Park Underwater Trail, conservationists have provided protected waters for sea turtles. The great sea turtles have nearly been killed off by human hunters, with the result that the 300-pound green turtle is now an endangered species. The leatherback or trunk turtle—for which Trunk Bay is said to be named—is also in grave danger of extermination. The trunk turtle, attaining a length of nine feet and a weight of three quarters of a ton, is the largest turtle in the world. There was a time when sea turtles arrived at Trunk Bay each breeding season in large numbers to deposit their eggs in the deep sand of the bay's beaches. After years of man's hunting them for meat and shell, and simply for sport, only a small number of the sea turtles return to the bay each year. Now, though, they are being bred in the protected preserve of the underwater park.

Marine biologists also welcome the new parks as seabed laboratories. Off Torrey Pines, a State Park at

San Diego, California, a recently established underwater trail runs eight miles along La Jolla Cove. In spring and fall, from the cliffs above the cove, one can watch the movements of the gray whales in their seasonal migrations along the coast.

During my visit, oceanographer Robert F. Dill and I donned snorkels and flippers to explore the head of a submarine canyon that starts in the shallow waters of the cove only 700 feet from shore. The winding, steep-walled valley runs 18 miles out to sea, where it merges with the San Diego trough 3600 feet below the surface of the Pacific Ocean.

Dr. Dill—a specialist in submarine geology—has studied this undersea canyon for a number of years. He believes the erosion that carved the canyon is still in progress, cutting it about an inch deeper each year. Dr. Dill's research suggests that weak ocean-bottom currents are the primary erosive forces involved. The question of what creates submarine canyons is of enormous interest to geologists and oceanographers. Some parts may be "drowned" terrestrial river canyons—carved by rushing rivers at a time when world sea level was lower than today and the shelves stood high and dry. Others may be the more recent work of undersea "turbidity currents"—massive streams of sand, mud and gravel that sometimes race down the shelves and slopes, possibly set in motion by earthquake shock waves. Still others may be the result of submarine rock slides, faulting or the sinking of crustal rock. Dr. Dill describes the submarine canyon at Torrey Pines underwater preserve as "a great natural laboratory for studying marine processes."

Going out to the canyon, I looked down through the unusually clear waters to the white, sandy bottom, as I swam easily and lazily above forests of kelp and grotesquely shaped sandstone formations. Swimming on the surface, I saw bright orange garibaldi, the "goldfish of the oceans," darting about among the swaying seaweed below. Lifting my face from the water, I saw a playful harbor seal only yards away. With his whiskery snout puckered as if in curiosity, the seal followed me for several minutes.

We headed toward the site of a 6000-year-old American Indian village that lies submerged at the canyon's mouth. Previously, the ocean here was 600 feet lower than today, and one of the ancient tribes had a major settlement near what was then the shoreline.

"We discovered this sunken city several years ago," Dr. Dill told me, "and already we have recovered more than 2000 stone bowls and other Indian artifacts. Unfortunately the site is within shouting distance of shore. It lies in water so shallow that anyone can wade almost all the way out to it. Since we first uncovered the settlement, many of its treasures have been looted for souvenirs. Now, with the park's protection, the site is being preserved for scientific exploration and study."

John Pennekamp Coral Reef State Park is a 75-mile preserve in the Atlantic Ocean off Key Largo, at the foot of the Florida peninsula. It is the largest and most popular of the public underwater recreation areas established in the United States and the Virgin Islands. Between 1963—when this 21- by 3½-mile offshore tract was opened to the public—and 1971, it drew more than two-and-a-half million visitors. Some 350,000 people now come each year. Only a small percentage of these use snorkeling gear or more sophisticated underwater breathing equipment. Most engage in underwater observation from large, glass-bottomed excursion boats, small motorboats or pedal-powered pontoon rafts. The waters, only a few feet deep near shore, reach a depth of 60 feet farther out at the edge of the Gulf Stream. The sea is generally so calm and clear that objects as small as a dime are visible on the bottom. Because of the refraction of light by the clear water, the deep reefs appear to lie close to the surface.

In Pennekamp's shoals are found most of the species of living coral found in the Atlantic and Caribbean. Ponce de León, sailing along the edges of these reefs, called them *Los Mártires*, or "the martyrs"—perhaps for the unlucky seamen who had lost their lives in shipwrecks here. Many more ships have met with disaster on the reefs since Ponce de León's day. One wreck is the *H.M.S. Winchester*, a British man-o'-war that foundered in a hurricane in 1695 while sailing home from Jamaica. It now lies half-buried on the sandy bottom only 30 feet below the surface. It is easily seen from boats, and can be explored at firsthand by skin

A graceful, flippered human interloper descends into a watery world of muted sunlight and coral forests to pick up Buck Island's underwater trail.

219

divers. Spearfishing and the removal of coral shells is prohibited, assuring the preservation of the natural marine environment.

In California's Big Sur region, the Pacific waters of the Julia Pfeiffer Burns State Park never get above 55° F. Divers must wear wet suits for protection against the cold. Whales, sea lions and sea otters are at home in the park's waters. On land, forests of oak and redwood stand right at the edges of the sheer rocky bluffs that drop straight into the sea—one of the most dramatic meetings of land and water in the world.

In coming years new underwater parks are certain to be established along more of the world's shores. As we become increasingly aware of the need to preserve the waters and the living things of the seas in their natural environments, the simple idea of declaring some of our remaining coastal areas "off limits" to all but sightseers becomes more and more appealing. It is one way to guarantee that our children—and their children—will inherit a world whose sea life has not been totally despoiled.

An alert photographer recorded this singular tableau at Buck Island's underwater park—a parrot fish lured to a plaque displaying its likeness.

We Must Stop Killing Our Oceans

By Gaylord Nelson, U. S. Senate

Throughout history, man has believed that at the sea's edge his power to destroy stopped and nature's invincibility began. Even Rachel Carson, in her 1951 book *The Sea Around Us,* saw the oceans as one last haven, safe forever. How could it be otherwise when the oceans are so vast that continents are mere islands in their midst, so deep that a Mount Everest could be lost beneath their surface? How does one pollute a volume of almost 320 million cubic miles? How poison an environment so rich that it harbors 200,000 species of life?

The vulnerability of the marine environment becomes dramatically clear when we stop to realize that even though the oceans blanket three fourths of the Earth, their productivity is limited mostly to the narrow bands of undersea land extending from coastlines which comprise the continental shelves. Eighty percent of the world's saltwater-fish catch is taken from these shallow coastal waters, which make up only a tiny fraction of the total sea area. In addition, almost 70 percent of all usable fish and shellfish spend a crucial part of their lives in the estuaries—the coastal bays, tidelands and river mouths that are 20 times more fertile than the open sea, seven times more fertile than a wheatfield. Cut the chain of life in these areas, destroy the myriad bottom organisms, pollute the continental-shelf waters and you will also eliminate the vital ocean fisheries.

Already pollution or overfishing, and sometimes both, have gouged fisheries around the world. Meanwhile, in a headlong rush to create more land, vital coastal tidelands are being filled for highways, industry, bridges and waterfront homes. At the same time the remaining estuaries are fed billions of gallons of sewage and industrial waste every day. These poison fish, choke out oyster and clam beds and render the bays and tidelands unfit for anything.

While the vise tightens on the critical inshore areas that lace the coastlines, pressure also builds on the ocean beyond. In 1968, for example, some 48 million tons of solid wastes were carried out by barge and ship and dumped in ocean waters off the United States. These wastes include garbage, waste oil, dredging spoils, industrial acids, caustics, cleaners and sludges, airplane parts, junked automobiles and spoiled food. During his two papyrus-boat trips across the Atlantic, author-explorer Thor Heyerdahl sighted plastic bottles, squeeze tubes, oil and other trash that had somehow been swept by the currents to mid-ocean. On some days the crew hesitated to wash because of the amount of pollution.

About 7000 feet off the sunshine-and-salt-spray won-

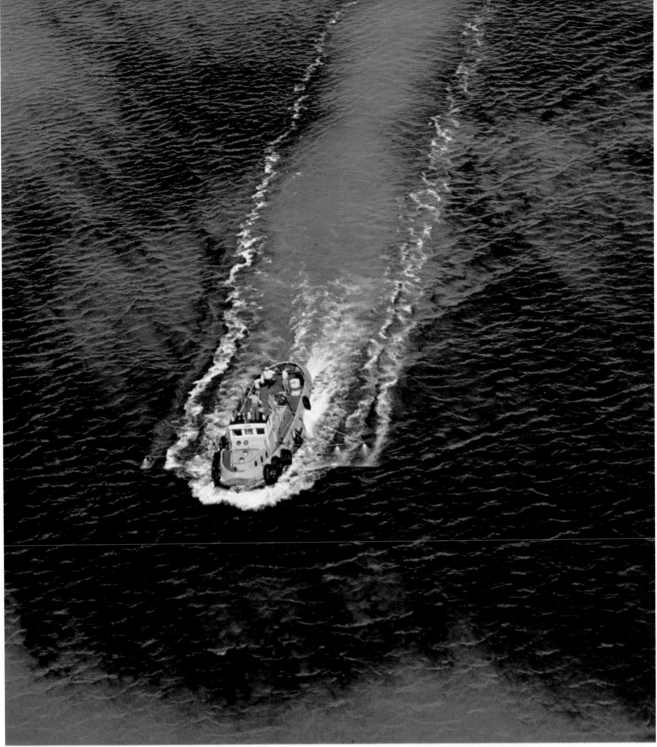

Cutting through a dark oil slick in the Caribbean Sea off Venezuela, a specially rigged pollution-control craft sprays the slick with a chemical that causes oil to congeal and sink to the bottom.

derland of Miami Beach, Florida, there is a man-made phenomenon mockingly known as the "Rose Bowl." The Rose Bowl is a large, bubbling splotch of ugly brown that sprawls over those blue-green waves. It is raw sewage piped out from Miami Beach and three other nearby communities. As yet, only rarely do wind and tide combine to wash the debris back onto the beaches.

Ordered more than ten years ago by Florida's health department to treat its sewage, Miami Beach only recently considered its first step—extending the discharge pipe one mile farther out to sea. But will this do any good? In fact, Durbin Tabb, marine biologist at the University of Miami, says that because of prevailing winds, extending the pipe means the sewage is just going to be blown back inshore on somebody else's beach.

Southeast Florida is booming, with a megalopolis of ten million people predicted within 20 years. But the Rose Bowl is an ominous sign of big trouble ahead for that supposedly limitless resource on which the Florida economy is built—the sea and the beaches.

Fishermen, divers and others whose lives are entwined with the sea, report similar situations all along U.S. coastlines and in many other parts of the world. Examples:

• Filter-cigarette butts, bandages and chewing gum have been found in stomachs of fish caught near New

At low tide the polluted broth of New York Harbor leaves behind a high-water line of refuse.

On a Cornish beach, workers heap black piles of seaweed soaked by 1967's Torrey Canyon oil slick.

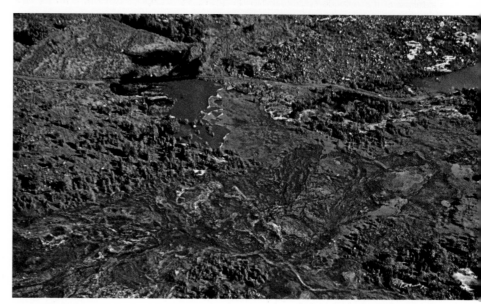

A slimy mass of industrial pollutants mixed with broken ice spreads toward the Baltic Sea from a river estuary below Hallstavik, Sweden.

Its shores under trash, its waters choked with sewage, Italy's Venice Lagoon—seen here near Chioggia—is fast becoming a biological wasteland.

Near Santa Barbara, California, a duck that was one of an oil slick's toll of shore creatures offers tragic testimony to the menace of pollution.

York City's sewage-sludge dumping ground, as close as eight miles out in the Atlantic. Meanwhile some beaches in northern New Jersey, near the Atlantic shipping lane into New York Harbor, have been turned into a nightmarish scene of plastic bottles, tar from oil slicks and even dead animals.

● Human and industrial pollutants pour into Oslo Fiord and many major harbors along the coast of Norway in such tremendous quantities that large areas are now devoid of marine organisms.

● Six to eight oil slicks *per week* are now reported along the coasts of Britain. Oil tanker accidents such as those involving *Torrey Canyon* in 1967, the *Pacific Glory* in 1970 and the *Panther* in 1971 have posed major threats to Britain's beaches, but meanwhile, the illegal dumping of oily waste and the improper cleansing of shipboard oil tanks continue to pollute Britain's busy coastal waterways.

● In 1965 five people died and 30 were stricken by mercury poisoning in the Minamata Bay area of Japan. The source of the poisoning: fish that had absorbed industrial pollutants. As far back as 1953 the sea had given warning. At that time, residents of the same area fell ill with the same disease. More than 100 were afflicted; 43 died.

The restricted seas of the world are in even worse shape. Deadly hydrogen sulfide has been building up in the depths of the Baltic Sea. If it spreads much more, experts say, the Baltic could become an oceanic desert. Throughout the Mediterranean, scores of its famed beaches are closed to the public because of pollution. In one stretch of the Italian Riviera alone, 67 open sewers discharge their effluent along the beaches, spoiling them for swimming and recreation.

Perhaps more than any other problem, the dramatic oil-well blowouts in the sea and the oil-tanker breakups have begun to awaken us to the total inadequacy of our present ocean policies. The list of places where oil has blackened beaches, killed thousands of birds and posed lingering threats to marine life is already long: Florida, Nova Scotia, the Bahama Islands, New England, New Jersey, Puerto Rico, southern California, southern England. We do not possess the technology to contain the oil from ocean disasters. Yet oil-carrying tankers are being built to gigantic scale, increasing the risks of monumental spills. And by 1980, if present trends continue, we shall be drilling 3000 to 5000 new undersea oil wells each year.

Radioactivity from the fallout of early nuclear tests can still be found in any 50-gallon sample of water taken anywhere in the sea. Investigators of a massive killing of sea birds in British waters found unusually high counts of toxic industrial chemicals used in making paints and plastics. The worldwide use of toxic, persistent pesticides is having serious adverse effects on many species of fish-eating birds and birds of prey over vast portions of the Earth, and there is evidence that

these poisons can also attack marine phytoplankton, a food fundamental in the chain of ocean life.

The conclusion is unavoidable. If tough, intelligent action is not taken now, we will make of the oceans the same mess we have made of the land. And the greatest losers of all will be the people of the world.

Although the day is late, there is still hope. But turning back the massive assault on the sea will be a tremendous task. All nations together must establish an international policy on the sea that sacrifices narrow self-interest for the protection of this vast domain that is a common heritage of all mankind. It is a task that, for the future of man, must be of the highest priority; a challenge that will test our intelligence as a species, our decency as human beings, our sense of moral responsibility to generations yet unborn.

Specifically, here are the major steps that must be taken:

1. *As soon as practicable we must end the dumping of wastes into the sea, the big lakes and the coastal areas of our rivers and bays—except for liquid wastes treated at least to levels equal to the natural quality of the ocean waters.*

We are running out of an 'away' in which to throw things away. Our only choice now is to put our technology to work at finding ways to recycle wastes back into the economy, and it is encouraging to note the recent progress of ocean pollution control in the United States Congress.

2. *In order to avoid the kind of chaos and destruction now apparent on the land, we must set tough controls before undertaking new ocean activities such as building offshore jetports and drilling offshore oil wells in new areas.* The public must be fully informed and consulted at every step, in deciding, for example, whether to allow marine industry, with all its paraphernalia, to create a new sea horizon, or whether huge supertankers will be permitted in coastal waters.

As for offshore oil wells, we should halt all drilling in ecologically sensitive areas. And we should prohibit new drilling *anywhere* until there is convincing evidence that it will not harm the marine environment, and until we have the technology to contain oil spills. Up to that point, untapped oil and mineral deposits under public jurisdiction in the sea should be held unexploited.

3. *We must halt the reckless dredging and filling of priceless tidelands and the carving up of ocean-front in the name of "progress."*

Some marine biologists say grimly that, unless we act now, the current accelerating pace of ocean pollution will put an end to significant life in the sea in 50 years or less. This would be a catastrophe, posing grave consequences to a world dependent on these vital resources for food, raw materials, recreation and, in the near future, probably even living space.

Once More, Holland Defies the Sea

*For ages the Dutch have battled the North Sea, and now, in the most costly and
ambitious assault so far on their ancient adversary, they hope to secure their land for good*

More than half of Holland's 13 million people live on land that shouldn't be under their feet at all. This is the land—called "polders"—that the Dutch have wrested from the sea over many generations. Now it comprises some 50 percent of Holland's 12,798 square miles. Most of the polders lie below sea level behind a 1200-mile bulwark of coastal dunes and manmade dams and dikes; thus modern pumping stations and the few remaining old-fashioned windmills must labor unceasingly to remove excess water produced by seepage and rain. Since

the Zuyder Zee was closed by the 20-mile Barrier Dam in 1932, the Dutch have converted 400,000 acres of former Zuyder Zee bottom into rich farmland—a tremendous boon to the most densely populated nation in Europe. But in 1953, savage North Sea storms drove the ocean over southwest Holland, killing 1835 people and ruining half a million acres of farmland. This grim reminder of Holland's vulnerability inspired the "Delta Plan," the biggest dam-building effort since the Zuyder Zee closing—and possibly Holland's last major project.

*Drainage canals intersect in an area of Dutch farmland that was once beneath the salt waters of the
Zuyder Zee. In the foreground, a pumping station's sluices direct excess water toward the sea.*

Holland's far north, like other regions, is also vulnerable to the sea. In 1969 the five-mile-wide Lauwers Sea was closed by a rock-buttressed dam.

With a superhighway on its back, Holland's Barrier Dam knifes 20 miles across the mouth of the Zuyder Zee (right), shutting out the North Sea's tides.

In an earlier stage of construction at Lauwers Sea Dam, rocks are dumped at the base of concrete caissons that were sunk across the mouth of the inlet.

Taming the Delta's Tides

With her northern coast well protected, Holland is now directing her age-old battle against the sea toward her low-lying southwestern Delta region. After recovery from the disastrous floods of 1953, the Dutch put into effect the Delta Plan—aiming to complete the project by 1978.

In the Delta, many fingerlike islands and peninsulas crowd the North Sea estuaries of the Rhine, Meuse and Scheldt rivers. Over the years the land in this area has been inundated by the sea time and again. Today, Holland's overriding concern is to shield these lowlands permanently from the ravages of the sea with an 18.5-mile network of new dams.

By dropping this weir's twin gates, the Dutch can restrict the flow of the Lower Rhine River and divert its water through canals into the Zuyder Zee.

This work island is one of the steppingstones for a six-mile-long barrier dam Holland is building to block the tides from its Eastern Scheldt estuary.

The causeway atop Grevelingen Dam is just one link in the new highway system that connects many previously isolated islands in Holland's Delta region.

When completed, the new dams will close all but two of the Delta's estuaries and will take the burden of coastal defense off the older and weaker dikes that now line bays and riverbanks farther inland. The waterways leading to the ports of Rotterdam and Antwerp, Belgium, will not be dammed, but existing dikes along these seaways will be raised and strengthened. When the works are finished, the extent of Holland's shoreline directly exposed to the North Sea will be reduced by 435 miles with a matching dip in dike maintenance costs. As a further benefit, the Delta Plan provides for diverting and damming fresh water from the Rhine and Meuse rivers to create large lakes and reservoirs. In addition to providing new recreational areas, the freshwater basins will provide a weapon to combat concentrations of ocean salt in farm areas. Because of the sea's recent inroads, Holland's soil salinity has been increasing. In building up freshwater levels, the Dutch will be able to head off threatened disaster to their farmlands by flushing salt water farther and farther back toward the North Sea.

Large freshwater lakes (light blue) are forming behind dams in the Delta and Zuyder Zee regions. Light green area shows total extent of reclaimed land.

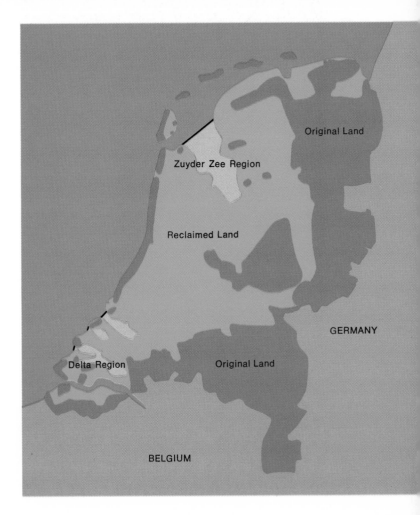

Tides drive 190 billion gallons of water through the 89-foot-deep main channel of Brouwershavense inlet every day. A cable was stretched across the channel so that foundation rock could be deposited.

Another section of Brouwershavense inlet was dammed by towing concrete caissons into position and sinking them across the channel, an enormous job that drew boatloads of sightseers to the final closing.

After the Brouwershavense tidal currents were blocked with a wall of rock, dredges heaped sand on either side of the cableway to form a broad dam capable of resisting the pounding of the ocean surf.

Dams Across the Open Sea

Years of sacrifice have only made the Dutch more resourceful in their latest offensive against the sea. Dikes are quickly heaped up by modern suction dredges, then waterproofed with asphalt or nylon carpeting to prevent washing until they stabilize. Where once Holland saw her defenses confined to riverbanks and shorelines, new dams are routinely laid in the open sea. Overhead cables stretch from point to point carrying gondolas that dump tons of rocks until a dam foundation rises above the water. Where the tide is too strong to build with rocks, immense prefabricated caissons are towed into place and sunk end to end until the gap is closed against the surging tide.

Although conceived primarily to end the threat of flooding, the Delta Plan is already beginning to yield a further dividend. The new dams carry superhighway extensions that shorten traffic links to the major cities, ending the isolation of the Delta islands and opening them to industry and tourism. But adverse effects have also appeared. With the Delta estuaries sealed off from the sea, the prosperous Delta shellfish industry will be destroyed. However, the Dutch government makes grants to assist fishermen in finding new livelihoods. And all Dutchmen know that many sacrifices must be made if they are to prevail against the sea.

Verdant and productive again after the floods of 1953, Holland's Delta farmlands will be secure when Delta Plan projects are completed in 1978.

Snug House
on the Ocean Floor

By Steven M. Spencer

The four young scientists woke each morning to sunlight dancing down through shuttling schools of silver-gray needlefish. From plastic portholes they looked out on yellow snappers grazing in the living coral gardens that surrounded their underwater home. They had breakfast, and after strapping on their scuba gear, the men emerged from the sharkproof vestibule to swim shoulder to fin with their neighbors—including mackerel and amberjacks, blue-green parrot fish and gaudily striped demoiselles.

For 60 days—from February 15 to April 15, 1969—the four American scientists lived and worked continuously under the sea, never coming to the surface. The aquanauts were involved in an underwater living experiment called Tektite I. The project was named for the tektites, or glassy pebbles, found on Earth and believed to be the debris of meteorite explosions on the Moon. Just as some of those mysterious objects from space have found their way into the oceans, so had the aquanauts of the Tektite project, which began as a study for the United States space program.

The Tektite habitat was anchored on the bottom of Lameshur Bay, off St. John Island in the Virgin Islands. Home for the Tektite scientists was a tight little duplex standing under about 50 feet of opalescent green water on a sandy area several hundred feet from the shore.

On the sea floor in the Virgin Islands an aquanaut displays a spiny lobster to a scientist inside a plastic porthole of the Tektite I ocean habitat.

The six-foot walls of coral surrounding the habitat were profusely overgrown with marine life—sea fans, sea whips, brightly colored sponges—forming an underwater rock garden. Beyond lay more reef formations, many rising as high as 15 feet above the bottom. By propelling themselves over these ridges with their flippered feet, the scientists could explore the plain that sloped away to the outer depths of the bay.

Their undersea house consisted of two air-filled cylindrical steel tanks, 18 feet high, connected by a crawlway. The cylinders were mounted on a base structure anchored to the ocean bottom with 87 tons of lead ballast. Painted white, the habitat attracted hundreds of fish, and night and day they swam up to peer into the portholes.

Each cylinder contained two rooms. In one of the lower rooms—comfortably furnished and carpeted—the men slept, ate, read and listened to tape-recorded music. An upper room, the bridge, was filled with communication controls and laboratory equipment. Another room housed transformers, air-conditioning compressors and a freezer locker for food. (The menu relied heavily on frozen dinners, but fresh fruits and vegetables were occasionally lowered from the surface in a sealed container at the end of a line.) The fourth chamber, the "wet room," opened directly into the sea through a vertical hatch. Here the men kept their scuba gear and diving suits, and here after each swim they showered with fresh water and changed into dry clothes.

The aquanauts' habitat was connected to the support barge by a thick "umbilical"—a bundle of hoses and cables. These brought down fresh water, air and electric power, as well as providing communications.

The Tektite crew chief was Richard A. Waller, an oceanographer in the United States Bureau of Commercial Fisheries. His teammates were Conrad V. W. Mahnken and John VanDerwalker—both Bureau biologists—and Dr. H. Edward Clifton, a geologist with the U.S. Geological Survey. All of the men were in their thirties.

The project was jointly sponsored by the U.S. Navy, the Department of the Interior, the National Aeronautics and Space Administration and the General Electric Company, which designed, built and equipped the undersea house. The main purpose of Tektite I was to establish guidelines for future submerged-laboratory research projects. The American space agency's interest was to observe the behavior of men confined for long periods in small, semi-isolated quarters—vital information for long space voyages.

The Tektite I experiment developed much new knowledge about "saturation diving." When a submerged diver breathes air or some other breathing mixture under pressure for any length of time, his blood stream and body tissues become saturated with the gases he is breathing. The greater the pressure, the

In a scene suggestive of science fiction, Tektite aquanauts set up their habitat for a 60-day sojourn on the ocean bottom. The cagelike structure is a movable shark refuge for aquanauts working outside.

more gas is absorbed. For descents of less than 150 feet, a breathing mixture of nitrogen and oxygen—ordinary air—can be used without endangering the diver. At greater depths and pressures, nitrogen has highly toxic effects; the first symptom of nitrogen poisoning is the narcotic state called "rapture of the deep." Since Tektite I was located at a depth of around 50 feet on the sea floor, ordinary air (at about 2½ times atmospheric pressure) could be used as the breathing mixture.

The advantage of living on the sea floor was that the men were required to "desaturate" only once—at the end of two months—instead of after each excursion into the water. This was because the pressure of the air they breathed in the habitat was kept equal to the pressure in the surrounding sea. Desaturation at the end of the two-month period required the aquanauts to spend almost a full day in the project's decompression chamber, where they were gradually brought back to normal atmospheric pressure. Too-rapid decompression allows the pressurized gases dissolved in the blood to "bubble" rapidly out of solution, causing the painful and often fatal affliction called "the bends."

Surprisingly, the men were not aware of any extra breathing effort in inhaling and exhaling air more than twice as heavy as normal air—which they did 24 hours a day, for two months. "What was important in the Tektite biomedical tests," said Dr. Christian J. Lambertsen, head of the Tektite team of doctors, "is that there were no important changes at all—pulmonary, cardiac or neurological."

The aquanauts gave themselves daily physical checkups—pulse, blood pressure and electrocardiograph readings. At night electrodes recorded their brain waves to show how well they slept. Meanwhile, topside in the control barge, doctors and "behavior monitors" sat in front of a bank of closed-circuit television screens and open microphones, keeping them under round-the-clock surveillance. (Some 40 aquanauts participating in later Tektite studies also came through the experience without significant physical problems, although a number of them couldn't take the psychological stress of isolation under water. Tragedy struck the man-in-the-sea experiments, however, when Sea Lab III Aquanaut B. L. Cannon died of carbon-dioxide poisoning at a depth of 600 feet on the continental shelf off California in February 1969).

The Tektite scientists made all kinds of observations and experiments. For example, much of the bottom sediment in Lameshur Bay consists of disintegrated coral and shellfish skeletons. Studies of sediment formation were made on carefully marked-off grid patterns. Coral reefs are known producers and potential reservoirs of petroleum, aquanaut-geologist Clifton explained, and aquanauts working on the continental shelves may one day help provide clues in the search for oil and other minerals.

In the lower part of the left chamber, Tektite aquanauts ate and slept; the room above held communications and laboratory equipment. The right cylinder housed the air compressor and a "wet room" that opened directly onto the ocean floor. Air pressure at 2.5 times normal atmosphere kept out the sea.

In the future, men living in the sea may play important parts in fish-farming and aquaculture—the farming of marine plants, such as kelp and other seaweed. Thus the Tektite scientists studied the habits of marine creatures and conducted a census of local marine life.

The aquanauts counted scores of species of fish, ranging from 2-inch anchovies to 12-foot sharks. Occasionally their finny friends played rough. Biologist Mahnken told of an encounter with a school of two dozen big amberjacks near one of the coral ridges. Each fish weighed 20 to 30 pounds. "They swam out of the darkness toward us in such tight formation that they looked like one huge fish," he said. His oceanographer colleague, Waller, added, "I think they were claiming their territory, because they started bumping us in the back and shoulders. Finally we took the hint and left."

As emergency refuges from sharks, five way stations had been set up on the floor of the bay. These were cage-like booths with plastic domes, a spare bottle of air and a telephone line to the habitat. For safety's sake, whenever the men swam out of the habitat they went in pairs. As a further safety measure, support divers hovered over them in boats, keeping track of their whereabouts by the trains of bubbles from their breathing gear.

One of the most active research projects was VanDerwalker's pursuit of the spiny lobster, an edible crustacean resembling the fresh-water crayfish. The Virgin Islands population of spiny lobster (which supplies the local market for lobster tails) had fallen off in recent years. The aquanauts snared 140 and tagged most of them. To some they attached thumb-size sonar transmitters, so that each could be tracked as it moved about in the sea in the vicinity of the habitat.

The lobsters are nocturnal, VanDerwalker discovered through the sonar-tracking devices. At night they would move out to the algae-covered sand flats, presumably foraging for mussels, clams and immature queen conchs. During the day most took refuge in naturally occurring shelters, or "hotels," among the coral. But many lobsters spent the day in caves 1000 feet seaward of the Tektite habitat. Possibly they hid out in the caves to avoid sharks and other predators.

Practical conclusions? "I think the reef could support more lobsters," VanDerwalker said. "A hatchery might be set up to introduce more young to the area, and perhaps something could be done by using low-frequency sound signals to draw the sharks away from the lobster grounds, and dispatch them."

Another inhabitant of the coral that became a Tektite study object was the curious "cleaning shrimp." This species, light blue and about an inch and a half long, lives among the poisonous tentacles of the sea anemones. "By flicking their antennae they attract passing fish," Mahnken said. "As the fish swim in, the shrimp hop aboard them and begin to pick off their external parasites. They clean the scales, fins and gills,

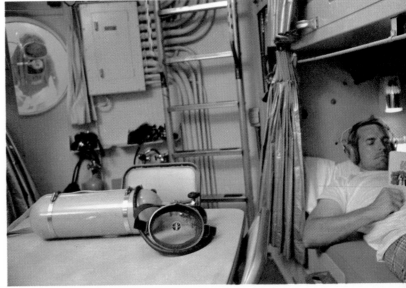

Cramped quarters required strict scheduling by the aquanauts; while one rests another swims in the sea (above), and a third prepares a meal (below).

and even eat away the dead flesh if the fish has a wound." Apparently this shrimp is immune to the stings of the anemone—a fact of interest to the pharmaceutical industry. Perhaps the secret of the shrimp's immunity will lead to chemicals of medical value to humans.

Today, scientists of many nations emphasize the need for further exploration of the sea floors. It seems increasingly certain that in the future man will turn to the continental shelves for new sources of food and minerals in an overcrowded world.

While the depths below 2000 feet are likely to be explored only with diving vehicles built to withstand tremendous pressures, the success of the Tektite program proves the feasibility of broadened exploration of the shallower reaches of the Earth's continental shelves.

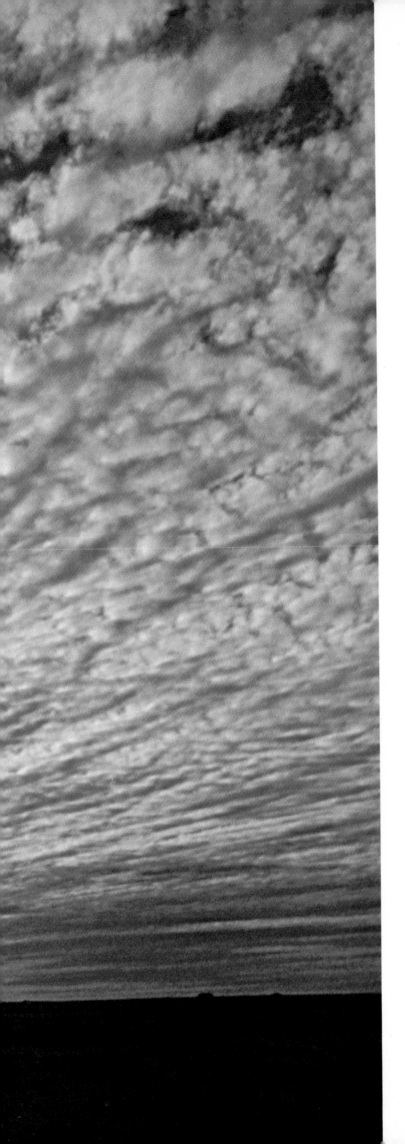

Part Five

OUR
MAJESTIC
ROOF
OF AIR

High above an African desert, a canopy of clouds spreads an ever-changing pattern across the sky.

Chapter One

THE AIRY DOMAIN

As our planet journeys through space, the only shields between us and the Sun's fierce radiation are our many-storied blanket of air and our outer planetary electron screen

The Deep Realm of the Atmosphere

By Theo Loebsack

The measureless sea of air above us surrounds the Earth like a gigantic husk to a height of several hundred miles. It reaches into all valleys and hollows and fills every man-made depression. It penetrates into the finest rock crevices, between the petals of a flower and between the pages of a book. It covers every inch of the 197 million square miles of the Earth's surface, and, if we calculate how much force it exerts on our planet, we have to write a 5 with 15 zeros to give the weight in tons. Every square inch of the Earth's surface at sea level bears a weight in air of almost 15 pounds. About 95 percent of this huge mass of air is concentrated in the first seven miles above the Earth's surface; the remaining 5 percent spreads farther upward to a height of several hundred miles, fading imperceptibly into space.

This sea of gases is not dead. It is as full of life as a cultivated garden. In the densest bottom layer live an incalculable number of plants and animals, including about 3.5 billion human beings. It is a zone teeming with growing, flowering and breathing individuals. And above this layer there is also an abundance of life. Up to a height of about six miles the air is full of suspended bacteria, fungal spores and pollen grains. Between 2500 and 4500 feet, a cube of air the size of a lump of sugar contains an average of about 70 microorganisms, among them mold spores, cocci and yeast species. Moreover, specially equipped balloons sent aloft by scientists in the past decade have captured the spores of molds at altitudes up to 14 miles.

The air is also full of inanimate matter such as dust, soot, volcanic ash, salt crystals from sea spray and fine sand grains raised by storms in the deserts. In addition there is cosmic dust. Each day a weight of two to three thousand tons of this dust falls from space onto the Earth.

Although the air at sea level, except for its water-vapor content and impurities, is made up of the same constituents as at great heights—roughly, 78 percent nitrogen, 21 percent oxygen, and small amounts of argon, carbon dioxide, neon, helium, krypton, hydrogen, ozone and traces of other gases—it becomes less dense at higher altitudes. At about six miles the atmosphere is already so thin that a human would suffocate there in a few minutes. At 12 miles a candle could not be lit because there is not enough oxygen to make it burn. In the thin air at even greater heights, the number of gas molecules per cubic inch declines dramatically. In other words, the distance between the air's particles grows larger with increasing height. As a general rule, air density diminishes tenfold for each 12½ miles of altitude. At 125 miles the atmosphere is thus only about one ten-billionth as dense as at sea level, comprising a near vacuum. Nonetheless, even at this height each cubic centimeter of space contains billions of molecules of gases—enough to make up a detectable atmosphere.

One can no more define the upper limit of the atmosphere than the edge of a puff of smoke. Somewhere far above the Earth the air gradually merges with the near vacuum of interplanetary space, but we cannot draw a clean line where the atmosphere may be said to end and space begin.

Although the air above us appears uniform, scientists have found that the atmosphere is divided into distinct layers. In the bottom layer, the troposphere (from the Greek *tropos:* to change), occur the effects we call "weather." This is the turbulent region where clouds and rainbows are formed, as well as gales, monsoons and sandstorms, where cloudbursts, hail, and thunder and lightning rage.

Miles
400 –
350 –
300 –
250 –
200 –
175 –
150 –
125 –
100 –
75 –
50 –
25 –
8 –

degrees F.
+2256° –
+2100° –
+1146° –
+324° –
–135° –
+28° –
–47° –
–68° –
+59° –

Ionosphere

Auroral
Displays

Noctilucent
Clouds
D Layer
Ozone Layer

Mesosphere

Stratosphere

Troposphere

*While probing the atmosphere, man has set these records with each category of air and space craft:
(bottom to top) turboprop plane, 46,214.5 feet; glider, 46,267 feet; piston-engined plane, 56,046
feet; balloon, 113,739.9 feet; jet plane, 113,890.8 feet; rocket plane, 314,750 feet; first manned
U.S. space flight (Glenn), 100-163 miles; first manned Soviet space flight (Gagarin), 112-203 miles;
first Soviet satellite (Sputnik I), 135-587 miles; first U.S. satellite (Explorer I), 224-1585 miles.*

237

As a rule the temperature gets colder as we rise higher in the troposphere. Paradoxically, the top of the troposphere is colder above the Equator than over the poles. Why? We must remember that the centrifugal force created by the Earth's rotation is greater at the Equator than at the poles. Quite naturally the troposphere is influenced by this latitudinal discrepancy in centrifugal force; in short, the air is not uniformly distributed about the Earth, but bellies markedly out around the Equator. Here the troposphere has a height of about 13 miles, whereas it measures only five miles at the poles. Above the Equator, the roof of the troposphere is thus higher and colder than at the poles; the two temperatures average −112° F (−80° C) and −58° F (−50° C) respectively.

What surrounds us several miles up in the troposphere? A cloudless, deep blue sky, with an intensity of color never seen from the surface of the Earth. Why is the sky up here so deep a blue, and why is it blue rather than yellow, green, pink or red?

The answer lies in the composition of white sunlight, which is actually a jumble of light waves comprising all the colors of the rainbow, and in the scattering effects of the gaseous molecules in the air upon these different-colored waves as they pass down through the atmosphere. The violet and blue rays are deflected farthest from the direction of the original white sunbeam. This is because the molecules' scattering effects are great on the violet and blue rays, much smaller on the yellow wavelengths and almost nonexistent on the red. Once deflected, the blue rays meet other air particles, which deflect them again. The scattering process is repeated endlessly, causing the blue waves to follow a zigzag course through the air. When they finally reach us, they do not seem to come directly from the Sun but, like a fine spray of rain, from all corners of the sky. Thus the sky appears blue and the Sun yellow, because the yellow and red rays undergo less scattering and reach us in a nearly direct line. In the evening, when the Sun is near the horizon, its rays have to traverse an exceptionally long path through the thick air of the troposphere; hence even the yellow light is widely scattered. The red rays are least affected, so that the orb of the setting Sun appears a deep red.

Diminishing atmospheric scattering of light explains why the sky grows darker with increasing height. At great heights the Sun's rays meet far fewer air particles than at sea level and are scattered and deflected less frequently. Therefore the light no longer seems to come from all around us, but to be concentrated mainly at its source: the Sun. Where there is no atmosphere there can be no colored sky. Because of this, the sky at great heights is black as night, while the Sun shines with a dazzling white intensity.

Ascending from the roof of the troposphere into the thin, crystal air of the stratosphere, weather balloons

record a marked rise in temperature. This increase, most dramatic between 16 and 32 miles, had long puzzled scientists. Today it is known that this warming occurs within layers of ozone which predominate in the stratosphere at these altitudes. Ozone is produced when intense ultraviolet light from the Sun splits the naturally occurring oxygen molecule (O_2) into two atoms. These then unite to make ozone (O_3). Later, when the ozone breaks down into normal oxygen again, the original energy of the ultraviolet light is liberated as heat, a process sufficient to raise the temperature to well above freezing at a height of 32 miles.

From humanity's perspective, this stratospheric ozone acts as a lifesaver: if the ultraviolet rays reached us at full strength, their powers of penetration and destruction would cause grave biological damage. The ozone layer, therefore, acts as an umbrella against the most dangerous of the Sun's rays.

In this unique portrait of the Earth's atmosphere photographed on a space flight more than 100 miles above the surface, refracted light from the setting Sun turns the dense troposphere into a band of color.

Fifty miles up, at the pitch-black, star-studded ceiling of the mesosphere, the thermometer falls to $-135°$ F ($-93°$ C). Up here the air molecules are too widely scattered to transmit sound, and a deathly silence prevails. Below, clouds drift by, looking like small pieces of cotton wool, their upper surfaces brilliantly reflecting the light of the Sun. Where visible, the horizon is misty, and for the first time the curvature of the Earth appears clearly against the black sky, giving us an idea of the immensity of our planet and the powerful gravitational forces of the Solar System.

At a height of about 50 miles the air is electrically charged; electrically neutral air particles have changed their nature and are capable of attracting or repelling each other. We have penetrated the first of the four (or occasionally five) electric layers of the atmosphere; it is the so-called "D" layer, which exists only when the Sun shines on it.

As with so many of the atmosphere's secrets, we lack exact knowledge about the electric layers. We do know, however, that the atom consists of a positively charged nucleus surrounded by a shell of negatively charged "electrons" in fixed orbits. All atoms, of which all matter, including the air, is composed, are constructed on this basic principle. But up here, cosmic rays and radiation from the Sun attack and change the structure of the atoms' electron shells, knocking out some of the electrons from the ordinary electrically neutral system. When a neutral atom loses an electron, the positive charge of its nucleus predominates and it becomes a positively charged "ion," while the electron is split off as a negatively charged particle.

All four ionization layers teem with positively and negatively charged atomic fragments; when named, this zone in the upper atmosphere—50 to 600 miles above sea level—was called the "ionosphere."

239

The ionosphere is one of the most intriguing of the Earth's atmospheric realms. In it appear the ghostly polar lights, the auroras. From this electrically charged zone low frequency radio waves are reflected to Earth as if from a mirror, enabling short-wave radio signals to be bounced great distances around the globe. As radio waves travel only in straight lines, one could not otherwise receive signals from a transmitter lying below the horizon.

At 58 miles the last clouds glide past us as faintly glowing gossamer veils. They are the "noctilucent" clouds—either fine ice crystals or dust—that can occasionally be seen from Earth at dawn or dusk when the Sun is on, or just below, the horizon. Only then, when lower cloud formations are still in shadow, do the Sun's rays rake these thin clouds at an angle that makes them visible from the Earth's surface.

The upper atmosphere, too, is the realm of flashing meteors. The air here is just dense enough to ignite these "bullets from outer space" by friction. Geophysicists have calculated that the atmosphere catches perhaps a hundred million of these small stone splinters daily, heating them to incandescence as shooting stars for a few seconds before they vaporize harmlessly. Only a small fraction ever reaches the ground.

Some 120 miles up, in the ionosphere, the temperature of the air reaches nearly 1500°F (800°C). And higher up it grows hotter still. The air temperatures at these great heights are deceptive, however. Air temperature is a measure of the speed of air particles set in motion by the Sun's rays or some other source of energy. From a height of 40 miles upward, however, there are not enough particles in a given volume of space to create much heat—regardless of their velocities. But neither is the air thick enough to shield an object from the direct rays of the Sun. As on the airless Moon, the temperature of an object in direct sunlight soars, while shade temperatures plunge to hundreds of degrees below zero.

Approximately 600 miles above the Earth's surface the air is incredibly thin, but there still is air. This is the exosphere (from the Greek *exo:* outside), where the gravitational pull of the Earth acts only very weakly on the air particles. Their number is very small compared to those in lower layers of the atmosphere, but they move much faster because of their direct exposure to solar energy. Under these circumstances some of the particles escape completely from the atmosphere, perhaps wandering in space until they reach the gravitational sphere of another star. Others may wander around the Earth like tiny satellites, while still others dive back again into lower layers of the air.

The exosphere, extending indefinitely into space, is the uppermost story of our atmosphere; above it lie the Van Allen radiation belts, and the near-empty reaches of the universe beyond.

Storms That Rage Beyond the Sky

By Ralph E. Lapp

For years, space was thought to begin just beyond the last thin wisps of gas in the outer reaches of the Earth's atmosphere. Space was by no means an empty void; astronomers had long been observing tremendous flares erupting from the surface of the Sun and they theorized, correctly, that particles of matter streamed from these flares far out into the Solar System. But when man began orbiting artificial satellites, we suddenly discovered there are enormous radiation "storms" in space which lash the Earth like an island caught in a hurricane.

We do not feel the effect of these cosmic storms because we live at the bottom of an ocean of air that provides a stout shield against the rays that bombard our planet. However, a thin drizzle of the most powerful rays penetrates to sea level. If more of this radiation reached the surface and if our atmosphere were not such an effective protective blanket, we could not survive, or at least a different form of life would have emerged on this planet.

What are these peculiar rays that strike the Earth's outer mantle of air and penetrate deeply into the atmosphere? Today, scientists know that cosmic rays are made up of atomic nuclei expelled by the Sun and stars. Traveling through space at velocities approaching the speed of light—some 186,000 miles per second—they possess enormous energies. Most of the cosmic rays that reach the roof of the Earth's atmosphere are the nuclei of hydrogen atoms. But nuclei of heavier elements are also among the rays reaching Earth. When these energetic particles collide with the gas molecules of the upper air, they trigger secondary cosmic ray showers, which consist of the debris of these collisions. Lower down in the atmosphere, this cascade effect finally peters out, so that only a small fraction of the cosmic rays from the upper atmosphere ever strike the Earth's surface.

Scientists have been studying cosmic rays for over half a century. In their quest for knowledge about the rays, pioneer researchers sent instruments aloft in free balloons to gather information that might tell of their origin. Using Geiger counters, they discovered that when the balloons ascended to higher altitudes, the clicking rates of the Geiger counters mounted. In other words, there were more cosmic rays at high altitudes.

The nature of the cosmic radiation both intrigued and puzzled many scientists—among them American physicist James A. Van Allen. During World War II Van Allen had worked on the design and construction

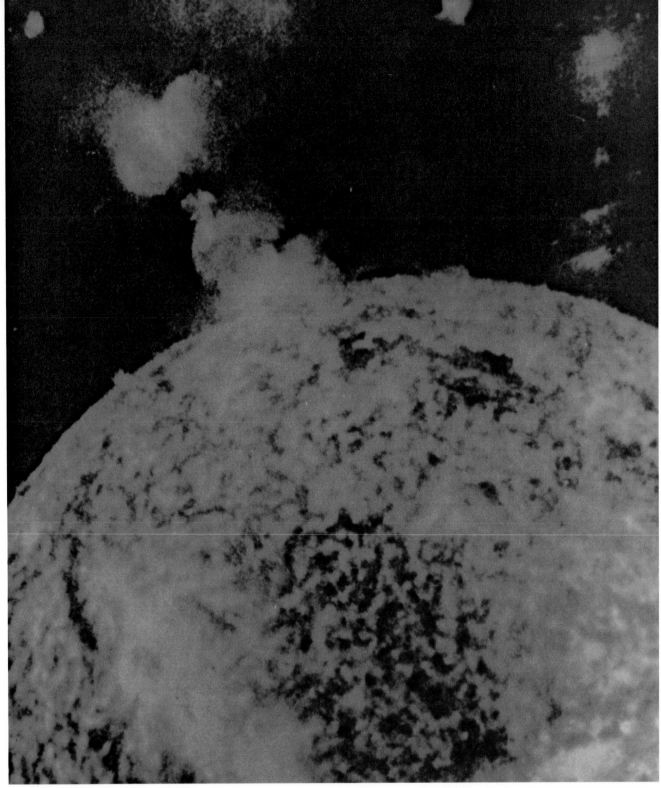

An infrared photograph records a solar flare as it bursts from the surface of the Sun. These flares are responsible for the "solar storms" that bombard the Earth with abnormal amounts of radiation.

of miniaturized instruments for the U.S. Navy. After the war he turned his attention to rockets and high-altitude research.

Van Allen designed instruments for nose cones of captured German V-2 rockets, and helped in the development of new research rockets such as the "rockoon"—a balloon-rocket combination employing a high-altitude balloon as the launching platform for a rocket sent straight up into the sky.

In 1956 Van Allen edited a book titled *The Scientific Uses of Earth Satellites*. He worked on tiny instru-

ments that could be crammed into a satellite and waited impatiently for the first American satellite to be launched.

When Explorer I finally streaked into orbit on January 31, 1958, a Van Allen instrument package was aboard. Van Allen had nestled a single Geiger counter tube about the size of a small cigar inside the 18-pound payload. Batteries supplied voltage to the tube, and a tiny amplifier magnified the electrical pulses produced by the Geiger counter whenever electrically charged particles, such as cosmic rays, passed through it.

241

Best observed during a total eclipse, the Sun's corona is a dense concentration of the particles that radiate to the outer regions of the Solar System.

A device was attached to scale down the rapid counting rate, so that every 32 electrical pulses were counted as one. This output was fed to twin radio transmitters, which in turn telemetered the information to Earth.

Below orbiting Explorer I a global network of receiving stations had been arrayed to pick up the satellite's radio signals when it swung overhead. Sixteen such receiving centers were scattered around the world. Explorer I described an elliptical path around the Earth. Its closest approach to the planet's surface was 224 miles above sea level. The most distant point in its orbit lay 1573 miles out in space.

There were no surprises in the first compilations of the satellite's transmissions. These were reported by stations within the United States above which Explorer I descended closest to the Earth. They indicated cosmic rays in numbers which agreed fairly well with estimates made from earlier rocket experiments.

Then, a few weeks after Explorer I had been launched, reports on the first orbits began to stream in from tracking stations in all parts of the globe: Australia, Singapore, Nigeria, Chile. On the other side of the planet from North America, the satellite had climbed to its highest altitude—higher than either of Russia's earlier satellites, Sputniks I and II, had gone.

When Van Allen and his colleagues collected and analyzed the data, they were perplexed. At very high altitudes the Geiger counter recorded *no* cosmic rays.

This was so unexpected that the scientists were tempted to conclude that at remote distances from Earth the instruments were malfunctioning. But this seemed unlikely, for they continued to work perfectly at lower altitudes. Certainly, scientists were reluctant to believe that there were no cosmic rays at the farther limits of Explorer I's orbit. In the light of what was known about the Solar System, such an explanation made no sense whatsoever, and the mystery only deepened.

Of course, there was another explanation for this strange behavior of the Geiger counter. Far out in space something was occurring that caused it to go dead. Then, after its high-altitude death, the counter came back to life as it was carried to lower altitudes again.

Suppose, the scientists hypothesized, that at high altitude the counter ran into a vast swarm of cosmic rays so intense that it could not cope with all the counting? Overwhelmed by the superabundance of rays, the counter would blank out.

Van Allen and his associates had not expected that this would happen, although they knew that it could. Van Allen checked out the new hypothesis by subjecting a duplicate Geiger counter to a powerful beam of X rays in his laboratory, and verified that it went dead when overloaded. He concluded that the dead counter in Explorer I must have been bombarded with many, many times the number of cosmic rays that struck it closer to Earth.

The temporary death of a Geiger counter, then, led to the discovery of a great zone of radiation encircling the Earth. The "Van Allen radiation belt"—as this zone was first called—rings the Earth like a slightly flattened inner tube. Beginning roughly 500 miles above the Earth's surface, this zone—about 1000 miles in width—extends far out into space.

Not long after the Explorer I mission, an American space shot was aimed at the Moon but fell far short of the mark, crashing back to Earth. While it was regarded as a flop by the U.S. lunar program, it was a resounding triumph for Van Allen and his colleagues.

Why? The satellite had gone far enough into space to relay back to Earth the exciting news that there was, in fact, a second great "Van Allen belt" in space. This second belt, beginning about 12,000 miles out, increases steadily in intensity for several thousand miles and then gradually tapers off—reaching more than 50,000 miles from Earth. Later space probes revealed that the heart of this second belt has a crescent shape

The Van Allen belts envelop the Earth in two overlapping zones of strong radiation. Buffeted by energy streaming from the Sun, the belts are compressed toward the Sun, elongated away from it.

with the crescent's tips, or cusps, pointing inward toward the Earth's magnetic poles.

Although the Earth's enveloping radiation belts were only discovered in 1958, their existence had been foreseen more than half a century ago by a Norwegian theoretical physicist, Prof. Carl Störmer. He had studied the effect of the Earth's magnetic field upon the paths of electrically charged particles and calculated how a charged atomic particle, such as an electron or proton, would behave as it approached the Earth.

The Norwegian scientist described a wide variety of curved and corkscrew paths that atomic particles arriving from space could follow. Störmer suggested that the Earth's magnetic field would trap cosmic radiation in belts like those that were discovered by Van Allen and his colleagues.

If we think of a huge bar magnet running through the center of the Earth with its ends slightly offset from the geographic North and South poles, we can project magnetic lines of force stretching from magnetic pole to magnetic pole, and swinging far out into space over the magnetic equator. Prior to satellite research, Earthbound scientists could only speculate about how far out in space the Earth's magnetism had significant strength. But where the Earth's magnetism did prevail, any charged particles caught between its lines of force would, in effect, be imprisoned in an invisible, curving tube stretching from one magnetic pole to the other.

These particles would be condemned to spiral endlessly back and forth far above the Earth along a corkscrew path, trapped by invisible, but nonetheless effective, magnetic walls. Hence the Van Allen belts are often called "trapped radiation" zones.

Yet this explanation of the manner in which cosmic radiation is trapped far out in space by the Earth's magnetic field still remained an unproven theory. A way to test the theory was proposed by Nicholas Christofilos, a Greek scientist doing research in the United States. He suggested that small A-bombs be shot into space and detonated at high altitudes. He reasoned that many atomic fragments from the explosions would be blown into the Earth's magnetic field. If they became "trapped" and were swept from pole to pole, this would settle the whole question.

Secret nuclear tests were made, in the summer of 1958, involving the detonation of small nuclear bombs at an altitude of about 500 miles. Rockets shot from other points on the Earth's surface identified the resulting atomic fragments as they raced around the world, trapped like bees in a bottle. These tests constituted a striking confirmation of the theory explaining the Van Allen belts and gave scientists conclusive evidence that the Earth's outer environment was far more complex than they had ever imagined.

The Stand-in That Succeeded

As the 1950s came to a close, the United States confidently expected to be the first nation to place an artificial satellite in orbit around the Earth. Plans were well underway to launch the first in a series of Vanguard scientific satellites in late 1957. Then suddenly, and with no advance notice whatsoever, the Soviet Union announced to a startled world that it had orbited Sputnik I on October 4. A month later, Sputnik II was launched with the dog Laika aboard. America's hopes for leadership in space received another major setback in December when the first full-scale Vanguard attempt lost thrust at liftoff, fell over and burned up on the launch pad. With the Vanguard program delayed indefinitely, U.S. scientists hurriedly converted a U.S. Army Jupiter rocket into a scientific spacecraft. It was launched successfully (right) on January 31, 1958, as Explorer I. In orbit more than ten years, Explorer I revealed the Van Allen radiation belts. But more importantly, the substitute satellite restored faith in the U.S. program and guaranteed that it would continue.

Lights go on (lower right) as a central Australian dust storm draws a dark curtain around the town of Alice Springs. Fine dust from storms like this can be carried thousands of miles before it settles.

What You Don't See on a Clear Day

By Clyde Orr, Jr.

Even when clear of clouds, palls of smoke and banks of fog, the air around us is not the purely clear gaseous realm it appears to be. It teems with invisible solid and semisolid bits of matter—viruses, spores, bacteria, pollen, smoke and dust particles that have a profound influence on life on this globe.

Physically, chemically and biologically there are great differences among the particles encountered in the air. Some wind-blown bits are so large they sting when they strike one's cheek, but particles of this size seldom travel far before falling to the ground. Finer particles may be carried many miles before settling during a lull in the wind, while still tinier specks may remain suspended in the air indefinitely. Jostled this way and that by moving air molecules, drifting with the slightest currents, the finest particles are washed out of the atmosphere only by rain and snow.

These finest particles are so small that scientists measure their dimension in microns (a micron is about one 25-thousandth of an inch). Included are pollen grains whose diameters sometimes measure less than 25 microns; bacteria, which range from about two to 30 microns across; individual virus particles, measuring a very small fraction of a micron; and carbon smoke particles, which may be as tiny as two hundredths of a micron.

Particles are frequently encountered in concentrations of more than a million per cubic inch of air. Since, with normal activity, a human being's daily intake of air is about 450,000 cubic inches, this means, of course, that each of us inhales astronomical numbers of foreign bodies.

Particles larger than about five microns are generally filtered from the air in the nasal passages. Other large particles are caught by hairlike protuberances in the air passages leading to the lungs and are swept back toward the mouth. Most of the extremely fine particles that do reach the lungs are exhaled again—although some of this matter is deposited in the minute air sacs within the lungs. From the air sacs, particles may go into solution and pass through the lung walls into the bloodstream. If the material is toxic, injurious reactions may follow as it enters the blood. Fine particles retained in the lungs may cause permanent tissue damage, as in "black lung" disease, which afflicts coal miners, and silicosis, caused by the buildup of insoluble silicon dusts in the lungs.

Given sufficient time and still air, all but the smallest airborne particles settle to the ground under their own weight. Their rate of fall is closely proportional to particle size and density. For example, vast amounts of fine volcanic ash were thrown into the air by the eruption of the East Indian volcano Krakatoa, in 1883, and again by the Alaskan volcano Katmai, in 1912. In both instances the finer dust reached the stratosphere and spread around the world high above the rains and storms that tend to cleanse the lower atmosphere. In fact, many years elapsed before these volcanic dusts entirely disappeared from the atmosphere. Since a two-micron dust particle may require about four years to fall 17 miles in the atmosphere, the lingering effect is not in the least surprising.

Dust storms are also prolific producers of airborne debris. Not infrequently Europe is showered with dust originating in the Sahara. In March 1901, for instance, an estimated total of two million tons of Sahara dust fell on North Africa and the European continent. Two years later, in February 1903, Britain received a deposit estimated at ten million tons. On many occasions Sahara dust has fallen in muddy rain and reddish snow over much of southwestern Europe. During North America's droughts of the 1930s, ten million tons of dust at a time were blown aloft in the heart of the continent. Occasionally, high winds swept the dust eastward 1800 miles to cloud skies along the continent's Atlantic coast.

Turning from volcanoes and dust storms to the waters of the Earth, we find still another potent source of airborne particles. When the wind whips off the crest of an ocean wave, or a calm sea is agitated by rain or by air bubbles bursting at the surface, the finer droplets that enter the air quickly evaporate, leaving tiny salt crystals suspended in the air. Winds carry these salt crystals over all the world.

Normally, airborne salt particles from the sea are

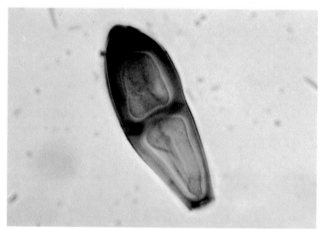

Wheat rust, a familiar threat to North American farmers, is caused by the fungus spore, enlarged above.

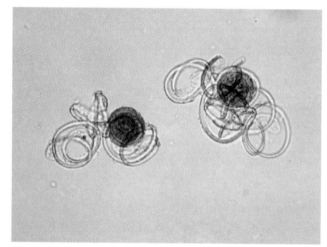

Humidity loosens hairlike appendages on spores of the horsetail rush, freeing them to drift in the air.

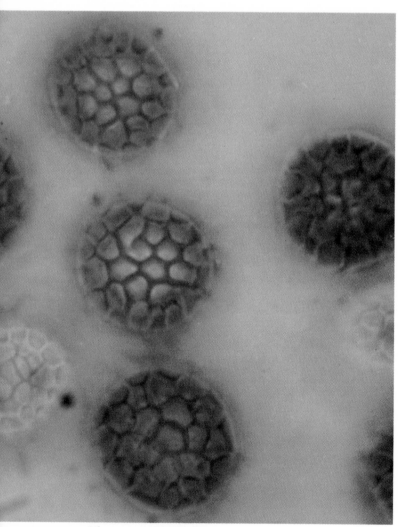

The corn-smut fungus, greatly magnified here, can infect crops great distances from its point of origin.

less than a micron in diameter. It would take a million billion of them to make a pound. Yet they play a special part in weather processes because they are hygroscopic—meaning they have a distinct ability to absorb water from the air. In a kitchen salt shaker this is an undesirable trait, but in the atmosphere it means rain for crops, rivers and waterfalls. Why? Because raindrops usually form about tiny particles that act as condensation nuclei. Generally, each fog and cloud droplet also collects around a particle of some type at its center. Tiny crystals of sea salt serve better as condensation nuclei than do other natural particles found in the air. Dust and smoke particles, however, also act as droplet nuclei.

Indeed, even dust from meteor showers may occasionally affect world rainfall. It has been observed that periods of maximum rainfall occur in both the Northern and Southern hemispheres at about the same time. This nearly uniform rainfall peak has not been explained on a climatological basis, but meteors may afford a plausible explanation.

When the Earth encounters a swarm of meteors, each meteor striking the upper reaches of the Earth's atmosphere is vaporized by frictional heating. The resulting debris is a fine smoke or powder. This "stardust" then floats down into the cloud system of the lower atmosphere, where such dust might readily serve as nuclei upon which ice crystals or raindrops would form. Confirmation that this does actually happen is found in the observed fact that increases in world rainfall come about a month after meteor systems are encountered in space by the Earth. The delay of a month allows sufficient time for the meteoric dust to fall through the upper atmosphere. Occasionally, large meteors leave visible trains of dust. Most often their trails disappear rapidly, but in a few witnessed cases a wake of dust has remained visible for an hour or so. In one extreme instance—the great meteor that broke up in the sky over Siberia in 1908—the dust cloud traveled all the way around the world before it dissipated.

Among the more spectacular producers of foreign

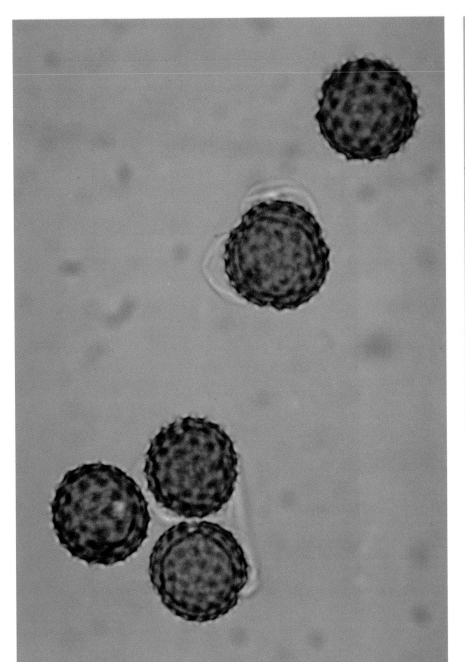

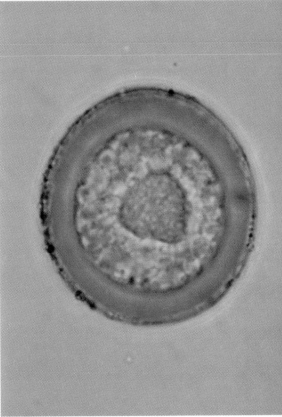

When junipers bloom, they fill the air with invisible clouds of these microscopic pollen grains.

Ragweed pollen, greatly enlarged at left, is a major cause of hay fever in late summer and autumn.

247

matter in the air are large forest fires. Because of the violent updrafts created by these conflagrations, smoke particles are carried to great heights, and, being small, are spread over vast distances by high-altitude winds. In the autumn of 1950 Canadian forest fires in the province of Alberta produced smoke that drifted east over North America on the prevailing wind and crossed the North Atlantic, reaching Britain and finally the continent of Europe. The light-scattering properties of this dense smoke made the Sun look indigo and the Moon blue to observers in Scotland and other northern lands.

Among the most prolific sources of foreign particles in the air are the wind-pollinated plants. Unhappily, pollens pose a serious problem for millions of people. When pollens are breathed by allergic individuals, a reaction occurs that irritates sensitive tissues of the eyes and the respiratory tract, the severity of the reaction depending upon the amount of pollen in the air and the length of exposure.

Closely related to pollens are spores, the reproductive bodies of fungi, including molds, yeasts, rusts, mildews, puffballs and mushrooms. These tiny bodies are adrift everywhere in the air, even over the oceans. Though similar to pollens in general appearance, spores are not fertilizing agents. Rather, they are akin to seeds in that they give rise to new organisms wherever they take hold. Fungal spores have been found as high as 14 miles in the air over the entire globe.

Most fungi depend on the wind for spore dissemination. One mold shoots a tiny pellet into the air, where it bursts, discharging spores like sparks from a sky-rocket. Another ejects spores in rapid succession like a machine gun. Still others loose broadsides of millions of spores that look like dense puffs of smoke. Once airborne, spores are carried easily by the slightest air currents.

A single spore of the wheat-stem rust fungus landing on a wheat stalk will reproduce 100,000 or more spores within a week or ten days. Corn smut is even more prolific—indeed, it is one of the most potent multipliers in the plant kingdom. In just two weeks of growth, a single spore can reproduce 240 billion new spores.

In many fungi, spore discharge continues from May to September. At this time the number of spores in the air is incalculable. During epidemic years, visible clouds of spores of wheat and oat rusts rise all through the threshing operation at the rate of nearly a million per square foot per day. Throughout the year, spores of one sort or another remain in the air. One has only to leave jelly uncapped, and in time it will be covered with a mold that has sprung from airborne fungus spores. The molds we see on stale bread and shoe leather also arrive via the air.

Long-distance dissemination by winds makes the control of spore-carried plant diseases extremely diffi-cult. For instance, the Canadian wheat crop may be infected by spores traveling with the wind more than a thousand miles from the United States or northern Mexico. New Zealand grains often suffer infections at nearly the same time as those in Australia, apparently from spores blown more than a thousand miles across the Tasman Sea.

Mountain ranges can act as high-altitude barriers to the wind, sharply decreasing the interchange of spores and pollens. The difficulties that pollens encounter in crossing from one mountain valley to another are exploited by plant geneticists to keep flower varieties pure. Thus it is California's fertile valleys separated by high mountains, more than her climate, that enable her to grow much commercial flower seed. Likewise, grain fields that are separated by mountains experience little spore exchange, limiting the chances of epidemics of fungal diseases.

Once, physicians were taught that infectious microorganisms quickly settle out of the air and die. Today, the droplets ejected, say, by a sneeze, are known to evaporate almost immediately, leaving whatever microorganisms they contain to drift through the air. Only a relatively small fraction of the microorganisms man breathes cause disease. In fact, most bacteria are actually servants. Some, for example, convert atmospheric nitrogen into usable plant food. Pathogenic or disease-producing microorganisms, however, can be insidious adversaries. Most propagate by subdivision—that is, each living cell splits into two cells. Each of the new cells then grows and divides again into two more cells. Provided with ideal conditions, populations thus multiply quickly.

Fortunately air is not a medium in which microorganisms thrive. Unless there is enough humidity in the air, many desiccate and die. Short exposure to the ultraviolet radiation of the Sun also kills most microorganisms. Low temperatures greatly decrease their activity, and elevated temperatures destroy them rapidly. Yet many microorganisms survive in the air, despite these hazards.

Among the tiniest of airborne particles are the viruses on the borderline between living matter and lifeless chemical substances. Viruses have diameters ranging from about three-tenths to one-hundredth of a micron. Under high magnification, individual strains may look like rods, fuzzy balls or pieces of string. Those infecting man or animals are often roughly spherical, but plant viruses tend to be characteristically elongated. Viruses are the cause of many diseases: the common cold, smallpox, influenza, polio and German measles, to name only a few.

With its populations of birds and insects, as well as its countless legions of microscopic particles, both living and nonliving, the Earth's atmosphere is a richly varied environment—one no less intriguing than the jungles and the seas, the mountains and the deserts.

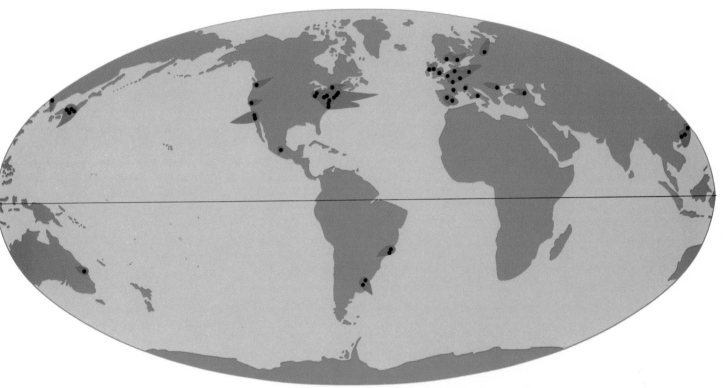

Air pollution—worse in densely populated, highly industrialized regions—is generally influenced by seasonal variations in wind circulation. The pollution pattern here is typical of January. In places like Japan and the United States, prevailing winds carry dirty air far out over the sea.

How Polluted
Is the Air Around Us?

By Wolfgang Langewiesche

People can have clean air if they demand it. The remedies for most forms of air pollution are known; they need only be applied. Already, in a number of the world's cities, air-pollution control programs have been initiated, usually with striking effect upon the quality of the urban environment.

A case in point is the steel-making city of Pittsburgh, long one of the dirtiest places in the United States because of sooty smoke from its coal-fired blast furnaces. One small measure of the success of the air-pollution control program adopted there after World War II is that housewives who were forced to launder soot-blackened curtains every week now need do so only about twice a year. In London, where a fog-and-pollution crisis in 1952 had taken some 4000 lives, a similar climatic situation a decade later reaped a fraction of this tragic toll, thanks largely to strong British air-pollution controls instituted in the intervening years.

But, while some of the world's most contaminated environments are being improved, other regions that were pollution free are now rapidly succumbing to a malaise extending far out from centers of industry. In Paris, where the air has turned corrosive, metal roofs that used to be good for 20 years now last only five, and lung cancer is on the increase. In Austria and Italy the beautiful valleys are filling with smoke. In Rome the picturesque pines are dying, their needles coated on the underside by an oily deposit that comes from the air. And everywhere the lungs of city dwellers, which should be pink, are black with dirt.

Pollution, of course, is not restricted to the cities and industrial areas where most of it is produced. Long clouds of industrial and automobile pollutants are found at sea, hundreds of miles from their origins in built-up coastal areas, and particles from factory smoke and automobile exhaust are carried miles up in the atmosphere. Indeed, many scientists fear that the broad dispersal of man-made pollutants throughout the atmosphere cannot help having unpredictable effects upon global climate, just as man's widespread use of pesticides and his dumping of untreated industrial wastes and raw sewage have upset the "balance of nature," making large stretches of land and water inhospitable to plant and animal life.

All taken from the same spot, these pictures show New York City struggling through a 1969 air-pollution crisis. Above is the first day, September 1.

Though the concentrations of air pollutants vary from region to region, the ingredients are usually much the same. In the United States, whose industrial areas are among the world's heaviest producers of air pollution, the National Air Pollution Control Administration is now closely monitoring the nation's air at some 7000 sampling stations located throughout the country. About 165 million tons of industrial and automobile pollutants are estimated to enter North America's air each year. Of these 165 million tons, about 17 million are small, solid particles of matter—including soot and ash, and cement, coal and other industrial dusts.

With the exception of pollutants suspended in the air as droplets, the remaining 148 million tons of contaminating matter are gases created in the combustion of fuels, including sulfur oxides from coal and heavy fuel oil with a high sulfur content, nitrogen oxides, and carbon monoxide and unburned hydrocarbons from automobile exhausts. Since carbon monoxide and hydrocarbons make up roughly 100 million tons of the total, the automobile is responsible for most of North America's air pollution. The Air Pollution Control Administration is also keeping a watch on environmental levels of scores of other harmful or potentially toxic substances, including mercury, asbestos, lead, nitrate, ammonia, cobalt, arsenic and radioactive materials.

Some of these things are poisonous while others are merely dirty. Some attack stone and metal, and others are known cancer agents. Some react chemically to

Second day, September 2: The air becomes dirtier.

Third day, September 3: Visibility is diminished.

form *new* poisons killing vegetation and destroying materials. And of course all of this dangerous matter is put into the air by nobody else but *us*.

Most of these pollutants can be prevented from entering the environment at their sources, given a willingness to invest in the necessary technology. For instance, much industrial air pollution can be eliminated by fitting factory chimneys with devices such as the electrostatic precipitator. This works by electrostatic attraction—the way a glass rod rubbed with silk picks up bits of paper. Mounted in a factory chimney, the precipitator picks soot and other particles out of the smoke as it passes up into the sky.

Another type of device mixes the gases with water in a whirlpool bath that causes the pollutants to dissolve in the water, which can then be purified. Other proposed or operational devices pass the gases through bags like those in vacuum cleaners, or chemically filter toxic compounds from industrial smoke. None of these industrial antipollution devices is inexpensive, however.

There are alternatives. New York, for example, has improved the quality of its air by attacking the problem at the other end—restricting the sulfur content of the coal and fuel oil burned in its power plants and regulating the burning of wastes in private and municipal incinerators.

Perhaps the most striking antipollution success story is that of London, which today enjoys 50 percent more sunshine in the winter months than in the year before Britain's passage of the Clean Air Act of 1956. Most of this dramatic gain has resulted from a diminution of coal smoke in London's air. Londoners, who almost universally had burned soft coal in their grates, now burn coke or "smokeless" coal. For most householders the act thus meant a more elaborate grate or an enclosed stove—if not a complete changeover to gas or electric heat. The cost of conversion was borne 30 percent by the householder, 30 percent by the local government, 40 percent by the national government.

Of the various gases liberated in the burning of coal, coke or heavy fuel oil, the most common pollutants are the oxides of sulfur, primarily sulfur dioxide and sulfur trioxide. In the United State alone nearly 50 million tons of sulfur oxides are pumped into the atmosphere each year.

In a humid atmosphere, some of the sulfur dioxide, which is soluble in water, is converted to sulfurous acid, which clings to the façades of buildings and automobiles, making stone crumble and eating into metals. The gas smells bad, makes the eyes sting and irritates the lungs. Sulfur trioxide dissolves in droplets of moisture in the air, producing sulfuric acid, far more corrosive than sulfurous acid.

As the world's demand for electricity increases, requiring the burning of more and more fossil fuels to generate power, the amount of sulfur oxides added to the atmosphere is almost certain to increase, unless steps are taken to stem the production of this pollutant. While Europe has ready access to the sulfur-free natural-gas deposits of the North Sea, Russia and

Fourth day, September 4: Buildings seem to vanish.

Fifth day, September 5: Conditions start to improve.

North Africa, North America's natural-gas reserves are not nearly so large. But even in Europe, natural gas cannot completely replace coal and petroleum as a fuel. Thus if sulfur dioxide levels in the world atmosphere are to be minimized, either sulfur will have to be removed from coal and petroleum prior to combustion, or processes must be employed to cleanse stack gases. Several types of technology have already been devised to perform these functions, but they are costly. Yet if we hope to maintain the atmosphere at its present levels of contamination, let alone reverse the pace at which the Earth is being poisoned, we will have to meet these costs.

As the number of automobiles in the industrialized world has multiplied in recent decades, the internal-combustion engine has emerged as a formidable polluter of the air. The bulk of automobile exhaust is carbon dioxide and water vapor—both of them harmless. Mixed with them, however, are carbon monoxide, a deadly poison; benzopyrene, a cancer agent; nitrogen oxides and unburned hydrocarbons produced by incomplete combustion of the gasoline. Add fumes from hot and half-burned oil and a sloppage of raw gasoline. All this becomes part of the air around us.

When automobile exhausts collect in still air, and the fumes are bathed in the rays of the Sun, complex photochemical reactions occur within the fumes, producing a heavy haze called "photochemical smog." Smog damages trees and crops, makes the eyes smart and irritates the nose and lungs; it causes the deterioration of many materials, such as rubber and synthetic fabrics.

More than exhaust fumes is required to make smog. No matter the number of cars gushing exhaust, the air will remain comparatively clear on days when upward currents can carry the fumes away to high levels and thin the gases out in the vast air ocean. But when this normal vertical circulation of the atmosphere ceases, pollutants are trapped near the ground where they are created.

What causes this blockage of atmospheric circulation? As each of us knows, the air at high altitudes is ordinarily colder than near the Earth's surface. Sometimes—and particularly over and downwind of major cities—the normal situation is reversed, and a layer of warmer air lies on top of cooler air near the ground. Such an "inversion" acts as a lid. Rising fumes and smoke are trapped by the warm-air layer and cannot rise farther. And as time passes, pollutants build up in the atmosphere beneath the inversion layer, and the

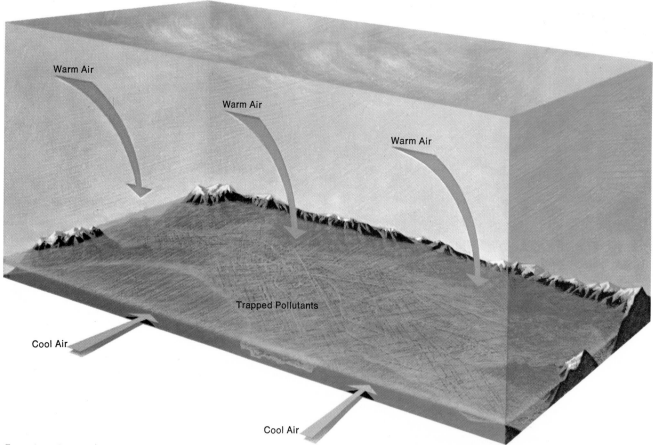

Los Angeles smog is created when warm-air currents trap cool air along the ground between the coast and the mountains. Stagnating lower air fills with pollutants which react with sunlight to form smog.

Pollution Dilemma: Change Color or Die

A British insect called the peppered moth has proved to be a survivor of industrial pollution, not one of its victims. The white peppered moth (upper left) long inhabited British woodlands, its black and white speckled wings acting as camouflage against resting places of pale gray tree bark. However, since the 19th century and the era of heavy industrial smoke, trees have darkened with soot. The light moth became conspicuous enough to be seen and decimated by hungry birds. Then, by chance, a mutation produced a dark strain of moths (lower left) which were better able to survive in the polluted environment. The first dark peppered moths were found in 1848. Since then, more and more of the dark variety have appeared until they now total 99 percent of the peppered moth population. The few remaining light peppered moths now flourish only in pollution-free corners of Britain.

Sun goes to work on them to produce photochemical smog.

A stagnant temperature inversion, abundant sunshine, a hilly location that walls in pollutants and, of course, its many automobiles make Los Angeles, California, the worst smog disaster area in the world. Recent U.S. laws requiring American automobile manufacturers to cut exhaust emissions from new cars by 90 percent within this decade, while posing technological problems, seem to hold out the best promise of less dense smog over Los Angeles and other American cities.

The classic London fog is also caused by a strong temperature inversion layer that forms at a low altitude —only 300 or 400 feet above the ground. Below this lid, in the past, the smoke of millions of coal-burning fireplaces was trapped along with the fog, making the air almost unbreathable. Even now, with smoke pollution greatly reduced, London's fog frequently cuts visibility to less than five yards—but it consists mostly of water droplets, not noxious smoke particles.

The inversion that brings on the Los Angeles smog is much higher—about 2500 to 4000 feet. It leaves much more room for smoke to dissipate. But, like the London inversion, it is persistent. Elsewhere in the world an inversion usually lasts a few days, then is blown away. The Southern California inversion sometimes continues day and night all summer long.

Unquestionably, if we wish to restore our planet to anything near its state before the world-wide mushrooming of population and industry in this century, we will find no miracle cures. Indeed, the price we pay in money and ingenuity to conserve the environment may prove almost as high as it was to promote the growth of industry in the first place. But we seem to have no choice other than to accept the need for anti-pollution controls, unless we are prepared to perish on a poisoned Earth.

253

Chapter Two

ILLUSIONS IN LIGHT

*Occasionally our atmosphere mixes physics with artistry, as when it paints halos around
the Sun and Moon, or when it creates rainbows, strange polar lights and fanciful mirages*

The Sky's Rarest Spectacles

By Richard Beidleman

It had been a chilly night for April throughout the Severn countryside of southern England. As the barley sowers left their homes at dawn to start their planting, the Sun rolled slowly up out of the eastern woodlands, bringing with it a never-to-be-forgotten spectacle.

Moving in consort with the rising Sun in the pale blue sky were four celestial spheres of equal brilliance. They flanked the Sun in pairs. Encircling the Sun were two colorful halos passing through the paired spheres. From sunrise until noon, this prodigy rode the spring sky above England in the year 1233, then vanished as mysteriously as it had appeared, leaving wild thoughts in the minds of hundreds of peasants, yeomen and gentry who had no ready explanation for the spectacle.

In the spring of 1912 and the autumn of 1913, brilliant halo systems were seen by many people in Europe, Britain and North America. In a letter to the British journal *Nature*, an observer described one of the more unusual displays:

The first thing noted was an object high over the setting Sun, just like a moustache brushed into a fierce upward curve. This had a metallic lustre like burnished brass, and marked the contact between two coloured circles, the top one, of which only about one-sixteenth was visible, showing two colours, silvery blue on the concave and rusty buff on the convex. The lower halo was complete down to the horizon, and showed all the colours, while from the Sun itself a long slender cone rose about halfway up to the moustache, and had exactly the same colour and lustre.

Through the centuries many such strange lights have been seen in the sky, ranging from simple halos to complex illuminations that vaulted the entire heavens. All such phenomena are caused by the bending or scattering of light rays in an unusually moisture-laden atmosphere.

There are two major types of celestial rings. True halos, most frequently observed in the winter, are caused when sunlight or moonlight is refracted, or bent, by ice crystals that float high in the upper atmosphere. Coronal rings, on the other hand, may appear in an overcast summer sky. They result from the distortion of light waves by many tiny water droplets of uniform size suspended in a thin cloud. Coronas—sometimes wrongly called halos—may occur either singly around the Sun or Moon, or appear in sets. This type of atmospheric effect has nothing to do with the solar corona—the luminous gas enveloping the surface of the Sun.

Most often, halos and coronas have an almost colorless appearance. However, both can be quite chromatic, especially those that form around the Sun. Smaller halos may possess vivid red interiors, changing to bluish white on their edges. The largest halos, if not pure white, show these same colors reversed, the blue being on the inside. Colored coronas are also blue on the inside. One sort of poorly developed corona, the aureole, ranges in color from bluish white on the inside to brown at its outer rim. Rings around the Sun are usually less evident than those around the Moon because the Sun's brilliance obscures them.

Although it is not always apparent, a haloed sky is overcast with high-altitude clouds consisting of tiny ice crystals—cirrus ("mares' tails"), cirrostratus or cirro-nebula formations. Coronas, on the other hand, appear in a lower cloud cover of suspended water droplets, also quite thin. Coronas vary greatly in size, but are usually much smaller than halos. Ordinarily halos and coronas do not appear together in the sky, and cer-

Complex solar displays such as this combination of halos, arcs, pillars and "mock suns" are exceedingly rare cold-weather phenomena. This one was photographed during an expedition to Byrd Land, Antarctica.

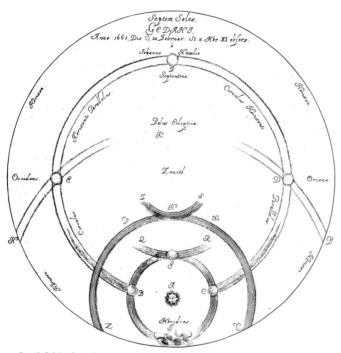

In 1661 the sky over Danzig, Germany, was filled with criss-crossing solar halos as recorded in this contemporary drawing by the astronomer Hevelius.

tainly not in a single cloud, because of the basic difference in the way in which they are formed.

There are variations on simple halos, some of which are extremely spectacular. The "pseudhelion," or false sun, for example, is visible only from aloft. It appears as a reflection of the Sun on the upper surface of a horizontal bank of clouds. At daybreak or dusk the low-lying Sun will occasionally project a person's shadow against a distant cloud bank. This enormously enlarged shadow forms what has been called the "specter of the Brocken" (after a mountain in Germany where the phenomenon was frequently witnessed). An individual looking at his shadow on a cloud may see a series of from one to five colored rings around the shadow's head, called "the glory" or "Ulloa's circle." These circles are formed, much like coronas, by the diffraction of light rays in the cloud. The same effect can often be seen by passengers in an airplane flying between the Sun and a cloud. The plane's shadow is usually surrounded by at least one colored ring.

The "halo of Bouguer," a large ring of light hanging in the sky opposite the Sun, is another variation. Colorless, it is often called a "white rainbow," because, like

Mock suns or "sun dogs" are optical tricks played by sunlight when it is refracted by clouds of ice crystals. Mock suns sometimes project vertical "pillars" high into the sky as in this Alaskan scene.

a rainbow, it appears in the heavens opposite the Sun. This halo is also caused by light reflected from ice crystals suspended in the atmosphere.

Strange vertical shafts of light often rise above the horizon at night. Caused by neither the Sun nor the Moon, these are manifestations of our artificially illuminated civilization. They occur on still nights. When flat snow crystals float in the air parallel to the Earth's surface, lights from the ground are reflected back from the surfaces of the snowflakes, causing the shafts of light to appear in the sky. The beams may be either white or colored, depending on the color of the light source on the Earth's surface.

Sometimes when the Sun is on the horizon a "mock sun" will appear just above it, occasionally with a "sun pillar" projecting from the mock sun high into the sky. First recorded by the German astronomer Hevelius on April 10, 1682, this phenomenon has since been attributed to reflection and refraction of the Sun's light by airborne ice crystals. When the Sun is slightly above the horizon, mock suns may appear both above and below it, with sun pillars extending both upward and downward.

The refracting angle of the ice crystals in the air determines the radius of the halo that appears around the Sun or Moon. Most commonly seen is the halo whose arc extends 22° on either side of a line between the observer and the center of the Sun. A 46° halo, called the Great Ring, is seen less often and has rarely been measured. Another halo, the halo of Hevelius, is an immense ring of 90° radius. But its appearance is an extremely rare event.

Given unusual atmospheric conditions, halo patterns may become strikingly complex. Both the 22° and 46° halos may form around the Sun at the same time, each shimmering with color. In contact with these halos, bright arcs of light may appear as if drawn tangent to the halos with a compass. Passing through the true Sun, parallel to the horizon, there is often a white, luminous band—the parhelic circle, along which brilliant secondary images of the Sun are formed. Technically these are known as "parhelia"; more commonly they are referred to as "sun dogs," "mock suns" or, by seafaring men, as "windgalls."

The most brilliant mock suns or sun dogs occur on the parhelic circle at or just outside the 22° halo; their distance from the halo increases as the Sun rises higher in the sky. Moon dogs or mock moons, known as parselenae, have also been observed.

Although these magnificent displays are occasionally to be seen toward sunset in winter, they occur more often on a chilly dawn. In the Arctic and Antarctic, of course, they are commonplace. As the Sun rises, the sun dogs move out of their halos and assume a comet-like appearance. By noon the vision has usually disappeared, leaving the true Sun to complete its course through the sky alone.

A pair of mock suns frames the real Sun over a flat Canadian horizon, near Regina, Saskatchewan. The scientific term for these odd lights is "parhelia."

High, thin clouds of ice produce the most commonly seen halo—a delicate circle of light such as this one crowning a tree in a Canadian forest.

Mock suns, sun dogs, mock moons and sun pillars are created by the refraction and reflection of light rays from atmospheric ice crystals that have a particular orientation with respect to the Earth's surface. The parhelic circle and ordinary sun dogs are produced when the Sun's rays strike the prismatic faces of hexagonal ice crystals floating in a vertical position in still air. The white parhelic light is reflected from the crystals, while the sun dogs are created by refracted light. Sun dogs that lie above and below the real Sun, vertical pillars through the Sun and tangential arcs are caused by refractions from those ice crystals that are horizontally oriented.

The brightness and distinctness of the sun dogs depend upon the number and arrangement of the ice crystals in the air. Frequently the sun dogs may be reddish in color on the side toward the Sun and may appear elongated where they intersect the parhelic circle. In rare cases a 22° sun dog may also be surrounded by its own halo, a portion of which passes through or near the mock image itself.

Complex halo displays never fail to arouse interest and excitement among witnesses, sometimes with historic consequences. Indeed, the cross that the Emperor Constantine reportedly saw in the sky in about A.D. 312, after which he converted to Christianity, may have been part of such a display. A parhelic circle, with its pillars above and below the Sun, would form such a cross, and it would appear to be flaming.

Solar halos are most common during the spring in the Northern Hemisphere, with March the most prolific month. In the Southern Hemisphere, solar halos are most common in late autumn. Lunar halos, on the other hand, are most often seen in the Northern Hemisphere during January. Some solar halos have persisted as long as ten hours, but on the average, both solar and lunar halos remain visible in the sky for less than two hours.

It is understandable that strange lights in the sky have led people to attribute a prophetic nature to their appearance. One of the most common of these prophecies has some basis in fact: clouds that produce coronas often do portend precipitation, and the adage "the moon with a circle brings water in his beak" reflects this. Equally reliable is the American Indian saying that predicts: "When the Sun is in his house [corona], it will rain soon."

Some of these sayings do not bear close scrutiny, however. For example, "the bigger the ring, the nearer the wet" is actually a misapprehension. A corona will become smaller as rain nears, because the water droplets in the cloud are growing larger at this time. The bigger drops have optical properties that produce a smaller, not a larger, ring.

The high cloud cover that causes halos often precedes bad winter weather. A number of scientific studies have examined the reliability of halos as harbingers of precipitation. One study done early in this century found that precipitation almost always followed a 22° halo within 12 to 18 hours and a 46° halo in 24 to 36 hours. Another set of observations conducted over a ten-year period revealed that 70 percent of winter solar halos were followed by precipitation within 31 hours.

Portentous or merely dramatic, these strange lights in the sky display endless variety. Human beings will continue to watch for them as they did in 13th-century England, "for the wonderful novelty of it."

How Light Creates the Rainbow's Colors

According to the Bible, Noah saw a rainbow as a sign from God after the Flood. In the 17th century, men sought a scientific explanation for this glorious natural phenomenon. The French scientist and philosopher René Descartes finally confirmed the explanation now known to be correct. The Sun's natural white light rays bend as they pass through dense concentrations of atmospheric moisture. Since white light is a composite of all visible colors, its bending—or refraction—makes it separate into bands of each of its main hues as it ribbons across the sky: red, orange, yellow, green, blue, indigo and violet.

High above Kiruna, Sweden, the northern lights swirl in a ghostly "folded ribbon." The rarest and strongest of auroral displays, it occurs when solar storms bombard the Earth with abnormal radiation.

Cold Fire in the Polar Night

Swishing down from polar skies in colorful, undulating sheets, the ethereal aurora long defied understanding, but now it is known to be a manifestation of solar radiation

"Sometimes on a fine night," wrote Aristotle long ago, "we see a variety of appearances in the sky: chasms . . . trenches . . . blood-red colors. . . ." Going on then to theorize that the air was turning into liquid fire, the Greek philosopher was actually puzzling over the spectacle of the northern lights, the Aurora Borealis, sometimes seen as far south as Singapore. The northern lights have an equally dazzling south-polar twin: the Aurora Australis.

Quietly shimmering almost constantly in polar nights, only rarely does the aurora intensify enough to be seen in the latitudes where most people live. Such auroral storms—when contorting, brilliant ribbons of light may be accompanied by eerie hissing and crackling in the air—usually follow within a day the eruption of a big solar flare on the Sun's surface. Solar flares, in turn, happen most frequently during periods of intense sunspot activity. Sunspots seem to follow an 11-year cycle (whose most recent peak came in 1969).

Scientists now believe that auroras are created in much the same way as pictures on a television screen. To produce a television image, electro-magnets focus a beam of electrons on a fluorescent screen. The Earth's magnetic field does the same thing with electrons and other particles streaming in from the Sun by focusing them on the "screen" of sky, high above the Earth's magnetic poles. At the poles, the Earth's magnetic fields dip earth-ward in a funnel-shaped pattern. As solar particles spiral down toward Earth in the magnetic funnels above the poles, they hit and excite atoms in the upper air. The excited atoms give off the flashing spectral light of the aurora. Oxygen atoms are known to emit both red and green light; nitrogen, violet, blue or red; and incoming solar protons themselves may be a weak source of yellow and red. Why do unusual auroral storms follow the eruption of solar flares? Because these intensify the solar wind with dense clouds of high-velocity particles.

Low in arctic Sweden's dawnless winter night, the aurora lights a warm flame against the total dark.

A blossom of rays ushers in the aurora's peak activity, which usually occurs after midnight.

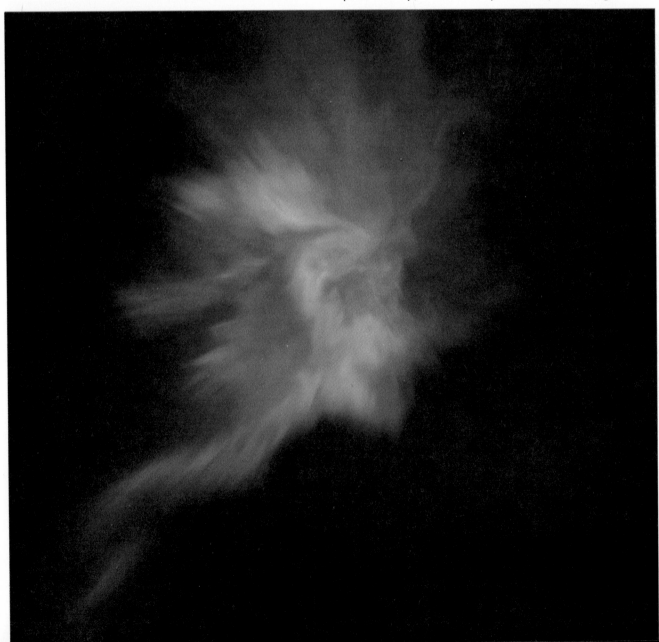

Blue-green rays of the northern lights pierce an October night at Whitehorse in Canada's Yukon Territory, creating a gala air for an old Yukon paddlewheeler dry-docked as a historic monument.

During a period of intense sunspot activity, dual streamers of auroral fire light up the sky over Alaska.

With mercurial moodiness, arctic Sweden's auroral arc flashes down from the sky in intense, rayed bands.

The World That Isn't There

By Theo Loebsack

Some years ago, on a windless day in the German North Sea port of Cuxhaven, a little boy ran home from play, shouting excitedly: "Mother, an island is falling from the sky!" His mother's disbelieving smile vanished when she looked out the window; for there, before her startled eyes, hung the offshore island of Helgoland upside down in the sky. Its red coastal cliffs were unmistakable, and the dunes along the shore and other details were clearly visible. There the island dangled as if suspended by a giant hand—about to come crashing down at any moment. Of course, the island did not fall—it was a mirage—and the child's fears disappeared when the atmospheric phantom finally vanished toward evening.

The apparition of Helgoland, which occasionally appears in the sky over Cuxhaven, is but one of a ghostly repertoire of atmospheric phantoms that have long dazzled and puzzled observers. One such specter, native to the Arctic, deluded men for nearly a century. In 1818 the Scottish explorer Sir John Ross set out from Britain to find the elusive "Northwest Passage," the sea route that was supposed to link the Atlantic with the Pacific along the northern coast of North America. North of Baffin Island in Canada, Ross entered unknown waters. When he went on deck one morning, he saw the channel blocked by huge mountains. Thinking he had sailed up a blind alley, he turned about and reported that the Northwest Passage did not exist.

Almost a hundred years later the American polar explorer Robert Peary also reported the existence of an unmapped Arctic mountain range. "We saw the mountains and called them Crockerland," he reported.

By now the mysterious Arctic mountains had aroused worldwide curiosity. What lay behind these peculiar mountains? And where exactly were they located? Might they conceal ore deposits, or perhaps gold? Did unknown tribes live there? Many adventurers and explorers set out for the Arctic, but they never found the range. Finally the American Museum of Natural History in New York City donated $300,000 to mount a scientific expedition into the area. Donald MacMillan, the leader of the expedition, became the hero of the hour in the world press.

But where Peary had seen mountains, MacMillian found only an icy waste. Where according to Peary there were deep and wide channels, floes of pack ice threatened MacMillan's ship. Finally Crockerland did appear—but, to the astonishment of the crew, it was some 200 miles farther west than where Peary had placed it.

MacMillan sailed as far as possible through the ice floes. Then he dropped anchor and set out on foot over the ice with a team of picked men. When they tried to approach the mountains, however, the mountains receded. If the men stood still, the mountains stood still as well. When the team started forward again, the mountains receded once more, their icy summits and snowfields beckoning in the polar sunlight, their dark valleys promising mineral wealth.

Redoubling their efforts, the men finally approached so near to a valley enclosed on three sides by mountains that they felt certain of success. But when the Sun sank below the horizon, the surrounding peaks and foothills dissolved as if by magic. Dumbfounded, the men could only stare mutely at their real surroundings. They were on a vast expanse of ice, ice in all directions, ice as far as the eye could see. Not a hill or a mountain was anywhere in sight. MacMillan's group stood there in the pale green Arctic twilight, the victims of one of the greatest tricks that Nature has ever played on man.

Mirages, the optical illusions created by special atmospheric conditions, occur when light rays are bent, or refracted, as they pass through adjacent layers of air of different temperatures, and hence of different densities. Let us, for instance, imagine a desert where the Sun has warmed the sand, and the sand, in turn, has heated up the layer of air immediately above it. Over the top of this hot thin layer of air on the desert floor will lie colder layers. Because hot air is less dense than cold air, light rays move through it more easily than through cold air. The change in the velocity of a light ray as it passes across the boundary between air layers of different density causes the light ray to bend out of its straight-line path.

Suppose, then, that an observer is standing on a sand dune somewhere in the desert. There is a group of palm trees on a dune several hundred yards away from the observer, and between these dunes lies a layer of hot air, warmed by the searing desert sands. In such a situation the observer will see two groups of palm trees where just one exists. One image will be received in the usual manner, the light rays following a direct route across the intervening air. A second image will appear upside-down below the first, created by rays that have arrived at the observer's eyes via a refracted path. These rays travel diagonally downward from the palms into the heated air blanketing the desert and are refracted upward, meeting the eye of the observer from below—as if a mirror were lying on the desert floor showing an inverted reflection of the palms. At the same time light from the sky is also refracted by the hot shimmering air, giving the impression of a sheet of water in which groups of palms are standing, the real ones upright, the mirages upside-down.

The "puddles" we sometimes see in the summer on highways and other hot surfaces are also mirages.

In this Antarctic mirage, the low mountains are real and the high ones are an illusion. Cold mirages produce images rightside up above the object, whereas desert mirages are inverted below the object.

They are light from patches of sky refracted by the heated air above the hot surface. Many tales are told of wanderers in the desert being driven mad by this type of mirage. Neither figments of the imagination nor hallucinations, they are shimmering, refracted images of the blue sky. "*Bahr el Shaitan*," the Arabs call them; "Lakes of Satan." Often, too, desert air creates mirages in which a far-off oasis, city or some other distant place appears close by, contributing further to the nomads' lore.

North America's southwestern deserts are also notorious for their mirages. A tragic story is told by the inhabitants of Cochise County in Arizona, where a ten-mile lake lies along a railway. The lake's water is visible summer and winter, although the lake is actually dry in the summer. During the summer, light from the sky refracted by the air above the sun-baked lake bed merely gives the illusion of water. A pilot who remembered the lake after seeing it in winter tried to land a seaplane on it during the summer. As he

began his landing approach, the phantom suddenly disappeared. The pilot crash-landed his plane, which disintegrated. He died later of his injuries.

When the boundary between the layers of hot and cold air is uneven, the refracted images are frequently distorted. The American explorer Roy Chapman Andrews once saw fabulous animals like huge swans wading in a lake in the Gobi Desert. From a few hundred yards away they seemed to be gigantic creatures from another world, wandering around on stiltlike legs nearly 15 feet long. Andrews immediately asked the expedition's artist to make drawings of the unusual beasts. He himself approached the lake stealthily. As he got closer, the water shrank and its inhabitants changed shape. The plump giant swans became slender antelopes grazing peacefully on sparse desert vegetation. A hot-air layer had produced the impression of water, and the unevenness of the layer had grotesquely distorted the images of the animals' bodies.

During the Anglo-Turkish battles of World War I,

a mirage forced the British artillery to stop firing. An illusory landscape appeared before the artillerymen's eyes, completely masking the enemy's position. The British commander's report on the shelling contained the remarkable sentence "The fighting had to be temporarily suspended owing to a mirage."

Napoleon's army encountered mirages in Egypt in 1798. Confronted with topsy-turvy landscapes, vanishing lakes and blades of grass that turned into palm trees, his troops at one point are reputed to have fallen on their knees and prayed to be saved from the impending end of the world. At least one member of the expedition, the French mathematician Gaspard Monge, kept his head and came up with a scientific explanation that dispelled the mystery.

A different sort of mirage is the polar type, which occurs when the air immediately above the ground is very cold and there is a warmer layer above it. The observer may then see images of distant objects displaced upward in the sky. These mirages often present double images. When, for instance, ships or icebergs are floating in a calm sea, they create a mirror image in the water below them. A distant observer often sees both the image of the object and its reflection in the warm upper layers of the air. One case of such "double ex-posure" was documented more than half a century ago in the Antarctic during Commander Robert Falcon Scott's South Polar Expedition of 1912. Men returning to the coast from an inland trek saw the support ship *Terra Nova* hanging in the sky in double form. There was an inverted picture of the sailing ship, with an upright image above it, and smoke was drifting in opposite directions from the respective cookhouse chimneys. Although the ship itself was hidden by intervening hills, its mirage showed that all was well on board.

Occasionally the air plays some amusing tricks. Conditions for the inverted type of mirage sometimes occur above Paris. Then the Eiffel Tower entertains the residents of the French capital by balancing an upside-down image of itself.

Near the North American coast during the First World War, a German submarine commander, on looking through his periscope, saw the skyscrapers of New York apparently hanging in the air above him. The city seemed about to plummet upside-down into the water, and the baffled submariner is said to have retreated hastily out to sea.

Mirages do not always show the true shape of an object. Images may be enlargements, contractions or

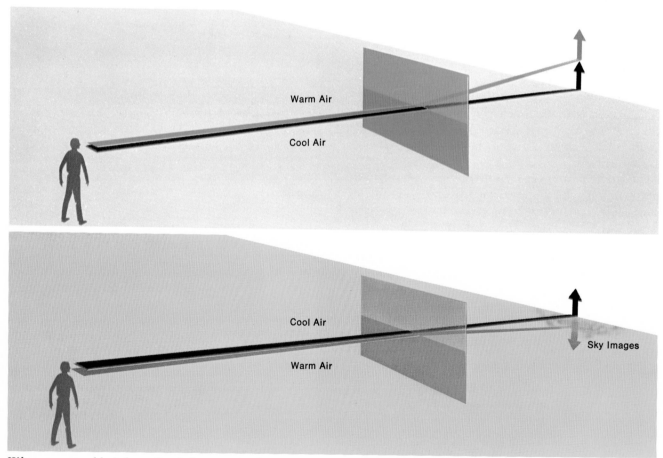

When a very cold air layer lies below a warm layer, mirage images appear above the real object (top).
When the layers are reversed (bottom), the image is inverted below, and reflected sky appears as water.

distortions like those seen in a hall of mirrors, according to the position and composition of the refracting air layers. In mirages, a drifting ice floe in the Arctic may become a dangerous iceberg, a palm tree shrink to a blade of grass and a fishing hut appear to be a colossal palace.

In certain situations one may see a mirage of an object that lies around a corner. This can happen, say, where a chill, refracting layer of air is created alongside a vertical surface, such as an icy cliff wall. Light rays may then be bent by refraction to curve around the base of the cliff. A wanderer in North America's Rockies reported having been warned of a bear lurking around a bend in a trail when he saw a mirage of the creature refracted in this way.

Perhaps the world's most intriguing mirage is the fata morgana, which occasionally appears in the Strait of Messina, between Italy and Sicily, and also in Japan's Toyama Bay. This apparition takes its name from the Italian fable about the magic undersea palace of the fairy Fata Morgana—or Morgan le Fay. The name has long been widely used to denote any form of mirage.

For the fata morgana to occur, the sea must be fairly warm, heating the layer of air in contact with it, and a second layer of warm air must exist at some higher altitude, so that a cool layer of air is sandwiched between two warm layers. The cold middle layer thus produces a double mirage and also acts as a cylindrical lens that magnifies vertical dimensions.

When the fata morgana appears, all imaginable mirages are combined—upright, inverted, enlarged, diminished and distorted multiple images. The fata morgana does not appear suddenly; rather, it is usually preceded by a ghostly cloud in the sky. When the air above the Strait of Messina is hot, and the water is calm, this strange cloud shimmers with the image of a splendid harbor town. Soon a second town may appear, piled on top of the first, and yet a third, all with shining towers and palaces. Sometimes it seems as if the houses lie below the water—where the fairy Morgana is said to dwell. One seems to see people strolling through the streets in billowing white garments.

The source of the refracted images of the fata morgana in the Strait of Messina is debated to this day. Some observers believe that the mirage is only a refracted image of the Sicilian harbor of Messina; others say the mirage is really the image of a strip of coastline, whose trees and stones have been magnified and distorted to suggest palaces and towers. Another theory settles on an obscure Italian fishing village, transformed into a beautiful town by the magic of the mirage.

Whatever its cause may be, the fata morgana remains an indescribably beautiful aerial effect. Though her phantom images greatly exceed all other types of mirage in variety, like the others the fata morgana results from the natural laws that govern the behavior of light waves traveling through the air.

Heat mirages may distort shapes on the horizon and make them seem to float (above). Many heat mirages create mirror images. Horses below are on dry land, but seem to be standing in a reflecting water pool.

Chapter Three

WEATHER'S MANY FACES

*Most of the time the air above us moves in well-known patterns and bathes the Earth
with the breath of life—but when its mood changes, it breeds storms of awesome violence*

The Passing Parade Above Us

By Richard M. Romin

The sky is like an enormous motion-picture screen. Passing across it almost every day are hundreds of images made of clouds. They may resemble nightmarish monsters, slender mares' tails, cream puffs, cauliflowers—indeed, they may appear to sketch anything from the outlines of continents to profiles of human faces. We say we are looking at a "mackerel sky" when high-flying clouds are arrayed like the pattern of wavy bars on a mackerel's back. Aviators talk of flying through "soup" and among "woolpacks." And the cloud show can become especially dramatic when it is accompanied by violent atmospheric phenomena, thunder and lightning and raging winds.

The imaginary pictures we see overhead result from the realities of physical chemistry. Every cloud is composed of either tiny water droplets or ice crystals. And just as motion pictures have titles, so do clouds.

Clouds were given the general names we use today by Luke Howard, a London pharmacist, in 1803. This astute observer of the skies divided all clouds into three basic groups—*cumulus, cirrus* and *stratus*—and his system is still used throughout the world.

Although his choices of names were intended to be technical, Luke Howard, too, must have been seeing the clouds as pictures. Cirrus, for instance, is Latin for "lock of hair," "a tuft" or "curl." Cumulus means "heap" or "mass," and stratus is derived from the Latin *sternere*, "to spread out." In other words, cumulus refers to puffed-up clouds; cirrus to the wispy kind; and stratus to those that take the form of layers, or sheets.

Later weather scientists modified Howard's system of classification and made it more specific. Thus the modern names for clouds also relate to the altitude at which they form and their method of formation. For example, the word "nimbus" is always added to the names of clouds that produce rain or snow.

Meteorologists classify all clouds whose bases are found at heights of less than 6500 feet as low-altitude clouds. Among these are the cottony puffs of cumulus that often decorate blue skies during fair weather and their stormy big brothers, the cumulonimbus clouds, named by combining cumulus and nimbus. A flattened low-altitude cumulus cloud is given its own distinguishing name, stratocumulus. In the low-altitude group are also layers of stratus clouds and the dense, murky rain providers called nimbostratus. When the air is saturated with humidity, stratus clouds may form at or slowly descend to the surface, encompassing regions in dense fog. Fog is really cloud touching the ground.

The middle domain for clouds lies between the altitudes of 6500 and 20,000 feet. Any cloud whose base rests here is distinguished by the prefix "alto-." Here white clumps of altocumulus roll across the sky, sometimes in parallel bands; here also, milky altostratus layers dim the Sun, harbingers of rainy weather.

The highest clouds are also the most gossamer in appearance. These thin veils are found drifting 20,000 to 30,000 feet or more above ground. Their feathery look results from their being composed of tiny ice crystals and the action of the strong upper winds, which comb them out. In this domain, all clouds belong to the cirrus family.

Of course, not every cirrus cloud resembles a lock of hair; cirrus can also take the form of hooks or perhaps stalks of wheat, while their wispy waves are frequently called "mares' tails." The cirrostratus variation, which invades the high sky with a thin veil, sometimes means bad weather ahead; at times, too, it refracts the Sun's rays, forming a bright ring of light—a halo—around the solar disk. Cirrocumulus clouds are less common; these give the pattern that is appropriately called a mackerel sky. Amateur forecasters have been aware for a long

Feet
—25,000

Cirrus

—20,000

Cirrocumulus

Cirrostratus

—15,000

Altostratus

Altocumulus

—10,000
—9000

Cumulonimbus

—8000

—7000

Cumulus

—6000

—5000

—4000

—3000

Stratocumulus

—2000

Nimbostratus

Stratus

—1000

Fog

From highflying cirrus to banks of fog on the surface, the appearance, structure and height of different types of clouds are visible evidence of the forces in the atmosphere that produce weather.

267

Sun tinted clouds mark a dissipating thunderstorm over Mexico. With tops that rise to 50,000 feet, the thunderstorm, or cumulonimbus, is the most dramatic and violent of all cloud formations.

time that cirrocumulus clouds are emissaries of storms, especially when they blend and thicken into cirrus, cirrostratus or altostratus. To these observations we owe such rhymes as:

> Mares' tails and mackerel sky,
> Not long wet, nor not long dry.
>
> Mackerel scales and mares' tails
> Make lofty ships carry low sails.

Whatever the type of cloud, its origin is the gaseous, invisible water vapor in the air. Water vapor is always present in the atmosphere to some degree. It evaporates into the air from the leaves of trees and other plants, and from the surfaces of the oceans, lakes, ponds, rivers and streams.

The vertical circulation of the atmosphere constantly carries this vapor upward. Rising to higher altitudes where atmospheric pressure is lower, the air expands; in doing so, it becomes colder. This cooling condenses some of the water vapor. The result is usually the production of the small droplets of water that make up a visible cloud.

Tiny particles already suspended in the atmosphere help to form these droplets from the condensing vapor. Some of these particles are microscopic bits of dust from the land; some are minute salt crystals that enter the air along with moisture from the sea. Many of the air's solid particles are too small to be seen through an optical microscope. Meteorologists term these makers of droplets, whatever their size, condensation nuclei.

When the cloud's droplets form around condensation nuclei at altitudes where the temperature is lower than water's freezing point, they may float in the air as supercooled water droplets or become ice crystals. Many such crystals may come together in clusters. When these ice-crystal aggregations grow too heavy to be supported by the air, they begin falling to the Earth, melting into raindrops as they descend through warmer layers of air at lower altitudes. If freezing temperatures prevail all the way to the ground, the ice crystals arrive at the Earth's surface as snow. If the ice melts into rain as it passes through warm air at an intermediate altitude, then refreezes, the precipitation arrives as sleet.

Clouds, then, are the visible manifestations of warm, vapor-laden air rising in the sky. But what makes the air ascend to high altitudes?

The Sun's rays heat the Earth's surface, which radiates this heat into the air. Generally, therefore, the air close to the surface is warmed more than air at higher altitudes, since it lies closer to the source of heat. Warm air is lighter, or less dense, than cold air. Thus the Earth's gravity pulls the colder, heavier air of higher altitudes down to the surface, where it displaces the relatively warm, lightweight air, pushing it upward.

Normally the vertical circulation of the atmosphere happens as a slow, steady process. Often, though, a strong upward air current, called a thermal, occurs over a section of the land or the sea that is heated intensely by the Sun. The heat of the surface is transferred to large parcels of air just above. These lightweight parcels pass quickly upward, rising through cooler, drier air in much the same manner as bubbles rising in a kettle of boiling water. At some higher altitudes the moisture in a thermal condenses, yielding a small puff of cumulus. And where the water vapor condenses, it liberates heat in changing from a gas to a liquid.

As the thermal convection continues to feed the cloud with moisture and heat, it creates a warm corridor through which new air readily rises to great heights. In this way a seemingly insignificant cumulus may build rapidly to a rain-bearing towering cumulus. The air parcels in thermals often have a rotary motion that makes the cloud appear to boil. Within 15 minutes, a cumulus that begins by looking like a small puffball may boil over into an enormous cumulonimbus monster—the most powerful and violent of all clouds. These are the ones responsible for heavy rain showers, thunderstorms, hail and tornadoes. The air currents within these giants create severe downdrafts and updrafts. Flying through them is extremely dangerous. In their most advanced stages they develop an anvil-shaped top. Their bases are frequently skirted by shaggy cloud remnants. Radar observations and pilots' reports have indicated that the tops of some cumulonimbus clouds extend up to 60,000 feet.

Another kind of lifting, characteristic of windy conditions, occurs above hills and mountains. Where the wind is forced upward by the contours of the terrain, it carries aloft its moisture, which may condense to form clouds above and beyond a mountain. Often the wind is thrown into a wavelike pattern after encountering mountains or ridges, creating a series of "lee-wave" clouds downwind from the mountain. These clouds appear to remain suspended within the wave crests of the air stream. While new moisture is being

A thick, dark-bottomed boundary of clouds defines the edge of a frontal system over the western United States and signals a sharp change in weather.

Seen from above, cumulus clouds appear as billowing white heaps of varying sizes. They are made up of water droplets, and when massed, often bring rain.

carried up into one side of the cloud and condensing into droplets, moisture is also being removed from the other side of the cloud by evaporation as the air stream descends again to warmer levels. Although appearing to be stationary, lee-wave clouds are actually composed of moisture on the move; they may resemble waves, lenses or shapes of fishes.

Most intriguing are the clouds along the boundary of a "warm front." When a warm-air mass approaches a cold-air mass, a few fine threads of cirrus form along the high leading edge of the warm front. Behind them trails a filmy sheet of cirrostratus at a somewhat lower altitude, which soon coats the entire sky. Where warm air meets cold, the denser cold air cannot retreat as rapidly as the warm air advances; thus the warm air moves up as it moves forward along the surface of the cold-air mass, producing a sloping front. Although the base of the front may still be as much as a thousand miles away from the observer, the high leading edge of the advancing weather system, indicated by the cirrostratus, is directly above, like the windblown crest of an approaching wave.

In a few hours the clouds slowly descend below 20,000 feet, becoming milky, translucent altostratus formations. While the clouds are thickening, the pale sky is turning a dull white. Soon most of the sky will be filled with a descending ceiling of opaque clouds.

Generally, the darkened underside of the lowering layer will begin to roll and heave. Ragged cloud masses usher in the first few raindrops or snowflakes. At this stage the base of the cloud cover rapidly drops lower. At about 6500 feet the clouds become gray, murky nimbostratus, and soon rain, drizzle or snow begins falling in quantity from the shapeless belly of the overcast. Now the precipitation may continue for a few hours—or for days, depending largely upon the size of the air mass behind the moving front.

The passage of the front is marked by a shift in the wind, a change in temperature and a clearing sky. The appearance of the sky as the front passes will vary with the seasons, the location on the Earth and the direction of the strong upper winds.

Of course, no cloud phenomena are new. A poet long ago observed:

> When clouds appear like rocks and towers,
> The Earth's refreshed by frequent showers.

But today meteorologists are finding new ways to study the clouds. They are observing them by means

Jet Streams: The Sky's Wind Tunnels

They were unknown before World War II and the era of highflying planes; even today they are often elusive. Nevertheless, the Earth's jet streams are very real, powerful rivers of air that move along west-to-east courses at speeds sometimes approaching 400 miles an hour. Located at altitudes between 20,000 and 35,000 feet, jet streams are often used by airline pilots as "super tailwinds" to save hours of flying time and large amounts of fuel. Created by air temperature differences where weather fronts meet, jet streams are pushed eastward by the Earth's rotation. Frequently—as in the photograph at right—a jet stream's location is marked by a ribbon of altocumulus clouds that float below it.

of radarscopes and satellite photographs. To further understand what is happening within clouds, some scientists have begun to visualize them as concentrated collections of electrical charges. Some study the effects of seeding clouds with artificial condensation nuclei, while others employ computers to obtain mathematical "pictures" of cloud structures.

These modern methods of studying clouds point the way to more accurate predictions of the weather and perhaps hold out the promise that man will eventually learn how to control weather and climate. In the meantime, these dramatic forms that pass across the screen of the sky will continue to fascinate amateur and professional cloud watchers alike.

The Year Without a Summer

By George S. Fichter

The winter of 1815–16 did not differ noticeably from any previous winter in southern Canada and the northeastern region of the United States, and spring arrived as blustery as ever. In late March, after several rainy, foggy days, ice began melting in the ponds and streams. Winter's tight, icy grip was broken.

April's days brightened, with birds flocking in from their wintering grounds to find places to nest. Buds swelled and burst; trees and shrubs let out new green leaves. Spring flowers added splashes of color to the winter brown of the woods and to the greening meadows.

Later, as the scattered records maintained in those days show, people reflected that after a normal start, spring did indeed proceed more slowly that year than in any other within memory. Some people complained about the cold spring, but no one was alarmed. Cold Aprils are not really uncommon in this part of North America. But when the weather had not warmed in May, the unseasonably cold weather became a major topic of conversation. After all, it was now time to put out the fires and forget about having to keep warm. It was also time for fresh greens from the garden, and for crops in the field to be well above the ground. But winter kept hanging on. Morning after morning heavy frosts whitened the ground, and ice had to be broken in the water buckets.

Still, people generally were patient and confident that the weather would soon right itself. Old-timers recalled many years when it had snowed in May, but summer had always arrived and almost precisely on schedule. No one had reason to suspect that 1816 would be different.

But in June it became obvious that this year *was* unlike any that anyone had ever known. The month began properly, with the temperature rising into the 80s during the day. Then, on Wednesday, June 5, a cold blustery wind swept out of Hudson Bay and drove down across the St. Lawrence Valley and on into New England. Heavy rains whipped by strong winds lashed the land all afternoon and night. And each hour the temperature dropped—down, down, down. By the next morning thermometers registered in the low 40s and were still going lower when the snow began. At Bennington, Vermont, snow fell that day from just after daylight until midafternoon. When at last the storm stopped, the snow was 12 inches deep in Quebec city, and many parts of New England lay under a six-inch blanket of snow. It was, as a farmer noted in his journal, "the most gloomy and extraordinary weather ever seen."

Day after day the weird winter weather gave no signs of waning. Rather, it got worse. No thermometer climbed above 50 degrees, and most were in the low 30s. Tender crops that the hopeful farmers had put out earlier in the month were killed by the unseasonable frost, and the whole land looked as though it had been seared by a scorching fire. As one New Hampshire minister wrote: "Frosts killed almost all the corn in New England . . . the prospects for fodder are most alarming."

If at this point the weather had changed and returned to normal, the now panicky people might have been calmed. But the weather got no warmer. Through most of July and August the days started with temperatures in the 40s. By late August, early morning temperatures were in the 30s. On the few successively warm days, people gamely put out their gardens again. Farmers planted corn and other crops, hoping that somehow they might still get a harvest before winter. But time after time their gardens and fields were devastated by frost and hidden by snow. The killing frost that came shortly after mid-September was the first of the new winter. It was slightly earlier than usual.

People faced the coming winter fearfully, for their land had produced little food. Fortunately some had slim supplies of staples from bountiful harvests of the season before. They could manage. But they wondered what had happened to the summer of 1816 and pondered how they would survive if such summers became common. That winter of 1816–17 was an especially severe one, but spring came as usual. And there was a normal summer in 1817, as there has been ever since.

What caused the year of no summer? Some scientists believe that the warmth from the Sun could have been blocked by a great shield of dust. Because of the position of dust clouds in the upper atmosphere, very little solar radiation may have reached the region during that strange season. There was an abnormal amount of dust in the atmosphere because of several large volcanic eruptions. These culminated in the explosion of Mount Tambora, a tremendous volcano on the island

The World's Hottest, Coldest, Wettest and Driest Places

Listed below are extreme temperature and rainfall world records as established by official meteorological observation stations. But many inhospitable regions have no weather stations. Thus the Earth's true climate extremes may have gone unrecorded.

HIGHEST RECORDED TEMPERATURES (°F.)

1	Africa	136	Azizia, Libya
2	North America	134	Death Valley, California
3	Asia	129	Tirat Tsvi, Israel
4	Australia	127.5	Cloncurry, Queensland
5	Europe	122	Seville, Spain
6	South America	120	Rivadavia, Argentina
7	Oceania	108	Tuguegarao, Philippines
8	Antarctica	58	Esperanza, Palmer Peninsula

LOWEST RECORDED TEMPERATURES (°F.)

9	Antarctica	—127	Vostok
10	Asia	— 90	Oimekon, U.S.S.R.
11	Greenland	— 87	Northice
12	North America	— 81	Snag, Yukon Territory
13	Europe	— 67	Ust' Shchugor, U.S.S.R.
14	South America	— 27	Sarmiento, Argentina
15	Africa	— 11	Ifrane, Morocco
16	Australia	— 8	Charlotte Pass, New South Wales
17	Oceania	14	Haleakala Summit, Maui, Hawaii

GREATEST AVERAGE ANNUAL RAINFALL (inches)

18	Oceania	460.0	Mt. Waialeale, Kauai, Hawaii
19	Asia	450.0	Cherrapunji, India
20	Africa	404.6	Debundscha, Cameroon
21	South America	353.9	Quibdo, Colombia
22	North America	262.1	Henderson Lake, British Columbia
23	Europe	182.8	Crkvice, Yugoslavia
24	Australia	177.0	Tully, Queensland

LEAST AVERAGE ANNUAL RAINFALL (inches)

25	South America	0.03	Arica, Chile
26	Africa	0.1	Wadi Halfa, Sudan
27	North America	1.2	Bataques, Mexico
28	Asia	1.8	Aden, Arabia
29	Australia	4.69	Millers Creek, South Australia
30	Europe	6.4	Astrakhan, U.S.S.R.
31	Oceania	8.93	Puako, Hawaii

GREATEST OBSERVED RAINFALLS (inches)

32	1 min.	1.23	Unionville, Maryland
33	15 min.	7.80	Plumb Point, Jamaica
34	42 min.	12.00	Holt, Missouri
35	12 hours	52.76	Belouve, Réunion
36	24 hours	73.62	Cilaos, Réunion
37	1 month	366.14	Cherrapunji, India
38	1 year	1041.78	Cherrapunji, India

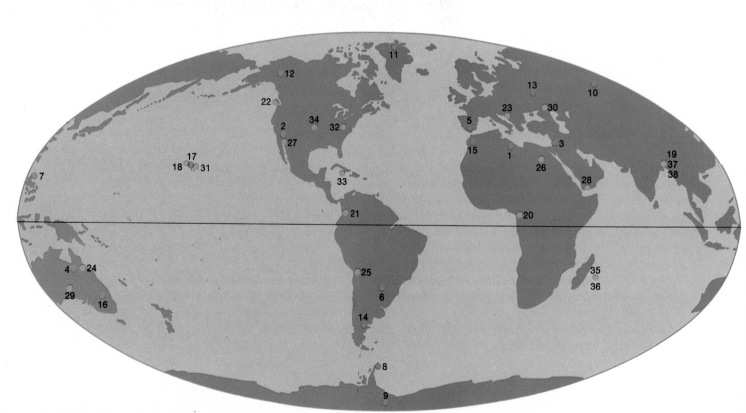

The numbers on this map indicate the locations of official weather stations where world records have been set for temperature (red) and rainfall (blue). The numbers coincide with those in the tables above.

of Sumbawa just east of Java, in 1815. This giant spewed an immense amount of dust into the sky, adding it to the large amounts thrown into the air by the earlier eruptions. The great blanket may have cast a screening cloak over much of the Northern Hemisphere that summer of 1816. This screen may have intensified local weather patterns in the New England–St. Lawrence Valley region and produced the bizarre summer.

Could it happen again?

Some scientists predict that it will—and perhaps soon, lasting even longer. They point out that there has been a general buildup of dust in the upper atmosphere for the past several decades. Some of the prognosticators say that if this trend continues for many years—say a century—the Earth could slide into another Ice Age, with glaciers forming on mountain peaks, spreading out from the poles and then riding out over the land. In the interval between the present and such a catastrophe, the summers would become increasingly cooler, the winters more severe.

Until recently volcanic activity throughout the world was largely responsible for the dust in the upper atmosphere. This dust does not disperse for many years. The cloud that resulted from the monumental explosion of Mount Krakatoa in 1883 affected weather in various parts of the world for more than ten years. For a good part of that time the dust caused the sunrises and sunsets in North America and in Europe to have a strange, rosy hue.

Dusts from volcanoes have caused a number of eccentricities in the weather over the years, but in the main, these have affected particular regions rather than altering the weather all over the world. Volcanic dusts are believed responsible, for example, for the especially cold winters in the southern United States in 1834 and again in 1895, when the hard-freeze lines extended to the very tip of Florida and stayed there day after day.

Now man has begun to add to the shrouds that reduce the amount of solar radiation reaching the Earth. The cumulative effect of pollutants added to the already existing dust can only be speculated about at this point, but there is general agreement that a gradual worldwide change in weather can be expected if present pollution trends continue. People will have to adjust to longer, colder winters and shorter, cooler summers. Some kinds of agriculture will have to be shifted southward. Crops that need a three-month growing season, for example, may not be possible in southern Canada, where the economy of the region is now built around 90-day wheat.

One thing is certain: Man's activities are now influencing a very basic natural phenomenon—the weather. If the direction were toward improvement, the world would be cheering. But if we are headed for worse weather, we must watch carefully.

Rainmaking Comes of Age

By Ben Funk

Over the Florida Everglades one day in the summer of 1970 a dazzling white cumulus cloud rose like a churning mass of whipped cream into a royal-blue sky. Flying at 21,000 feet, a reliable old propeller-driven airliner, which had been converted into a flying laboratory, flew into the top of the cloud, riding its turbulent drafts like a model-T Ford bouncing along a country road. Monitoring an instrument panel inside the aircraft, Dr. Joanne Simpson, director of the Experimental Meteorology Laboratory of the U. S. Oceanic and Atmospheric Administration, ordered a button to be pushed. Flares dropped from racks under the plane's wingtips into the heart of the cloud, filling it with smoke containing trillions of microscopic silver iodide crystals.

Suddenly the writhing cumulus cloud mass exploded; its creamy white head billowed higher and higher until it towered far above neighboring clouds.

What had caused the cloud's explosion? The cold-water droplets within it had begun rapidly collecting on the silver iodide crystals and freezing. As the water droplets froze, the heat they gave up powered the cloud's expansion, raising the cloud's temperature, approximately doubling its buoyancy and speeding its updrafts. As it mushroomed upward, it became a thunderstorm 40,000 feet tall, sucking in great quantities of additional moisture-laden air at its base. The cloud turned gray, then black. Thousands of tons of ice crystals formed, fell and thawed in the humid air below—and sheets of rain cascaded down on the swamps.

Miles away at the University of Miami's computer center, technicians used radar to measure the deluge from the "seeded" cloud against rain from unseeded clouds in the area. The difference was 2000 "acre-feet," enough water to spread a foot-deep layer of water over three square miles of land.

This particular cloud's artificially induced convulsion was one of the most spectacular in a long series of experiments that have moved the science of rainmaking toward solid, predictable accomplishment after a quarter of a century of controversy. The findings of the cloud-seeding experiments conducted over Florida suggest that days of good "seedability" can be diagnosed by computer, and on such days rainfall from local cumulus clouds can be as much as quadrupled through cloud seeding.

The first successful attempt to make rain by artificially seeding clouds was carried out in the United States in 1946 by Vincent J. Schaefer, a scientist working in the General Electric Research Laboratory. Why precipitation falls to Earth had long puzzled scientists, for they knew that the tiny ice and water particles in

When silver iodide crystals are released into the tops of clouds, they produce ice crystals. The heat released when the ice forms causes updrafts which make the clouds grow in size and shed rain.

the clouds are actually too light to reach the ground under their own weight. Some scientists suggested that the droplets in clouds might first be drawn together by electrical attraction, forming heavier drops, but calculations soon ruled this out as a possibility in most clouds. Others reasoned that airborne particles, such as windblown dusts and tiny salt crystals from the sea, might serve as condensation nuclei on which fine droplets could collect, building up into heavy drops.

Another line of reasoning suggested that ice crystals would automatically grow in size within a humid cloud, since the water vapor and droplets with which they came in contact would freeze onto the crystals. To explore these possibilities, Schaefer studied clouds

of water droplets supercooled to well below the freezing point of water in a refrigerated laboratory cloud chamber. He found what many studies have since verified—that even when temperatures fall as low as −30°F (−34.4°C), droplets within supercooled clouds may not freeze. Schaefer sprinkled a pinch of powdered dry ice (frozen carbon dioxide) onto such a supercooled cloud of droplets inside his chamber, watching through his observation window; lo and behold, a miniature snowstorm issued from the cloud! Soon after, he flew a plane over a cloud bank above a mountain in Massachusetts, spilling a few pounds of crushed dry ice into the cloud mass as seed. Almost at once snow began falling on the land below.

Schaefer had shown that precipitation could be artificially produced from a supercooled cloud by the introduction of dry ice. Then Bernard Vonnegut, a colleague, set out on a new line of attack, experimenting with particles that might act as condensation nuclei around which supercooled water droplets could collect and freeze. Vonnegut found that microscopic crystals of lead iodide and silver iodide worked much better than other salts, apparently because their crystal structures resemble that of ice.

Schaefer's and Vonnegut's discoveries were employed by amateur rainmakers who began recklessly seeding the clouds. Results were unpredictable, to say the least. Cloud seeding was like a new game for which the rulebook had not been written. When rain fell for some 60 days in a three-month period in the Abitibi region of Quebec outraged Canadians charged that rainmaking operations of the forest-protection services were the cause. Though the responsibility was never established, the Canadian government was beset with complaints from businesses and property holders on whom the heavy rains had inflicted great damage. Elsewhere, exaggerated claims by commercial seeders were not backed up by consistent results, and customers canceled contracts. Three states in the U.S. banned cloud seeding, and lawsuits were filed against the rainmakers.

Many scientists, meanwhile, insisted that there was no proof that cloud seeding worked any better than a Hopi Indian rain dance. They demanded proof—demonstrable, irrefutable evidence of cause and effect—obtained through controlled experiments. But no such evidence was forthcoming.

Then in 1966 the U. S. Academy of Sciences reviewed results in 18 commercial rainmaking operations and concluded that the rainmakers had increased precipitation by 10 to 20 percent. It advocated a government-sponsored program of controlled experiments and evaluation of uncoordinated research projects.

To meteorologists, one thing was clear: before precipitation patterns could be modified, far more had to be learned about clouds. Already deep into such work was meteorologist Joanne Simpson. In the mid-1960s she participated in Project Stormfury, a U.S. experimental hurricane-modification program. Stormfury researchers flew into a hurricane and seeded the cloud walls around the hurricane "eye" with silver iodide crystals to see whether the heat explosively liberated by the condensing drops within the supercooled cloud might be enough to break up and scatter the storm. The Stormfury tests achieved some success in breaking up hurricanes and provided insight into the nature of these violent storms. Working in the Caribbean with a technique of "dynamic" or massive seeding of clouds, the Stormfury team had proved conclusively that injections of silver iodide a thousand times greater than earlier experimenters had tried brought about fantastic cloud expansion under certain predictable conditions.

Now it was time to test the technique over land, where rainfall could easily be collected and accurately measured. The tests began in Florida in 1968. Each morning, atmospheric soundings were taken and fed into a specially programmed computer. If the computer showed that the clouds would grow to great heights naturally, whether seeded or not, that day's operation was canceled.

Eventually 14 clouds were picked, and each was hit with 20 flares containing a total of 2.2 pounds of silver iodide. Because of the vast number of silver iodide crystals released, most of the clouds appeared literally to explode in their sudden transformation into ice crystals. As the ice fell earthward into warmer air, it melted, forming rain. All but one of the selected clouds grew to thunderhead stature, and the yield in rainfall was doubled or tripled in comparison with untreated clouds in the area. Here at last was encouraging evidence from controlled experiments.

After the conclusion of the experiments in 1968, there remained a possibility that some of the heavy rainfall was solely attributable to nature. Therefore, Dr. Simpson wanted one more round of tests. These were conducted in 1970 and proved the effectiveness of cloud seeding beyond reasonable doubt. On days when the computer gave the go-ahead, average rainfall from all seeded clouds exceeded that of control clouds by well over 100 percent.

But for cloud seeding to provide rainfall, there must first be suitable moisture-laden clouds in the sky. There is optimism among rainmakers that the Simpson method will work over arid regions of the tropics where the air is often loaded with moisture that never falls as rain. But the methods that succeeded for Dr. Simpson in Florida cannot necessarily be transposed intact to areas with different meteorological conditions. Although experiments using the Simpson system have already been successful in Arizona, Pennsylvania and South Dakota, she says, "We just don't know yet the degree of seedability in other places. We can figure out, however, which areas are amenable to weather modification and which are not."

Nowhere will it be possible, she contends, "just to go out and put water on Farmer Brown's fields on a particular day. On each watershed, researchers will have to make a preliminary study. Then, when and if conditions are favorable, rain can be produced and stored in reservoirs until needed. This is the only sensible way seeding can be used to alleviate drought."

A local study of the type specified by Dr. Simpson has recently been carried out in a part of North America's Rocky Mountains by Dr. Lewis O. Grant of Colorado State University. Dr. Grant studied air and cloud flows to develop an effective method of controlling snowfall from certain clouds. He found that on days when water-filled cloud tops were relatively warm (ranging between −4° and 14°F), seeding could double or triple snowfall; when cloud tops were colder than −15°F, seeding could *decrease* snowfall by 30 percent or more. This information provided the basis for a model describing how seeding should be carried out under various cloud conditions.

In October 1970, largely because of Grant's efforts, a four-year pilot project was begun in the region to manage precipitation by systematic cloud seeding. The test site is a 1300-square-mile area of the San Juan Mountains in southwest Colorado, one of eight principal snow basins that together contribute about 85 percent of the average annual flow of the Colorado River. Records show that over 20-year periods, annual precipitation in this area has varied enormously. The pilot program has been seeding clouds in an attempt to maintain the vital snowpack at a size that can guarantee the optimum spring rush of melted snow into downstream irrigation systems.

Dr. Simpson notes that such operational cloud-seeding programs must be undertaken only with great caution. If nations are to move toward operational programs for improving rainfall and modifying weather, scientists must consider the consequences of weather modification in the light of competing interests within society. For instance, doubling the snowpack on a mountain watershed might help the farmer in the valley but hit the rancher with a stock-killing blizzard. More rain could benefit a power company, while depressing a tourist industry. Spreading out snow might ease traffic problems in a city but ruin a ski resort.

As cloud seeding and other weather-modification techniques are perfected, we may not just find ourselves with the talents to make rain or deflate a hurricane; mankind may develop the means to alter global climate. "We have come to that point where the scientific fraternity and the people must sit down and plan together the wise use of any new ability we create," remarked former U. S. Assistant Secretary of Commerce for Science and Technology Myron Tribus. "A scientist can bombard a nucleus with neutrons without asking the permission of the nucleus," Tribus commented. "But he cannot engineer the environment without consulting the people who will be affected."

With its fascinating promise and its not unreal perils, the science that was born a quarter of a century ago from a daring pioneer rainmaking experiment has finally come of age.

Heavy, moisture-swollen storm clouds lower as rain becomes imminent. Cloud seeding can produce results like this when conditions are right, but no rain-making techniques work where clouds are absent.

With the advent of space photography, we have learned much about how weather develops. This view from Apollo 11 shows the entire Pacific Ocean, with North America's west coast visible at upper right.

A New Look at the World's Weather

From a vantage point mentioned only in science-fiction tales a generation ago, man now gazes down upon his planet and gains unprecedented knowledge of the weather

Within the last 50 years, weather forecasting has progressed from the quaint, erratic generalizations of the almanac on a farmer's shelf to the precision of a camera eye boosted by rocket power into space, where it peers down at our planet from its perch on a satellite. The weather satellite is only one tool in the modern meteorologist's complex array of equipment, but to date it is his best. The United States orbited the first weather satellite, Tiros I, in 1960. It was equipped with television cameras that gave man his first global view of the formation and movements of the Earth's clouds. Manned satellites supplemented Tiros achievements as astronauts took color photographs of the Earth's weather patterns. Comparing these pictures with black-and-white Tiros shots, meteorologists were better able to interpret the views being relayed to them by Tiros. More sophisticated world weather information has since come from U.S. Nimbus satellites. This latest series of weather watchers takes both day and night pictures while infrared detectors collect temperature and humidity data from all over the world.

Probably the best way to forecast weather is to see it coming, which is exactly what weather satellites do. These pictures, although taken for the most part by astronauts, give the nonprofessional reader a good idea of what unmanned weather satellites see. Large cyclonic storm systems—like those shown below and at upper right—are among the phenomena that interest modern weathermen most, since they are the atmosphere's most serious disturbances. Satellite tracking of cyclonic storms began in 1969, the year in which several international meteorological organizations joined together to launch the World Weather Watch program. Supplementing satellites with specially equipped planes and ships, Weather Watch comprehensively charted Atlantic hurricanes, then expanded operations in 1972 to include typhoons spawned in the Pacific. With this greater flow of information from increasingly sophisticated instruments, meteorologists believe they can soon make accurate predictions of the seemingly capricious twists and turns in the courses of major storms.

A cloud eddy (above) reveals an area where prevailing winds stall just off Morocco's Atlantic coast.

On the way to man's first moon landing, Apollo 11 astronauts documented the circular sprawl of Typhoon Bernice as they passed over the Pacific at sunrise.

Clearly defined ranks of clouds spiral hundreds of miles inward toward the eye of a Pacific typhoon, giving Apollo 9 astronauts an unusual view of the typical structure of a major cyclonic disturbance.

Cirrus clouds, normally too high to be seen clearly from the ground, assume striking, feathery patterns when viewed from a spacecraft transiting the Atlantic.

From 100 miles up, the east coast of the United States south of Savannah, Georgia, lies blanketed under streamers of small, puffy cumulus clouds.

World's Most Powerful Storms

By David I. Blumenstock

What is a hurricane? It is air in motion—trillions of tons of air. Spread this across a circular area with a radius, say, of 300 miles. Start this mass rotating, with the air near the center swirling around at 75 to 150 miles an hour. Fill this system with clouds towering to altitudes of 40,000 to 60,000 feet. Dump water from those clouds at a rate of up to 20 inches of rainfall per day. Set the whole system in motion along some course. Picture the circling winds whipping up the ocean waves, piling the water onto the adjacent land. Such are the specifications for a typical major hurricane, whose effect when it hits a populated coastal area is seldom less than disastrous.

The hurricane has several names. Technically, it is termed a tropical cyclone. In the western Pacific it is called a typhoon; in Australia, a willy-willy. Whatever its local name, the hurricane is a spinning wind system born above the seas in tropical latitudes. Its substance is warm, moist, tropical air. The source of its strength is the Sun, which causes water to evaporate from the surface of the sea, forming the clouds of the storm. As this vapor condenses into water droplets high in the storm clouds, its heat is released; in this way, a medium-sized hurricane liberates as much energy through condensation in a single hour as would come from the explosion of some sixteen 20-megaton hydrogen bombs. About three percent of this enormous energy is harnessed in the hurricane winds.

Typically the hurricane season begins in late spring and extends into early autumn, with the most powerful storms coming toward the end of summer. Then, tropical seas have been basking in the Sun's near direct rays for weeks on end; the rate of evaporation from the ocean's surface is high, filling the air above the ocean with the vapor that fuels the hurricane "heat engine."

Weather reconnaissance satellites, including the United States Tiros and Nimbus satellites, have today largely taken over the role of hurricane tracking from the weather reconnaissance planes that were used to report on tropical cyclones in the past. Satellite cameras photograph the storms by day, while their infrared equipment takes up the task at night. The satellites transmit their information to receiving stations located at various spots around the globe, providing a world-wide network that can monitor a hurricane's birth, growth, movements and finally its decay. Planes, ships at sea and ground-based weather observation posts also gather information on these violent storms.

If all hurricanes traveled the same paths, forecasting would be greatly simplified. Unfortunately, a hurricane may do erratic things—steer zigzag paths, dip toward the Equator but then swing away, sometimes halt

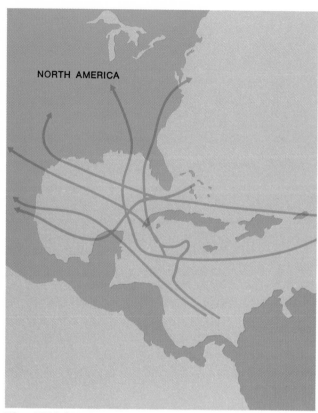

The 1970 hurricane season in the Atlantic (above) produced a below-average number of storms. But where their paths touched land, more than 85 people died.

In 1970 the Pacific—always more active than the Atlantic—produced a dozen major typhoons whose paths (below) tangled erratically over a wide area.

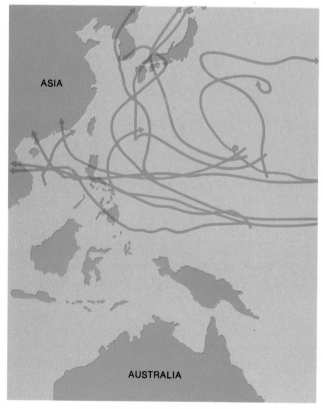

Battered by the double fury of high winds and storm-driven surf, this U.S. beach resort near Narragan-sett Bay took the full force of Hurricane Carol in 1954, one of the century's most destructive storms.

suddenly to retrace a path back in the direction from which it has come. Sometimes a hurricane executes loops, as one did early in the century. Moving northward in the Caribbean in October 1910, this storm scored a direct hit on the town of San Juan y Martínez in western Cuba; two days later it circled back in from the west and hit the town again.

To ships at sea, no storm poses a greater threat than a well-developed hurricane. The violent winds and the confused, mountainous seas inflict a beating upon even large aircraft carriers and ocean liners. During World War II the U. S. Navy's Third Fleet, commanded by Admiral William F. Halsey, was trapped in a full-blown typhoon about 500 miles east of the island of Luzon in the Philippines. The fleet's meteorologists had been warned well in advance that a typhoon was somewhere in the area, as had Fleet Weather Central 4000 miles away at Pearl Harbor in the Hawaiian Islands. Yet even at Weather Central, where coded weather reports streamed in continually from ships at sea, from aircraft in flight and from American-held bases on scores of Pacific islands, no one was certain just what the course of the typhoon was.

As fate would have it, Halsey's Third Fleet blun-

dered into the center of the typhoon. Destroyers and cruisers, battleships and carriers were tossed about like corks. "Proceed at will" was the order from the flagship, and the fleet scattered wildly before the wind and waves. Not a ship escaped heavy damage, and three sank with few survivors.

Though any ship is vulnerable when trapped in a typhoon at sea, by careful navigation a vessel can be kept clear of the regions where wind and wave conditions are most extreme. Less fortunate are the inhabitants of an island that happens to lie in the path of such a storm. If the hurricane strikes the island head-on, all they can do is seek what shelter exists and, if there is warning far enough in advance, evacuate coastal areas. If the island has a low relief, as is the case with a Pacific atoll, the devastation of a typhoon or hurricane may be catastrophic.

Nothing can protect these slight bits of land from the furious winds and high seas of a typhoon. A typical atoll is a ring-shaped coral reef rising above the surface of the sea in one or more low-lying islets that partially or wholly enclose a lagoon. Most of the coral formations that create atolls rest on the summits of dead volcanoes that reach up from the deep-sea floor into

Hurricane Gladys was stalled in the Gulf of Mexico, off the coast of Florida, when this photograph was taken from Apollo 7 in 1968. Highest winds, located near the storm's center, were about 65 knots.

The approach of a typhoon is dramatic on South Pacific islands such as Moorea. The sea rolls in, the air fills with spray and the wind grows stronger and stronger until everything bends before it.

warm, sunlit upper layers of the ocean where coral organisms can take hold and grow. The larger islets are not more than several hundred yards wide and a few hundred yards to a few miles long. At their highest points they seldom rise more than 10 or 20 feet above the sea. Characteristically, the coral reef upon which the islets are perched descends abruptly on the ocean side, forming a sheer submarine cliff. The shores of the atoll's circular lagoon slope down far more gently. The deepest parts of these lagoons rarely lie more than 200 feet below sea level.

An awesome demonstration of the impact of a typhoon on unprotected atolls was given on January 7, 1958, when an enormous storm struck the Marshall Islands in the western Pacific. Hardest hit was Jaluit atoll. The 1200 inhabitants of the atoll received their first warning of an approaching typhoon during the morning, when huge waves rolling in from the sea began to bang against the outside of the reef on the east. The tide was still extremely low, so that the atoll was not immediately submerged. But as the tide rose and the fury of the waves increased, the water surged past the outer edge of the reef and directly onto the atoll. Abruptly a strong wind began blowing from the northeast. The wind increased to hurricane force—some 75 miles per hour—and waves began rolling clear over the eastern islets into the lagoon.

By now it was afternoon. With the tide nearly at full height, the wind pushed the water across the land; upon the surface of the floodwaters huge confused waves were piled up by the wind, creating a soup of uprooted trees and thousands of coral fragments, some weighing 100 pounds or more. A survivor of the disaster recalls: "We half-swam, half-waded from one big tree to another, but since the trees were going down everywhere we weren't sure what was safe. I saw a whole family drown. They'd tied themselves to a huge tree, high above the water. Then the tree went over and they were swept away."

When the typhoon passed, it left coral debris spread almost everywhere to depths of three feet. Trees were often either uprooted and carried away or snapped off above the roots. In the lagoon were outwash deposits in the shapes of fans and deltas, as if left by small rivers over a period of many years. Native huts and food crops such as coconut, pandanus and breadfruit were destroyed wholesale. Astonishingly, only 14 persons were killed during the storm. Two more died from exhaustion afterward. The low death toll was due to the refuge provided by World War II, foot-thick Japanese blockhouses sunk deeply into the coral, and to the happy coincidence that many natives from the most populous islands happened to be visiting another islet that was never completely awash.

A snapped-off palm tree and the rubble of a beachside hotel testify to the tremendous power of a hurricane that swept over the low-lying island of Plantation Key, Florida, in September 1960.

When hurricanes strike coastal lands anywhere, the greatest danger is in zones where the winds are on-shore, sweeping water from the sea onto the land in the form of storm surges. The devastating effect of storm surges in densely populated coastal areas has been demonstrated far too frequently.

In the monstrous hurricane that smashed the New England coast of North America in 1938, some 600 lives were lost and the damage to property exceeded a quarter of a billion dollars. Most of the havoc occurred in coastal lowlands that lay to the east of the storm center, where the winds blew unimpeded directly off the sea. As the hurricane moved in from the south, these areas were submerged by a surge driven onto the land by the tremendous winds. Areas to the west of the center of the storm experienced heavy northerly winds and suffered little or no inundation.

Storm surges generated by hurricanes have taken a huge toll in human life. One that rolled in from the Caribbean Sea onto the shores of Texas in 1900 drowned 6000 persons. In 1864 a surge produced by a typhoon in the Bay of Bengal took a toll of some 70,000 lives in the low, flat ricelands of the densely populated coastal regions. In the same area, on October 7, 1737, there had occurred one of the greatest natural catastrophes ever known to mankind when a typhoon

struck at the same time as an earthquake, creating a 40-foot wave that rushed far inland, killing 300,000 people. In November 1970, tragedy was visited upon the region yet again when typhoon-generated storm waves rushed in over the coast of what was then East Pakistan, killing upward of 200,000 by drowning, with more perishing in the aftermath of famine and epidemic disease. This strip of coastline is especially vulnerable to the typhoon's storm surge: the coastal plain lies barely above sea level, being an expanse of delta land deposited over many millennia by the Ganges and Brahmaputra river systems where they flow into the Bay of Bengal. Moreover, the bay's U shape acts as a funnel concentrating along this coast the fury of typhoons moving north from the Indian Ocean.

Once it moves over land, a hurricane begins to lose power; its winds are slowed by friction with hills, mountains, forests and man's structures, and it can no longer draw warm water vapor from the sea to power its condensation "heat engine." Three hundred miles inland it may still generate high winds, blowing down power lines and disgorging heavy rains that produce flooding. But the farther inland a hurricane moves, the weaker it becomes. Some 600 to 800 miles inland it may be indistinguishable from a heavy rainstorm. The hurricane is truly a daughter of the Sun and the sea.

Inside the Tornado

By Bernard Vonnegut

Of all weather phenomena the most extraordinary is a tornado—the terrifying whirlwind that descends from giant thunderstorms. In its brief lifetime, seldom more than an hour, the tornado demonstrates that it is unquestionably the most violent of all winds.

Scientists now have some clues to why this is so, and to what produces a tornado in the first place. But we are sadly lacking in some of the more fundamental facts, and continue to work with little more than guesses. This is why the best experts on tornadoes are the few eyewitnesses who have seen them close up and lived to tell the tale.

People who have experienced a tornado often say its approach is preceded by hot, humid air under an overcast sky, with scarcely any wind. Some have seen swirling currents at the bottom of the storm cloud. Others have observed two slowly rotating clouds approaching one another. Most accounts mention frightening noise as the funnel forms and begins drawing near. In bygone days observers likened the roar to that of carriages rolling rapidly over rough cobblestones, while today most analogies favor a squadron of low-flying jet planes.

The roar probably is related to that also produced by hurricanes and typhoons. But observers sometimes report a puzzling sound "like the humming of a million bees" apparently peculiar to the approaching tornado. Then, when it strikes, those very near its path often hear what sounds like the explosion of a bomb.

The bomb analogy is apt, for after the tornado has passed, the landscape looks like the target of a saturation-bombing attack. Roofs are torn off, buildings ripped open, houses left askew on their foundations, while some structures have disappeared entirely, leaving only cellars or footings. Automobiles, airplanes and house trailers have been picked up, carried away, then dropped and smashed like toys. Wire fencing is twisted into rope or rolled into wads. Railroad locomotives are sometimes blown off the tracks and overturned, and the terrific force may even tear up the steel tracks.

In addition to feats of strength, tornadoes have performed clever tricks, such as leaving a flock of chickens barely living and plucked of their feathers. Tornadoes have wrapped sheet-metal roofing tightly around utility poles, flung lengths of lumber like spears through buildings, farm animals and, in one instance, through the tires of a truck. Equally impressive, although more delicate, is the tornado's ability to drive splinters or pieces of straw into people like nails.

In view of this ferocity it is understandable why the human toll is high, even though no single tornado lasts long. Around the world, hundreds of people are killed and thousands injured annually by tornadoes, and damage is on the order of millions of dollars.

The tornado's violence is so much greater than that of other storms that a fascinating question is often raised: How fast is the wind in the tornado's funnel? No measurements have ever been taken there; hence scientists must rely on indirect estimates, based on the damage done by the wind. One of the more reliable estimates was made by Dr. Edward Brooks, an American authority on wind-speed measurement. He calculated that wind velocities near 300 miles per hour were necessary to collapse high-tension cable towers during a particularly devastating tornado that tore through Worcester, Massachusetts, in June 1953. Other scientists have speculated that the whirlwind's velocity may reach 500 to 600 miles per hour—and some have estimated speeds greater than that of sound! However, these high speeds are controversial and some scientists believe the wind velocities go no higher than 200 miles per hour.

Wind speed, however, is only part of a bigger question: What makes the twister? Let us consider some of the facts already pieced together and also look at the remaining problems:

The tornado belongs to a family of rotational winds whose members range in size from the tiny dust devil to the giant winter storm that covers as much as half a continent. The tornado is one of the smaller members, and unlike some fairly energetic dust devils and mild waterspouts, which can form during fair weather, it apparently builds up only during a severe thunderstorm. Fortunately, only about one out of a thousand thunderstorms breeds a tornado.

Even before the funnel-shaped vortex makes its appearance, a swirling can be seen in the base of the mother cloud. Here, rising air is forming the tornado. As the swirl increases in size and speed, the funnel drops out of the cloud like an elephant's trunk reaching groundward. At times, instead of extending all the way to the ground, it pulls back into the cloud, where it is a hazard only to aircraft.

If it approaches the ground, the disturbance raises a dense cloud of dust. Observers have often reported seeing, within this dust cloud, flocks of birds circling the funnel. Actually, of course, the "birds" are roofs, bits and pieces of buildings and other debris traveling in the whirlwind. Damage is mainly limited to the path of the funnel. Thus one house may be totally destroyed, while neighboring buildings escape unscathed.

Although twisters are known to accompany thunderstorms in the tropics, they are more likely in the Temperate Zone—probably because these provide the greatest atmospheric temperature contrasts and consequently produce severe thunderstorms when warm and cold air masses meet. Central North America has the dubious honor of being the most tornado-ravaged part

These unusual pictures record the development of a U.S. tornado near Freeman, South Dakota, in 1965. Above, the funnel begins to drop from a massive cloud. Below, it swells in size as it gains power.

of the world. Thunderstorms in this area, especially in the spring, are also unusually intense. The tornado season begins in February in regions along the Gulf of Mexico, moving north in spring and summer.

Most tornadoes form in midafternoon, when solar energy has heated the air to its maximum temperature for the day. But they can also occur at night. Whatever the hour, the funnel usually moves with the velocity of its mother cloud—from as slowly as a man walks up to 70 miles per hour. Generally its havoc at any one place is over in a minute or less, sometimes in only a few seconds.

The most widely accepted explanation of the tornado today is that it results from the great temperature contrast between the warm, humid air in the lower levels of the atmosphere and the cold, dry air higher up. As the warm air rises into the cool air, its moisture condenses, liberating heat and causing a swirling updraft of sufficient intensity to create a tornado. But this explains the thundercloud far better than the tornado that drops out of the cloud. Temperature contrasts alone would not seem to account for winds of tornado strength—or even winds less than half as fast. Temperature differences also do not explain why only the largest, most vigorous thunderstorms ever breed tornadoes. Another idea attributes tornadoes to the falling of large hailstones, often observed in a tornado's vicinity. But most meteorologists find it hard to see how the hail could cause the tornado updraft; furthermore, tornadoes are not always accompanied by hail.

What, then, causes a tornado? For a while many years ago an electrical explanation was popular. In the United States, in the late 1830s, an American scientist, Robert Hare, stated: "After maturely considering all the facts, I am led to suggest that a tornado is the effect of an electrified current of air superseding the usual means of discharge between the Earth and clouds in those vivid sparks which we call lightning." At about the same time, the French physicist J. C. A. Peltier wrote an entire book giving arguments for the electrical nature of the tornado.

The idea of an electrical cause for tornadoes apparently enjoyed considerable acceptance until 1887, when Col. J. P. Findlay, of the U. S. Signal Corps, enumerated no fewer than 143 reasons why tornadoes could not possibly be of electrical origin. From then on the electrical idea lost favor in scientific circles and became an almost completely forgotten concept.

As an atmospheric research scientist whose experience has covered more than a quarter of a century, my personal interest in an electrical explanation for tornadoes began with the one that devastated Worcester in 1953. That evening, upon returning home from my work at a research laboratory in Cambridge, I received a telephone call from a friend. He alerted me to an un-

usual lightning display over the Atlantic Ocean to the southeast. Going to the shoreline, I found this was the most spectacular lightning I had ever seen. It was originating in the same storm that had produced the Worcester tornado only six hours earlier, and I saw it now in almost constant illumination—at least 20 lightning flashes per second. From photographs I took of the storm, then about 100 miles away, I was able to calculate that the flashes originated in a bank of thunderclouds whose tops were 12 or 13 miles above sea level. I then estimated the electrical power necessary to produce such a display and arrived at a fantastic figure—about 100 million kilowatts, which at that time was roughly equivalent to the generating capacity of the entire United States.

The bizarre idea then occurred to me that perhaps this electrical energy was in some way connected with the formation of tornadoes. A few rough calculations indicated this amount of power might possibly be enough to drive the winds of a tornado. I began looking through sources on tornadoes to check on two things: Was lightning uncommonly frequent in these storms? Had eyewitnesses observed any other unusual electrical effects? The answer was yes to both.

In one source, *Tornadoes of the United States*, I found a statement by a former airlines meteorologist that lightning of a peculiar and intense type is almost always associated with tornadoes. In the same book an

The funnel hits the ground, cutting a path eight miles long and 300 yards wide through the farmland.

After 20 minutes on the ground the tornado funnel begins to rise and dissipate. The storm is over.

287

Tornadoes Born of Fire

To learn more about why Australian bushfires are so vicious, a team of scientists set a carefully controlled 50-acre fire near Rockhampton, Queensland, in 1969 called "Operation Euroka." They may also have created the world's first man-made tornadoes. To get the blaze going, 6000 tons of gum tree scrub were cut and stacked around the 50 acres and simultaneously ignited at 900 separate points. Meteorologists noted weather conditions before the fire, then continued to monitor wind and temperature readings as the blaze mounted up. Their work produced some surprising results. They found that their mammoth inferno markedly changed the direction of surface winds and nearly tripled wind speeds near the fire's edges. But the most dramatic activity occurred at the center of the fire, where scorching winds, howling at speeds of more than 60 miles per hour, produced miniature tornadoes that swirled as high as 2000 feet. In having proved his ability to construct a tornado, man may have come a step closer to understanding one of Nature's most mysterious and destructive occurrences.

eyewitness related: "There was a screaming, hissing sound coming directly from the end of the funnel. I looked up, and to my astonishment I saw right into the heart of the tornado. There was a circular opening in the center of the funnel, about 50 to 100 feet in diameter, and extending straight upward for a distance of at least half a mile, as best I could judge under the circumstances. The walls of this opening were rotating clouds, and the hole was brilliantly lighted with constant flashes of lightning which zigzagged from side to side. Had it not been for the lightning, I could not have seen the opening or any distance into it."

In the course of my readings I found that the association between luminous activity and the tornado was so well known in ancient times that the Latin language provided a word, *prester*, for such twisters, meaning "a fiery whirlwind that descends to the Earth in the form of a pillar of fire." Obviously the electrical explanation I had fondly considered a new theory might properly be called the classical explanation.

Still, I had to have more evidence. So, whenever a tornado was reported in the United States, I wrote to the editor of the newspaper in the area, asking to hear from anyone who had seen the twister close at hand.

As is usual with tornado experiences, some testimony had a warm personal appeal. One reply, by a man in his sixties, went back to when he was eight. He remem-

bered others in his family, taking refuge themselves, shouting, "Willie, get under the bed!" But Willie knew nobody would dare come and get him. So he pulled a chair to the kitchen window and took in the "most wonderful sight of my life—a large electrical ball of fire that traveled in front of the twister." Only when the funnel came near enough to remove the barn roof nearby did he tell himself, "Willie, you better get off this chair."

My project brought much evidence that confirmed previous details and added valuable new ones. Where earlier reports mentioned a sulfurous or gassy smell, modern observers likened it to the smell from electrical machinery. Or, if they knew some chemistry, they alluded to the smell of ozone or oxides of nitrogen, which are produced by electrical discharges. People who had sought safety in storm cellars remembered that they were almost suffocated by a greenish gas. Others reported their faces and arms showed a severe sunburn, possibly because of close exposure to ultraviolet rays from the electrical activity. The funnel interior that people in previous generations had reported as a luminous tube was now described as "like a giant neon tube."

Such additional information was valuable. Yet most of it still did not answer professional skeptics who doubted "hearsay" reports and wanted "objective" supporting evidence or scientific measurements. However, two pieces of evidence with good, objective credentials have turned up in the past few years, and both may lead to wider acceptance of electricity's role in the whirlwind.

The first concerns a tornado that passed within six miles of a geophysical observatory near Tulsa, Oklahoma. As the twister passed, disturbances were recorded at the observatory. These included abrupt fluctuations in the electric current flowing through the Earth and a change in the intensity of the Earth's magnetic field. Such change of intensity indicates the sudden flow of a large electrical current. From the data, Prof. Marx Brook of the New Mexico Institute of Mining and Technology calculated that a current of over 100 amperes was flowing for a period of ten minutes while the tornado passed by. From this amperage and the voltages known to exist in thunderstorms, it seems possible that the electrical power being released might have been large enough to run the tornado.

The other piece of exciting objective evidence is a photograph taken by James Weyer, on Sunday night, April 11, 1965. From his home near Toledo, Ohio, he was taking shots of a thunderstorm. As the center of lightning activity moved eastward toward Toledo, his camera followed. Hail had been falling around Weyer's home. The hailstones now were much larger, and he turned experimental—was the lightning bright enough to show hailstones lying in the foreground? He took a series of pictures in Toledo's direction, unaware that the city was having a tornado experience.

In his studio the next morning, glancing at the developed negatives, he would have discarded one that appeared "fogged" by bright bands of light. But something teased his professional curiosity and he examined the negative more closely. Now he saw that the two somewhat vague vertical bands of brightness didn't extend past the picture area, as they would on a fogged negative. So he made a print. Since then, scientific analysis as well as reports by Toledo eyewitnesses indicate that what Weyer had recorded was a pair of genuine luminous pillars—possibly tornado funnels illuminated by some kind of electrical discharge within the vortex of each twister.

Such evidence goes a long way to show that in at least some tornadoes there is either energetic luminous activity or a strong electrical current. Yet, this evidence leaves more questions than it answers. For example, is the electrical activity common to all tornadoes? What is its nature? Is it a cause or a result of the tornado? If we could answer these questions, our meteorological knowledge would be greatly advanced. Furthermore, it is conceivable that scientists could devise tornado-control techniques that would eliminate the peril to humanity that still rides the whirlwind.

The Great Waterspout of 1896

By Michael J. Mooney

Thousands lined the shore, transfixed by what they saw. Less than six miles away a gigantic black column stood poised above the waters of Nantucket Sound, Massachusetts—a living apparition 144 feet in diameter and more than half a mile high.

Not for 27 years had a waterspout been seen off this section of the American coast. And never had one of such awesome proportions performed where so many people could view it in safety. For 40 minutes on the afternoon of August 19, 1896, the residents of Martha's Vineyard watched in awe. Experienced sea captains among the spectators called it the largest, most perfectly formed waterspout they had ever seen.

But what is a waterspout? How do they form? Where do they occur—and why?

A waterspout is basically a whirling vortex of wind and water, similar to a tornado but occurring over water. There are two distinct types. The tornadic or "stormspout" is usually a land-born tornado or tornado cloud that goes to sea. Like its land counterpart, a tornadic waterspout has violent winds, sinister black funnels and angry-looking clouds overhead. The second, more common type, is the fair-weather spout. Unlike the stormspout, it begins at sea level and swirls skyward like the familiar "dust devils" on land. Fair-weather spouts are usually small, of short duration and virtually harmless, being more curious than spectacular. The waterspout that visited the waters off Martha's Vineyard *was* spectacular—and definitely tornadic.

The morning of August 19 was partly cloudy and warm with a light northerly breeze. By noon an ominous line of thunderclouds had risen out of the northwest, heralding an approaching cold front. As the residents of the summer resort enjoyed their lunch, a menacing cumulonimbus cloud reached the junction of Vineyard and Nantucket sounds.

At 12:45 P.M. someone shouted, "A waterspout! A waterspout!" Within minutes scores of people were rushing toward the shore.

Only those near the beach saw much of this display, which lasted about 12 minutes. One witness, E. H. Garrett, later gave this impression of the scene:

"We were out on the beach and saw an odd-looking cloud in the sky. It seemed to have a curious appendage, at first, which one of the party described as looking like 'an icicle.' We turned to go home, when one of the group looking back saw the 'icicle' changing, and we all watched. It grew larger, then looked like a long, thin gray veil of mist, and as it descended the water from the sound began to rise."

It turned from luminous gray to black, remaining nearly perpendicular while flaring out, funnel-like, at the base of the cloud. By the time local photographers had set up their cumbersome equipment, the spout had dissipated.

By then a great throng lined the beach. Though no spout was visible, the sea beneath the huge cloud continued to seethe. Apparently a powerful vortex was still present, which, though invisible, could re-form again at any moment.

Sure enough, at precisely one o'clock the great cloud spawned another black funnel, much broader than the first. This time the column steadily swelled in size. In awesome majesty it swept slowly and sedately to the southeast, following the parent cloud above. Vacationing Rev. Crandall J. North was deeply affected by its visual magnetism:

"The sea was perfectly calm, the air almost motionless, the Sun shining brightly, light summer clouds hanging here and there over the deep blue sky; and in strange contrast with all the rest was this lofty mass of black vapor with its absolutely perpendicular support. To add to the weird effect, occasional livid streaks of lightning shot athwart the black monster cloud above. The column was slightly funnel-shaped just where it joined the cloud, and was of equal diameter the remainder of its length. At its base the sea was lashed

This picture, taken from Martha's Vineyard in 1896, shows one of the waterspouts which amazed summer vacationers as it rose high over Nantucket Sound.

into a mass of white foam and spray that mounted as high as the masts of a large schooner."

Those on shore were too far away to hear the roar of the colossus. Not so for those aboard the schooner *Avalon*, caught between the spout and shore. Not only did the monster seem to them to be twice as large and twice as terrifying, but the tumult of its tornadic winds was positively stupefying.

At 1:18 P.M. the great column began to dissolve. But two minutes later the astonished thousands beheld the *third* appearance of this fantastic spectacle. This time the spout was no longer columnar, but tapered to a point barely visible at sea level, with the entire trunk curving gracefully toward the southeast.

After five minutes its rotating winds lessened, condensation decreased and final dissipation set in. Prof. William B. Dwight, a summer resident of Martha's Vineyard who was a professor of geology at Vassar College, described the startling visual effects attending the final display:

"Toward the close of this phenomenon the eastern half of the sky became quite black with clouds, while the entire western half was brilliant with sunlight, which at this hour glanced easterly beneath the blackness. The chromatic effects were of an indescribably rare and beautiful kind. The surface of the sound for several miles out was lighted up with weird hues of bright blue, green, yellow and gray, in patches, according to the nature of the variable weedy and sandy bottom, greatly intensified by the solemn, black storm clouds and waterspout overhead. Thousands of spectators, crowding the beach, gazed on the sight with mingled admiration and awe."

With the disappearance of the vortex, the waters finally calmed and the triple display was over at 1:28 P.M., a little more than 40 minutes after the first excited calls echoed along the beach.

As the spout vanished, a cold front swept over Martha's Vineyard. At 1:45 P.M. the first rumbles of thunder were heard, followed by torrential rains after three o'clock. Dr. S. W. Abbott, secretary of the State Board of Health, described what happened next:

"The waterspout was soon followed by marked atmospheric disturbances. Thunder, lightning, hail and rain in abundance fell within an hour or more. A dense, dark cloud formed in the northwest, followed by a squall from the southwest, and the wind shifted in a short time from northeast to southeast, and then by southwest to northwest. The thermometer at 2:00 P.M. indicated 56°, a very low reading for a place where it has varied but little from 70° all summer."

Prof. Frank H. Bigelow, a U. S. Weather Bureau meteorologist, made an exhaustive study of the entire phenomenon and interviewed many witnesses, including several professional Weather Bureau observers who were on hand that memorable day. From this and a meticulous inspection of all available photographs, he

Rare twin waterspouts kick up clouds of spray as they touch down off New Providence Island in the Bahamas. Although waterspouts are the maritime counterpart of tornados, they are seldom as violent.

came to the following conclusions about the spout's dimensions:

Diameter of spout at the water	240 ft.
Diameter of vortex tube at middle . .	144 ft.
Diameter of tube at base of cloud . .	840 ft.
Approximate length of tube	3,600 ft.
Approximate height of top of cloud . .	16,000 ft.
Average forward speed of spout . . .	1.1 mph

When natural phenomena occur on a spectacular scale, they usually bring widespread destruction and heavy loss of life. How rare it is when a phenomenon on the scale of the 1896 waterspout causes no physical damage whatever and no casualties. In fact, aside from those aboard the *Avalon*, the only other persons remotely exposed to danger were the shaken crew of a catboat becalmed near the spout itself—a small price to pay for the privilege of viewing this dramatic marine phenomenon up close.

In the words of the Reverend Mr. North: "The scene presented to view was such as not one in a thousand had ever witnessed before or would ever see again." Observer W. W. Neifert aptly summed up the feelings of all those present that day: "It was a sight long to be remembered, and when the weather cleared, about 4 P.M., each expressed himself as being most fortunate in having escaped some dreadful calamity."

Part Six

EARTH'S GREAT NATURAL TREASURES

Our planet's crust provides many riches, but none more tantalizing than the flashing fire of gems.

The Mineral Monarchs

The Earth's crust has long been a treasury of great riches for mankind. From its top layer, the soil, come the timber and produce of forest and field; its deeper strata yield the stone, ores, coal and oil on which modern industrial civilization thrives. Yet today there are many urgent signals warning that industry's increasingly voracious appetite for raw materials may drive the crust's once-full treasury into early bankruptcy. If the mineral wealth described on the following pages is to enrich the lives of future generations, it seems certain that present trends will have to be reversed and replaced with worldwide plans for making the best use of our dwindling natural riches.

Any accounting of the Earth's natural treasury must begin with those royal minerals, the gemstones. With the new uses for gems in specialized industries, today's gem cutters are kept far busier than their predecessors —whose secret arts of cutting and polishing were restricted to producing jewelry. Every continent contains in its rock the raw crystals of at least some gemstones. South Africa, of course, is renowned for its fabulous diamond mines, but India, Brazil, Siberia, Tanzania and West Africa also supply diamonds to the world market. Other highly-prized gemstones are emeralds, mined chiefly in Peru, Colombia, South Africa and Russia; rubies, mainly from Burma and Ceylon; and sapphires, now primarily obtained in Australia and Asia. But these and myriad others—running the alphabet from aquamarine to zircon—are found wherever the natural circumstances responsible for their creation have occurred.

Diamonds are made up of carbon atoms packed in tight crystal structures under intense heat and pressure deep within the Earth. Most of the other precious stones, however, are situated in special geological formations called pegmatites, which originate from molten rock magma rising inside the Earth. The magma contains fluid, acidic solutions of minerals which move upward through fissures in crustal rock. As these solutions cool, their dissolved minerals precipitate as crystals. Pegmatites sometimes contain giant crystals of quartz, tourmaline and other gemstones—a single emerald crystal was once discovered that weighed several hundred pounds. But the most valuable crystals rarely grow very big. The largest known diamond, the Cullinan, found in South Africa in 1905, measured just four inches on a side and weighed 3601 carats—about 1.3 pounds. When cut it yielded 105 gems.

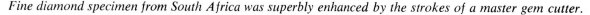

Fine diamond specimen from South Africa was superbly enhanced by the strokes of a master gem cutter.

White (top) and two black opals are from Australia.

Brazilian quartz produced this 7000-carat gem.

Coarse rock is the matrix for a brilliant ruby.

It is their near-perfect crystal structures that make gemstones the royalty of the mineral world. A gem's particular atomic arrangement within its crystal structure gives it the hardness, brilliance and transparency for which it is valued. A diamond's great hardness and its celebrated "fire" when held in direct light are both results of the carbon's crystal structure. A sapphire's asterism, or starlike gleam, is caused by an intersecting pattern of gas bubbles trapped within the crystal during its formation. Chemically identical oxygen-and-aluminum combinations, ruby and sapphire, owe their different colors—blood red and deep blue, respectively—to small traces of impurities: chromium atoms in the ruby, and iron and titanium in the sapphire. The emerald—consisting of oxygen, silicon, aluminum and beryllium—gets its beautiful green shades from titanium and chromium atoms present as impurities in the gem.

Artistry by Accident

Although those regal minerals, the gemstones, owe their finished beauty to the skilled hand of a gem cutter, Nature's own artistry with more lowly minerals is often equally breathtaking. The colors of a polished piece of jasper (opposite page, top) are like a dazzling Oriental landscape painting. A fine-grained form of quartz, jasper forms under intense heat deep in the Earth. This sample's delicate brushstroke-like swirls of color come about from varicolored iron and manganese impurities trapped in it at the time of its formation.

Nature's artistry sometimes takes unexpected and deceptive turns. At first glance, even an expert might be deceived by the lacy, branching pattern on the cracked slab of shale (below, left), so close is its resemblance to a fossil impression of an ancient fern. But this "fossil" is a dendritic (treelike) crystal growth of the mineral, manganese oxide. Moisture containing dissolved manganese oxide seeped into a fissure in the shale, crystallized there and was exposed only when the shale was broken up into slabs. Another example of mineral mimicry of living things occurs in the formation of barium sulfate crystals (opposite page, bottom). When crystals of this mineral develop in sand, the crystal structure becomes rounded and softened into petal shapes. As cleverly conceived as a piece of contemporary metal sculpture is the pyrite crystal formation (below, right). Better known as "fool's gold," pyrite's crystal faces develop chiefly in the shape of cubes. When the faces of the growing crystals intersect, as here, complex geometric forms are born that would delight the most sophisticated connoisseur of modernistic sculpture.

Pyrite from Spain resembles a cubist sculpture.

Branching crystals etched this slab of German shale.

Swirling impurities in polished jasper—a form of quartz—create a mineral landscape painting.

A large cluster of barium sulfate crystals forms a reddish American "rose" discovered in Oklahoma.

Mercury, the liquid metal, is extracted from cinnabar, a pink mineral prevalent in California.

The Rarest of Metals

Silver makes up only about .00001 of one percent of the Earth's crust, and platinum and gold each are a minute .0000005 of one percent. While gold and silver were man's measure of worldly wealth for a long time, the 16th-century discovery of platinum by Spanish explorers resulted in its use in counterfeit gold ingots—creating disdain for it as an adulterant. Ironically, platinum today commands twice the market price of gold—its resistance to corrosion and temperature distortion making it indispensable in many modern industries. But gold remains one of the foundations of international finance. Some $40 billion in gold reserves are held by the nations of the world.

Some 40 percent of the world's annual gold production comes from South Africa (see page 174). Mexico, the United States, Canada and Peru together mine more than half the world's silver, while Canada, South Africa and Russia are the major platinum producers. Rarely forming chemical compounds, gold is usually found in the Earth's crust as a relatively pure metal. Gold in some South African mines was liberated from the rocks of an ancient mountain range by weathering and now forms placer deposits in stream-bed gravel. Today it is mined from narrow "reefs." The California and Australian "gold rushes" of the mid-19th century began with the recovery of placer nuggets, flakes and dust. This was followed by the mining of the "mother lodes" upstream in the mountains. Platinum and silver also occur as almost pure metals. Although silver often forms compounds, such as silver chloride, its extraction from compounds is generally unprofitable.

A gold nugget typically contains in alloy about eight percent silver and traces of platinum, copper and other metals. The more silver present, the whiter the look of the gold. To refine gold, antimony (from the mineral stibnite) and mercury, or "quicksilver" (from cinnabar), were once widely used. Today electrolytic refining is the rule. An electric current is sent through an acid bath holding unrefined gold, causing pure gold to be deposited on cathode plates; the silver and other metals sink to the bottom in solution and are recovered later by means of chemical treatment.

Antimony ore, stibnite, comes from Bolivia, Mexico.

A California vein yielded the chunk of gold at right.

Platinum nugget weighs 17 ounces.

Silver (left) is from Norway.

299

Bedrock for Building

Most great civilizations are distinguished by their architecture, for which the crust's bedrock has long proved a vault of varied riches. As early as the 26th century B.C., the Egyptians were quarrying nearby limestone deposits for the first of the great pyramids, the Pyramid of Khufu. A sedimentary rock, limestone consists of the pressed and cemented fossil shells of trillions of long-dead marine creatures that flourished in ancient seas, their skeletons collecting layer upon layer on the sea floor over the eons. If Britain's chalk cliffs of Dover are the most famous limestone deposits, abundant supplies are widely available throughout the world, and have gone into cathedrals as readily as office buildings. Marble, the beautiful monumental stone of Greek and Roman builders, is a metamorphic form of limestone—its highly polishable crystalline

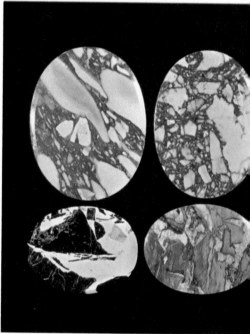

Renowned for their color and variety, Italian marbles and limestones are widely used for decorative work on modern buildings.

The crystals at right are gypsum, the basis of plaster of Paris, so called because of big gypsum deposits formerly mined in the Montmartre section of the city.

An American quarry in Barre, Vermont, is the world's largest producer of granite building blocks.

structure the result of the limestone's subjection to intense heat and pressure within the Earth. Sandstone, the basic building rock in Australia and the source of handsome church and "brownstone" house facades elsewhere, is a mixture of beach or desert sands cemented with silica, iron oxides or other minerals. The hardest and most plentiful rock of the continents is granite, an igneous rock—one that crystallizes as hot magma from the Earth's interior rises and cools. Because its hardness makes it costly to quarry and work with, granite's use (like that of marble) has more recently been limited to facings for major public edifices and monuments.

Chief among today's low-cost, highly adaptable building materials is, of course, cement—created by mixing powdered limestone and clay at high temperatures. Combined with water, sand and gravel, and poured into molds, it forms concrete. Reinforced with steel rods, concrete's unequaled flexibility makes it the architect's choice for an incredibly broad spectrum of projects ranging from suspension bridges to skyscrapers. Asbestos—the fibrous form of the heavy green mineral serpentine—is widely employed in tile, insulation and wallboard as a heat-resistant substitute for plaster of Paris. The latter is made by dehydrating gypsum, a mineral residue left over from the evaporation of water from salt lakes. Thus, as one material outlives its usefulness, the Earth provides another.

Asbestos—the remarkable rock that makes fireproof cloth and building materials—comes from a mineral family with tough, easily separated fibrous crystals.

Uniform fragments of coarse rock—for highway beds and building foundations—pile high around a U.S. crushing plant in the Shenandoah Mountains.

301

Basic Metals
for Industry

When Stone Age man entered first the Bronze and then the Iron Age, civilization was inexorably headed along the road toward giant skyscrapers and moon rockets. In the past, the great economic centers of our present industrial civilization grew up around accessible, easily worked sources of the basic metal ores. Today the world's prosperity hinges largely on man's continuing utilization of the basic metals that are found in the Earth's crust.

Although iron is Earth's most plentiful metal, most of it lies in the planet's core. Aluminum makes up about eight percent of the crust, iron just five percent, magnesium two percent, copper a mere five thousandths of one percent, with metals such as zinc and lead scarcer still. How, then, did sufficient concentrations of metals develop in the crust to the point where man can work them easily? Geologists tell us that within a cooling magma chamber the most common rocks, such as granite, are the first to crystallize. Among the last minerals to precipitate within the cooling lavas are the metals. Expelled upward and outward through fissures in the crust, they finally crystallize and form veins of ore. In North America, near Lake Superior and at Butte, Montana, for example, millions of tons of high-grade copper have been extracted from such veins. Because different metals crystallize at different temperatures within the rising, cooling magma, a lead-zinc deposit may lie over a copper deposit—with, say, tungsten found at lower levels in the vein.

But not all metals are mined from primary veins of this sort. Chile's rich deposits of copper sulphate and other famous nitrates, for instance, are the residue of groundwater that slowly percolated upward to evaporate in the dry air of the Atacama Desert. The world's greatest iron-ore reserves—in Australia, Venezuela, China, Russia, India and North America—were washed out of long-vanished mountain ranges and laid down as sediments on ancient lake and ocean floors. Bauxite—the ore of aluminum—results from the weathering of feldspar, a major constituent of granite. Canada's position as the world's leading aluminum producer stems not from its bauxite—it has none—but from its cheap hydroelectric power, which is used to refine the aluminum by electrolytic means. (The ore is shipped to Canada from Africa and South America.) Rarer metals, such as nickel, vanadium and molybdenum, are now also widely sought throughout the world because of their importance in making alloys of steel, copper and other basic industrial metals.

Fiery cascades of molten steel resemble a volcano's lava flow. The similarity is more than superficial—both are mixtures of many elements at extreme temperatures.

The United States—the world's leading producer of copper—extracts almost all its copper ore from open pits such as the immense Mission Mine near Tucson, Arizona.

Modern steel mills are highly automated, but many phases of production still require close-up work.

Using a long rod, a worker breaks a crust of slag to allow the molten steel to pour more easily.

A structural steel beam nears the end of its journey through a U.S. rolling mill in Pennsylvania.

Energy from the Earth

With most of its great forests fallen before the ax by the late 18th century, Europe turned to the Earth's crust for a new source of energy—coal. A "fossil fuel" consisting of the compressed remains of ancient swamp forests, coal is generally found buried beneath sedimentary strata that were deposited over the land by repeated floodings in past ages. With hard coal (anthracite) now largely exhausted, world mining operations today typically concentrate on vast, shallow seams of soft coal and lignite, which—though smokier and less efficient than hard coal—exist in vast reserves in many nations. Most governments have adopted laws regulating the flagrant abuse of the environment by soft-coal strip-mining operations, which in the past often left barren wastelands in their wake.

The Age of Coal was succeeded in the 20th century by The Age of Petroleum. Another fossil fuel, petroleum consists of organic matter that originated in tiny organisms which flourished in ancient oceans. By the 1970s the world's petroleum-producing nations—led by the United States, Russia, Venezuela, Iran, Saudi Arabia, Kuwait, Libya, Canada and Algeria—were turning out more than 13 billion barrels (42 gallons to the barrel) of oil annually, but were still finding it difficult to meet the world demand. In spite of the huge remaining Middle Eastern reserves (where more than half the world's known petroleum reserves are located), newly discovered deposits in the Soviet Arctic and Alaska and the prospect of extracting oil from shales and sand tars, geologists warn that all of the Earth's oil may be gone in about 100 years at expected rates of consumption. Hence the 21st century may become The Age of Nuclear Power—particularly in light of the promise of "breeder reactors" to create almost limitless amounts of radioactive fuels from natural uranium and thorium found throughout the world.

A concentration of oil rigs in California attests to the Earth's bountiful supply of this essential resource. But present rates of consumption may soon deplete the world's more accessible reserves.

With a splatter of drilling mud
a pipe is raised from a promising
oil bore in an Australian field.

Natural gas, a common by-product
of oil drilling operations, must
be burned off when there is no
way of storing and marketing it.

When coal lies near the surface, as at this American site in New Mexico, giant excavators strip
away layers of soil and rock to expose the seam.

Bleak, eroded hills, devoid of life, are the aftermath of uncontrolled strip mining for coal in the
Appalachian region of the eastern United States.

A Chemical Cornucopia

In the 20th century, human welfare has become increasingly linked to a cornucopia of chemical products, ranging from agricultural fertilizers to medicines. Some organic substances are used to synthesize plastics and pharmaceuticals, while others yield ammonia, synthetic fibers and other products as necessary to industry as to human existence. One of the greatest troves of inorganic chemical compounds is found in salt beds, from which the world now mines some 100 million tons of salts each year. No continent is without major salt deposits, which are nothing more than the mineral residues left behind by the evaporation of salt-rich bodies of water in past eras. The accumulation of salts in an evaporating body of water is a phenomenon seen today at the Dead Sea and Utah's Great Salt Lake. Here man is shortcutting the natural process by drawing off water heavily laden with dissolved minerals into evaporation beds, then mining the beds when they dry out (see page 150). In addition to the table salt so vital to human existence, salt beds are rich in potash, an agricultural fertilizer; sodium salts used in manufacturing soaps and paper; gypsum, the basis of plaster of Paris, and a wide range of rarer chemicals of great value to modern industry.

Important salt deposits are frequent in world desert areas. Among these are the rich nitrates of Chile's high Atacama Desert—the basis of explosives and fertilizers—and the borates of the western North American desert —critical to the glass and enamel industries. Concentrated in the desert floor by the upward percolation of mineral springs over the ages, the deposits have remained in place because there is little or no rain to wash them away. Sulfur deposits are common around the sites of extinct volcanoes, since sulfur wells up in its native state in volcanic vents. But sulfur is also found, along with natural gas and petroleum, in salt domes where it is extracted by forcing hot water down to the sulfur and back again to the surface.

Caustic soda is produced in El Caracol—"the snail"—a large freshwater pond with spiraling levees not far from Mexico City.

As salt precipitates in Colombian evaporation ponds, it is heaped up for further drying, then hauled away on the backs of Indian laborers.

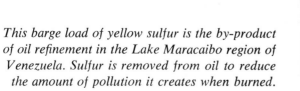

Along the Caribbean coast of Colombia, Guajiro Indians harvest salt by the ancient method of damming sea water in shallow, quick-drying ponds.

This barge load of yellow sulfur is the by-product of oil refinement in the Lake Maracaibo region of Venezuela. Sulfur is removed from oil to reduce the amount of pollution it creates when burned.

Soil's Vital Harvest

No precious metal, gemstone or ore extracted from deep inside the Earth is so valuable to humanity as the crust's top few inches of soil, the basis of agriculture and forestry. Soils result from the weathering of rock, lavas and volcanic ash into small grains and particles. Wind, water and temperature change play leading roles in breaking down solid rock. In tropical climates, hot days alternate with cool nights, inducing expansion and contraction in rocks. This causes them to crack. In cold climates, moisture seeps in, then freezes and expands, generating further cracking pressures. As the rocks split they are decomposed by the chemical action of weak acids and alkalis in rain and groundwater and are broken down by the abrasive action of wind erosion. Some minerals within rock undergo slow chemical change and form clay. More stable minerals, such as quartz, simply fracture into smaller and smaller sand grains.

The most bountiful agricultural soils are those rich in minerals and humus—decayed organic matter. But where heavy year-round rains occur, as in equatorial zones, the minerals and humus wash away, leaving the soil leached of nutrients. In semiarid regions marked by dry seasons, groundwaters seep upward from springs and dissolve the soil's minerals. At the surface, the water evaporates, leaving behind its dissolved minerals as a crust or residue. Where iron compounds collect in the soil in this way, the soil has a rusty red color. Calcium sometimes forms surface crusts and nodules through a similar process. Desert sands and

Rice, the staff of life for millions of people, grows to maturity in irrigated paddies. About 95 percent of the world's rice comes from continental Asia, the Philippines, Indonesia and Japan.

Golden stalks of rice crowd the terraced slopes of an Indonesian hillside at harvesttime. Despite great advances in agricultural technology, hand labor still prevails in major rice-growing areas.

soils are often extremely salty because of the accumulation of residues over many centuries. But careful irrigation and anti-saline chemical fertilizers can often make desert regions extremely productive.

Most of the world's successful civilizations developed in lands with good arable soils. South Asia, central Europe and North America's upper Mississippi basin are blanketed with a yellow subsoil called "loess"—dusts laid down by the winds along the margins of the massive continental glaciers of the last Ice Age. These regions today support most of the world's wheat, rice and corn agriculture. Egypt's Nile delta, India's Ganges, Indus and Brahmaputra basins, the Yangtse valley of China and the "fertile crescent" along the Tigris and Euphrates rivers have supported civilizations over the millennia because of soils built up and replenished year after year from mineral-rich sediments washed down from distant mountains.

When man respects the soil, he often brings forth both beauty and bounty as in this serene hillside farm scene in the Tuscany region of Italy.

In precise patterns, a well-organized corps of harvesting machines sweeps up and down a wheat field in California's fertile Sacramento Valley.

Gifts to Grace
Our Table

For some 10,000 years—the time since humanity first settled down from a hunting and gathering to a farming way of life—the soil has provided mankind with increasingly rich harvests from field, orchard and vineyard. Although hunger still sometimes stalks the family of man, more people are well fed today than ever before. Modern methods of food preparation, refrigeration, canning and shipping now put onto millions of family dinner and restaurant tables a diverse and palatable menu of meats, vegetables, fruits and beverages originating far from local farms and plantations. Between 1963 and 1968, when world population was increasing by 10 percent, combined world production of the staples wheat, rice, milk and potatoes scored a much greater increase—from 1.13 billion to 1.33 billion metric tons, around 18 percent. No authority expects agriculture always to prove capable of outdistancing uncontrolled population growth in this dramatic way. But there is every reason to believe that if population can be stabilized, mankind's tables will never lack for fine food and drink.

Freshly cut celery awaits shipment from a field in California to markets as far away as New York and thence to the salad bowls in thousands of homes.

Pruned close to the ground so they can be harvested easily, these grapevines in the Rhône Valley of France produce the famed Châteauneuf-du-Pape wine.

A golden stream of olive oil is pressed from the fruit of sturdy trees (left) thriving on rocky terraces near Florence, Italy.

A gray day in Amsterdam, Holland, is brightened (left) by a street stall's gay profusion of flowers.

At a market in Tunisia (right), fancy farm-ripened vegetables are arranged in a colorful sample of the soil's great productivity.

Once a rare and expensive delicacy, tropical fruits like these from the Philippines are made available throughout the world by fast refrigerated transport and efficient distribution systems.

Abbott, S. W., 290
Aborigines, 164–67
Aden, Gulf of, **21**
Aegean Sea, **64**
Afghanistan, **131**
Afifi, Hafez, 129–30
Africa, 96, **96–98**
   climate extremes, 272
   continental drifting, 22, 23, **96, 98**
   early man in, 42–46
   Golden Crescent map, **176**
   rain forests, 120
Agassiz, Louis, 84
agriculture, 130, 273, 308–10
Ahaggar Mountains, 96–98
Aichilik River, 138
air, 236, 245–48
   pollution, 249–53, **249**, 273, **307**
   see also atmosphere
air travel
   jet streams, 192, 270, **270**
   records, **237**
   tornadoes, 285
Alaska, 99, **141**, 304
   tidal bore, 187
   tundra, 138
Alaska Pilot, The, 187
Aleutian Islands, 23, 79, 99, 198
   navigation, 187
Aleutian Trench, 23, 79, 198
algae, **16, 27,** 117, 150, 207, 217
   in geysers, **158,** 159
Algeria, 304
Alice Springs, Australia, 164, **245**
Allosaurus, **32**
Almaden Wine Company, 81
Almannagja fissure, 161
Alps, 23, **85,** 103, 104
   elevation, 130, 131
   marine fossils, 24
   snow avalanches, 91–95
Altamira cave, 51, 55
altocumulus clouds, 266, 270, **270**
aluminum, 302
Ama Dablam, Mount, **134**
Amazon River, 144–45, **145**
   tidal bores, **185,** 187
amberjacks, 233
Ameca, Sierra de, 168–71
amino acids, 15
ammonia, 13, 15, 306
amphibians, 144
anaconda, 145
Anchorage, Alaska, **74, 76**
Andes Mountains, **125, 144,** 193, 207
   earthquakes, 79
   formation, 23, **101,** 103
   marine fossils, 24
Andrews, Roy Chapman, 263
anemones, sea, 233
Angel Fall, 90, 146
animals
   in the Arctic, 137, 139, 140
   in asphalt swamp, 38
   beginnings, 16–17
   in deserts, 126
   in fresh water, 144–45
   in Great Salt Lake, 150
   microscopic marine, 190
   mutation from pollution, 253
   in the oceans, 143–44, 190, 207–9,
                                      212–14
   quick-frozen, 34–37
   in rain forests, 124–25

   in Yellowstone National Park, 157–58
Ankylosaurus, 32–33, **33**
Anning, Mary, 27
Antarctica, 22, 23, **22–23,** 257
   climate extremes, 272
   glaciers, 84–85, **85**
   icebergs, 87
   mirages, **263,** 264
antimony, 298, **299**
ants
   army, 125
   in deserts, 126
   in rain forests, 123, 125
Apennines, 23
Apollo program
   Apollo 15, **18**
   weather photography, **277–79, 282**
aquaculture, 233
aquanauts, 230–33
Arabian Desert, 125
Arabian Peninsula, **21**
Arabian Sea, 132
arapaima, 144
Archaeopteryx, **25,** 27
Archelon, **32**
Arches National Monument, **111**
Arctic, 136–41
   halos, 257
   minerals, 304
   mirages, 262
   quick-frozen mammoths, 34–37
Aristotle, 73, 259
asbestos, 301, **301**
Ascension Island, 160, **160,** 208
ashab, 126
Asia, 272, 294
asphalt swamps, 38–41
Atacama Desert, **125,** 306, 312
Atlantic Ocean, 160, **280**
   abyssal plains, 208
   currents, 189, 190, 192
   lightning display, 287
   Mid-Atlantic Ridge, 160, **160,** 161,
                                 206, 208
   pollution, 223
   profile, **206–7**
   waves, 203, 204, **205**
   widening of, 20, 23, **23,** 99, 160
   see also hurricanes
Atlantic Seamount, **206**
Atlantis, 67, **68**
Atlas Mountains, **109**
atmosphere, 142, 236–40, **238–39**
   air and space craft records, **237**
   coronas and halos, 254–58
   formation, 13
   jet streams, 192, 270, **270**
   thermals, 269
   vertical circulation, 269
   volcanic ash, 246
atomic energy, 130, 304
Augrabies Falls, **147**
Augustine, St., 17
aureoles, 254
auroras, 240, 259–60, **259–61**
Australia, 22, 23, **22–23**
   Ayers Rock, 164–68, **165–68**
   climate extremes, 272
   fire tornadoes, 288, **288**
   minerals, 294, 298, 302
   monsoon forests, 120, **121**
   Pinnacles, 128, **128**
Australopithecines, 28
Australopithecus, **43**

Austria, 249
automobiles, 249, 250, 252, 253
avalanches, snow, 91–95, **92–94**
Avalon, 290, 291
Ayers Rock, 164–68, **165–68**
Azores Islands, 160, **160,** 208

B
bacteria, 16, **27,** 117, 236, 245, 248
Bagnold, R. A., 178, 179
Baikal, Lake, 143, 145
Baja California, 23, **209**
Baltic Sea, 97, **222,** 223
Balzac, Honoré de, 26
Banff National Park, **83**
Barberton, South Africa, 16
barium sulfate, 296, **297**
Barrier Dam, 224, 225
Barter Island, **138,** 141
bathyscaph, 210–12, **213,** 214–16
bauxite, 302
Beardmore Glacier, **85**
bears, 157
   short-faced, 41
Beaumont, Peter, 42–46
Bedouins, 126–27
belly plant, 126
Ben Franklin, 190, **190, 191**
Bengal, Bay of, 132, 284
Benguela Current, 190
Bering Strait, 40
Bermuda Rise, **206**
Bezymyannaya, Mount, 58
Bigelow, Frank H., 290
Bihar, 134
bioluminescence, 209
birds, 144, 190, 194, 248
   fossil, **25,** 27
   pollution affecting, 223
bison, 40
Black Hills, 103
"black lung" disease, 246
Blanc, Mont, 130–31
Bolivia, 120
Bolt, Bruce, 81
Boltwood, Bertram, 19
Bonneville, Lake, 151, **152**
Bonneville Salt Flats, **150,** 151, **151**
Border Cave, 42–46
Borneo, 120
Boshier, Adrian, 42–46
Boule, Marcellin, 47
Bourke's Luck potholes, **111**
Brahmaputra River, 132–34, 284, 309
Brahmins, 17
Brazil, 294
   rain forests, 120
Breuil, Henri, 51
Brewster, Sir David, 178
Bridal Veil Falls, 88, 89
Britain, 188, 223
bromeliads, 122–23
Bromo volcano, **62**
Brontosaurus, **33**
Brook, Marx, 289
Brooks, Edward, 285
Brooks Range, 136, 137, 138, 139, 141
Broom, Robert, 28
Brouwershavense, **228, 229**
Brower, Kenneth, 136
Brown, A. E., 179
Buck Island Reef National Monument,
                                 217–20
Buono, Giuseppe, 210, 211

Burma, 120, **131**, 294
Burns (Julia Pfeiffer) State Park, 220
Butte, Montana, 302
butterflies, 124–25

C

cactus
  in deserts, 126
  in rain forests, 122
calcium, 128, 308
California, 23, **23**, 76–81
  Yosemite Valley, 87–91, **88–90**
California, Gulf of, 187
California Current, 189
Cambodia, 120
camel, 39, **40**, 126
Campbell, W. A., 179
Canada
  climate of, 37, 273, 275
  forest fires, 248
  minerals, 298, 302, 304
  summer of 1816, 271
Canaries Current, 189
Cannon, B. L., 232
Caribbean Sea, 23, **204, 221**, 284
caribou, Barren Ground, **139**, 140
Carlsbad Caverns, 172, **172, 173**
Carson, Rachel, 220
Cascata delle Marmore, **147**
Caspian Sea, 145
Casiquiare Canal, 144
Castellani, Vittorio, 154, 155
Cauca River, **101**
caustic soda, **306**
cave paintings, 51–55, **52–53, 55**, 165–67
caverns, 172–73, **172–73**
  subterranean river, 153–56
Cecropia tree, 123
Celebes Islands, 120
cement, 301
Ceylon, 294
Challenger Deep, 210–14
Charybdis, 201
Chile, 190, 302, 306
  earthquakes, 79
China, 302
chlorophyll, 16
Christofilos, Nicholas, 244
cinnabar, 298, **298**
cirrus clouds, 266, **267**, 270, **279**
clay, 308
Clifton, H. Edward, 230, 232
climate and weather, 209, 266–76, 308
  Antarctic ice cap affecting, 84–85
  changes in, 31, 35–37
  climatic optimum, 36, 37
  erosion, 108
  forecasting, 271, 277–80
  salt particles, 247
  troposphere, 236
  weather modification, 276, 289
  see also clouds; hurricanes; rain;
                   snow; tornados
clouds, 142, **188, 234–35**, 236, 239, 247,
           266–71, **267–69, 279**
  cloud-seeding, 273–76, **274, 276**
  haloed sky, 254, **257**
  hurricanes, 280
  noctilucent, 240
  space photography, 277–78, **277–79**
  tornadoes, 285–88, **286**
  waterspouts, 289–91
coal, 250–52, 294, 304, **305**

coconut palm, 190
Colombia, 120, 294
  earthquakes, 74
Colorado Plateaus, 112, **113**, 114
Colorado River, **112**, 114, **115**, 276
Compositae, 122
conch, queen, 217, 233
condensation nuclei, 247, 267, 274, 275
  artificial, 271
Conner, Mount, 164
Constantine, Emperor, 258
continental shelf, 206–7, **206**, 220, 233
continental slope, 206–8
continents
  climate extremes, 272
  drifting, 19, 22–23, **22–23**, 31, 33, 36,
        78, 95, 99, 131, 132, 207–8
  formation, 19, 20, 22–23
  Pangaea, 22, **22**
  super-continents, 19, 22, 132
  see also specific continent
Cook, Captain James, 187
Cook Inlet, 187
copper, 302, **303**
Copiapó, Chile, 177
coral, 216, 219
  reefs, 186, 198–99, 217, **217, 219**, 230,
             232, 282–83
corn smut, **246**, 248
coronas, 254–58
Corythosaurus, **32**
cosmic dust, 236
cosmic rays, 240–44
cosmic storms, 240
coyotes, 39, 41
Cretaceous time, 29–31
Crete, 64–67
Cro-Magnon man, **43**, 54, 55
Cromwell, Townsend, 192
Cromwell Current, 192
Cullumber, Mrs. Alan, 81
cumulonimbus clouds, 266, **268**, 269
cumulus clouds, 266, 269, **269**, 273, **279**
Cuvier, Georges Léopold, 26–28, **26**
Cuxhaven, Germany, 262
cyclonic storm systems, 278, **278–79**

D

daisy family, 122
Daly City, California, 76, **77**, 81
dams, 224–29
Dart, Raymond, 28
Darwin, Charles, 18, 24, 28, 42, 177
Davos, Switzerland, **92**, 94, **94**
Dead Sea, 143, 151, 153, **153**, 306
Death Valley, 107
Delta Plan, 224, 226–29, **227**
Descartes, René, 258
desert(s), 108, 125–30, **125–30, 193–95**
  mirages, 262–63
  salt deposits, 151–52, 306, 309
  sand dunes, **105**, 106–7, **106–7**, 111,
            127, 128, 178
  soil, 308–9
diamonds, 294–95, **294**
diastrophism, 102
Dietz, Robert S., 22
dikes, 224, 227, 229
Dill, Robert F., 219
dinoflagellates, 207
dinosaurs, 29–33, **32–33**, 40
  fossil tracks, 27, 29
Diplodocus, **33**

dire wolves, 38, 39, **39**, 41
"D" layer, 239
Dubois, Eugène, 28
ducks
  eider, 201
  pollution affecting, **222**
dunes, **105**, 106–7, **106–7**, 111, **127**,
               128, 178
dust devils, 285, 289
dust storms, 111, **245**, 246
Dwight, William B., 290

E

Earth
  age, 17–19, 24, 115
  beginnings, 9, 10–14, **11**
  core, 13, 102, 302
  crust, **9**, 12–14, 19–20, 100–4, 112–13,
         294, 298, 302
  curvature, 239
  magnetic field, 244, 260, 289
  magnetic poles, 20, 31, 35, 95, 96,
            99, 260
  mantle, 100, 102
  mass, 13
  rainfall extremes, 272
  rotation, 189, 238, 270
  temperature extremes, 272
Earth Clocks, **12, 17, 28, 37, 46**
earthquakes, 72–81, 174, 284
  epicenter, 73, 80
  geyser's clock, 159
  near Himalayas, 134
  marine, 75, 208, 209
  prediction, 81
  Richter scale, 74
  turbidity currents in ocean, 219
  see also tsunamis
Eaton, Jerry P., 81
Ecuador, 74, 75
eels
  electric, 144
  European, 190–92, 208
Egypt, 67–68, 129, 300, 309
Eigg Island, 177, 178
Elburz Mountains, 91
El Capitan Mountain, 87–88
electric energy
  metal refining, 298, 302
  pollution, 251
  from tides, 187
  tornadoes, 287–89
electrostatic precipitator, 251
El Niño Current, 193–94, **194**
emeralds, 294, 295
Enchanted City of Spain, **110**
Eobacterium isolatum, 27
epicenter, 73, 80
epiphytes, 122–23, 125
Equatorial Current, 189
erosion, 108–11, **108–11**, 144, 148, **149**
Etna, Mount, **56–57**, 58, 69–71, **69–71**
Euphrates River, 129, 309
Europe, 246
  climate extremes, 272
  continental drifting, 20, 22, 23, **22–23**,
                      99
  minerals, 304
  natural gas, 251–52
Everest, Sir George, 130
Everest, Mount, 60, 100, 104, **104**,
      130, **132, 133**, 134–36, **136**, 220
Ewing, Maurice, 19

exosphere, 240
Explorer I, **237**, 241–42, 244, **244**
Eyre, Lake, **130**

F

Faeroe Islands, **111**
fata morgana, 265
faulting, 103
feldspar, 302
ferns, 122, **124**
Field, William O., 85
fig, strangler, **124**, 125
Findlay, Col. J. P., 287
fiords, 82
 defined, 91
 tides, 186, 187
 whirlpools, 199–200
fire,
 forest, 248
 tornadoes, 288
fish, 117, 143–44, 200–1
 cleaning shrimp, 233
 as food, 220, 221, 223
 food for, 188–90, 192, 200
Florida, 188, 221
fogs, 245, 247, 266
Folger, Timothy, 189
Font-de-Gaume, France, 55
food, 308–10, **308–10**
 fish as, 220, 221, 223
 shortage, 130, 192, 233, 310
Fornazzo, Sicily, **70**, 71
fossil fuel, 304
 see also coal; oil
fossils, 24–28, **25**
 age of Earth from, **17**, 18, 20, 24
 man, 28, 42–46
 marine, 19, 24, 26, 97, **99**, 102, 114,
 115, 131, 134–36
 oldest known, **27**
 as records of organisms, 24, 115–17
 rigid body parts, 17
 South Pole, 96
Franklin, Benjamin, 189
fruits, **311**
Fujiyama, Mount, 60, **63**
fumaroles, 157
Fundy, Bay of, 184, 186, **186**, 187, 197
fungi, 123
 atmosphere, 236, 245, 248

G

Galanopoulos, Angelos, 64, 67, 68
gammas, 216
Ganges oceanic trench, 134
Ganges River, 132–34, 284, 309
Gansser, Augusto, 134
garibaldi, 219
Garrett, E. H., 289
garúa, 194
gems, **292**, 294–96, **294–95**
Genesis rock, **18**
Genghis Khan, 129
geomorphology, 144
Gettysburg Seamount, **206**
geysers, **9**, 157–60, **157, 159**, 163
Gibraltar, **206**
glaciers, **63**, 82–86, **83**, 91, 147, 273
 flooding from, 162
 Ice Age, 36, **84**, 85, 309
 icebergs, 87, **87**
 lateral moraines, 91

medial moraines, 91
 in Sahara Desert, 95–99
 valley, 87, 90–91
Glomar Challenger, 19–22, 28
Gobi Desert, **126**, 177, 263
gold, 174–76, **174–76**, 298, **299**
Goldthwait, Richard P., 86
Gondwana supercontinent, 132
Gordon, Ernest, 168
Grand Banks, 190
Grand Canyon, 112–17, **112–17, 149**
granite, 300, **301**, 302
Grant, Lewis O., 275–76
grapevines, **310**
Great Lakes, 82
Great Rift Valley, 103, 153
Great Salt Lake, 143, 150–53, 306
Great Salt Lake Desert, 151, **152**
Greeks, 24–26, 64–67
Green Grotto, **173**
Greenland, 22
 glaciers, 84, 86, 87, 91
 icebergs, 87, **87**
 temperature, 272
Grevelingen Dam, **227**
Grímsvötn, Mount, 162
ground sloths, 40
groundwater, 129, 142, 157, 159, 308
guano fertilizer, 190
Gulf Stream, 188–90, **188, 191**
guyots, 209
gypsum, 152, **300**, 301, 306

H

Hagen, Tom, 134
hail, 269, 286–87, 289
Half Dome Mountain, 88–89, **89**, 91
Halley, Edmund, 17–18
halos, 254–58, **255, 256**, 266
Halsey, Admiral William F., 282
Hancock, G. Allan, 41
Hare, Robert, 287
Harrison, George, 174
Hawaiian Islands, 60–61, 203
 tidal waves, 196–99
Hawkes, Jacquetta, 50
Hebrews (Israelites), 17, 67–68
Heezen, Bruce C., 66
Heintz, Anatol, 29
Heintz, Natascha, 29
Hekla, Mount, 162, 163
Helgoland Island, 262
hematite, 42
herbivores, 39
Hernández, Miguel, 170
Hevelius, 257
 drawing of solar halos, **256**
Heyerdahl, Thor, 220
Hillary, Sir Edmund, 135
Himalayas, 19, 23, 103, 130–36, **131–36**
Hindu Kush Mountains, **129**
Holden, John C., 22
Holland, 224–29
holoplankton, 207, 208
Homo erectus, 42, **43**, 47
Homo habilis, 45
Homo sapiens, 28, **43**, 45, 47, 50–51
horse, **25**
Horseshoe Falls, **149**
horsetail rush, **246**
hot springs, 157
Howard, Lake, 266
Hudson Bay, 97

Humboldt (Peru) Current, 190, 193–94
Hurricane Carol, **281**
Hurricane Gladys, **282**
hurricanes, 205, 280–84, **280–82, 284**
 modification, 275, 276
 roar, 285
 space photography, 278, **278–79**
Huxley, Thomas, 28
hydrosphere, 142
hydrothermal energy, 163

I

Ice Age, 36, **84**, 85, 273, 309
 glaciers, 36, **84**, 85, 309
 Sahara Desert, 95–99
 Yosemite Valley, 87, 90–91
icebergs, 87, **87**
Iceland, 62, 97, 157, 160–63, **160–63**
Icthyosaurus, 32
Iguaçu Falls, **146**
Iguanodon, 29–31, **29–31**
India, 22, 23, **22–23, 131**, 132, **132**, 309
 earthquakes, 75
 minerals, 294, 302
Indo-Malayan rain forests, 120
Indus River, 132–34, 309
insects, 248
 fossil paths, 27
 in rain forests, 123–25
ionization, 208
ionosphere, 239–40
Iran, 304
Iraq, 129
iron, 302, 308
Iron Mountain, **86**
Irwin, Jim, **18**
islands
 formation, 19, 22
 hurricanes, 282, **283**
Italy, 249
 underground river, 153–56

J

Jacobshaven Fiord, **87**
Jago River, 139
Jaluit atoll, 283
Japan, 22, 23
 earthquakes, 74–76
 marine observatory, 216–17
 whirlpools, 201, **201**
jasper, 296, **297**
Java man, 28, 44
jet streams, 192, 270, **270**
Johannesburg, South Africa, 174
jökulhlaup, 162, **162**
Jonas, Cpl. Roger, 180
Jones, J. M., 179
Josephine Seamount, **206**

K

K-2, Mount, 130
Kalahari Desert, 179
Kamchatka, Siberia, 58
Kanchenjunga, Mount, **131**
Kaneohe Bay, 198
Kant, Immanuel, 10
Karakoram Mountains, 130
Katla, Mount, 162
Katmai volcano, 246
Kauai Island, **178**, 199
Kawela Bay, 196, 197, 199

Kelvin, Lord, 18–19, 100
Kelvin Seamount, **206**
Khumbu Glacier, 135
Kilauea, Mount, 60–61, **63**
Kilimanjaro, Mount, **63**
Klyuchevskaya Mountains, 58
Krakatoa Island, 61, 62, 64
Krakatoa volcano, 246, 273
Krichauff Range, **104**
Kurdistan Mountains, 47
Kuroshio Current, 189
Kuwait, 304

L

Labrador Current, 189, 190
Lafitte, Robert, 29
La Jolla Cove, 219
Lake District Mountains, 103, **103**
lakes, **81**, 142–45, 157
    natural asphalt, **41**
    underground, 153
Lambertsen, Christian J., 232
Lameshur Bay, 230, 232
Lanao, Lake, 145
landslides, 75, 76, **76**
Laplace, Pierre Simon, 10
Lapparent, Albert F. de, 29
Lascaux cave, **52–53**, 53, 54
Last Lake, 136–38, **138**
Lauwers Sea Dam, **225**
lava, 20, 104, **108, 110**, 161, **178**, 302
lead, 302
Leakey, Louis S. B., 45
Leakey, Mary, 45
Leakey, Richard, 42–46
Leroi-Gourham, Arlette, 49
Lewis, U.S.S., 210–14
lianas, 120, 122
Libya, 129, 304
life
    atmosphere, 236
    beginnings, 13–17, **17**
    deserts, 105, 193–94, **194, 195**
    Great Salt Lake, 150
    Gulf Stream, 190
    ocean, 190, 192–94, 207–9, 216–23
    rock calendars, 115–17
lightning, 287–89
    waterspouts, 290
limestone, 113, 300–1, **300**
Lisbon, Portugal, 72–73, **72**, 75
lithium, 152
Llanoria, 114
lobster, spiny, **230**, 233
loess, 309
Logan Gardens, **189**
London, England, 249, 251, 253
Los Angeles, California, 78, **252**, 253
Los Angeles basin, 39
Lyell, Sir Charles, 18, 75

M

Macchia, Sicily, 71
mackerel sky, 266–67
McKinley, Mount, 130
McMahon, Sir Henry, 105
MacMillan, Donald, 262
maelstroms, 199–201
magnesium, 152, 302
magnetic poles, 260
    polar wandering, 20, 31, 35, 95, 96, 99
Mahnken, Conrad, V. W., 230, 233

Mallet, Robert, 73
Mallory, George, 135–36
mammals
    in rain forests, 124
    sea-dwelling, 144
Mammoth Hot Springs, **158**, 159
mammoths
    great imperial, 38, **38**, 40
    quick-frozen, 34–35, **35, 36**, 37
man
    ancestors, 28, 42–51, **43, 44**
    in deserts, 126–30
    fossils, 28, 42–46
manganese, 208, **208**
Mantell, Gideon, 31
marble, 300–1, **300**
Marco Polo, 177
mare's tails, 266–67
Mariana Trench, 206, 210
    Challenger Deep, 210–14
Marshack, Alexander, 51
Marshall Islands, 283
marsupials, 41
    in rain forests, 125
Martha's Vineyard, 289, **290**
Martínez, Antonio, 170
Martinique Island, 58
Mártires, Los, 219
Matterhorn, 82
Matthews, D. H., 20
Mauna Kea, Mount, 60, 104, **104**
Mauna Loa, Mount, 60, 61
Maury, Lt. Matthew Fontaine, 189
Mediterranean Sea, **64**, 223
Mehlert, Charles, 218
Merced River, 87–90, **88**
mercury, 298, **298**
mercury poisoning, 223
Mesopotamia, 129
Mesosaurus, **32**
mesosphere, 239
Messina, Strait of, 201, 265
metals, 298, **298–99**, 302, **302–3**
meteors and meteorites, 240, 247
methane, 13, 15
Meuse River, 226, 227
Mexico, 168–71, 298
Mexico, Gulf of, 114, **282**, 286
Miami Beach, Florida, 221
Michelangelo, 203
Mid-Atlantic Ridge, 160, **160**, 161, **206**
Mid-Ocean Ridge, 208
Mid-Pacific Ridge, 209
Milford Sound, **148**
Milky Way, **11**
Miller, Hugh, 177
Miller, Stanley L., 15
Milton, John P., 136
Minamata Bay, 223
Minas Basin, 184
Mindanao Trench, 206
minerals, 152, 208, 294–311
mirages, 262–65, **263–65**
Mississippi River, 114
Moeraki Beach, **171**
molybdenum, 302
monadnock, 164
Monge, Gaspard, 264
Mono Lake, **108**
monsoon forests, 120, **121, 122**
Monte Soratte caves, **173**
Mont-Saint-Michel, 197
Moon
    coronas, 254–58

    geyser's clock, 159
    halos, 254–58
    lunar rocks, **18**
    origin, 13, **18**
    parselenae (moon dogs), 257
    temperatures, 240
    tides, 184
Mooney Falls, **149**
Morning Glory Pool, **158**
Morpho rhetenor, 125
mosquitoes, 123–24
moths
    fossil, **25**
    peppered, 253, **253**
mountains, 23, 100–4, 114
    block, **102**, 103
    chert, **27**
    dome, **102**, 103, **103**
    fold, **101, 102**, 103
    Grand Canyon, 114, **114–15**
    in oceans, 208, 209
    see also specific mountains
Mountford, C. P., 166
Mycenaean culture, 67
Mýrdalsjökul Glacier, 162

N

Nakous, Gebel, 178
Nameless, Mount, 58
Nantucket Sound, 289, **290**
Naruto Strait, 201
National Air Pollution Control
                Administration, 250
National Canyon, **113**
natural gas, 251–52, **305**, 306
navigation, 184–87
    whirlpools, 200, 201
    hurricanes, 282
Negro, Rio, 144
Neifert, W. W., 291
Nepal, **131**, 134, 135
New Guinea, **23**, 120
New York City, 223, **250–51**, 251
New York Harbor, 186, **222**, 223
New Zealand, 157, 163, 171
Niagara Falls, 146, 148, **149**
nickel, 302
Nimbus, 277, 280
nimbus clouds, 266, 270
Ninkovich, Dragoslav, 66
nitrates, 302, 306
nitrogen, 248
nitrogen poisoning, 231
Norkay, Tenzing, 135
North, Rev. Crandall J., 290, 291
North America, 114
    animals, 39, 40
    climate extremes, 272
    continental drifting, 20, 22, 23,
                22–23, 79
    desert arches, 110
    minerals, 302
northern lights, 259–60, **259–61**
    see also auroras
North Pole *see* magnetic poles
North Sea, 224, 226–27
nuclear power (atomic energy), 130, 304
nuclear tests, 223, 244
nucleic acids, 15

O

Oahu Island, 196, 203

Oceania, 272
oceans and seas, 143
  abyssal plains, 208–9
  age of Earth, 17–18
  basins, 184, 206, 208
  beginning of life in, 15–17
  continental shelf, 206–7, **206**, 220, 233
  continental slope, 206–8
  currents, 188–95, **192**, 202, 212, 219
  deep-ocean basin, 206, 208, **208**
  depth, 145, 206, 207, 213
  earthquakes, 75, 208, 209
  food shortage, 192, 233
  Grand Canyon, 114, **114**
  icebergs, 87, **87**
  life in, 190, 192–94, 207–9, 216–23,
          233
  marine observatories, 216–20, 230–33
  mineral gardens, 208, **208**
  ocean floor, 206–9
  pollution, 220–23, **221–22**
  ridges in, 20, 78, 100, 199, 206
  "Rose Bowl," 221
  sand dunes near, 106, 107
  seamounts, **206**, 209
  Sea-states One—Seven, **204–5**
  sediments, 20, 24, 28, 111, 171,
          209, 232
  submarine canyons, 208, 209, 215–19
  submarine landslides, 198, **209**, 215,
          215, 216
  submarine valleys, 199
  tidelands, 220, 223, 229
  trenches, 78, 206–9
  turbidity currents, 208, 209, 215,
          216, 219
  undertow, 205
  volcanoes, 18, 63, 207, 209, 282–83
  waves, **182–83**, 202–5, **202–5**, 246
  see also navigation; tidal waves; tides
Ochrida, Lake, 143, 145
octopus, brown, 216
Odell, Noel, 135
Okeechobee, Lake, 179
oil, 294, 304, **304–5**, 306, **307**
oil pollution, **221–22**, 223
Old Faithful, **157**, 158–60, **159**
Olga, Mount, **109**, 168
Oosthuizen, Petronella Francina, 174
opals, **295**
opossums, 125
Orange River, **147**
Orinoco River, 120, 144, 145
Oslo Fiord, 223
Osterberg, Jorj, 181
Ouachita River, 180
ozone, 238

P

Pacific Glory, 223
Pacific Ocean, **277**, **280**
  currents, 189, 192
  deep-ocean basins, 208
  guyots, 209
  Mid-Pacific Ridge, 209
  Ring of Fire, 58, **60**, 75, 207
  shrinking of, 23, **23**, 207
  trenches, 206, 207
  waves, 203, 204
  see also typhoons
Pakistan, **131**

Pakistan, East, 284
Pangaea, 22, **22**
Parasaurolophus, **32**
parhelia, 257
Paricutín, Mexico, 62
Paris, France, 249, 264
parrot fish, 216, **220**
Pastena, Cave of, 153–56, **154–55**
Pearson, Steven, 136
Peary, Robert, 262
pegmatites, 294
Peking man, 28, **43**, 44
Pelée, Mount, 58, 60–62
Peltier, J. C. A., 287
Pennekamp (John) Coral Reef State
             Park, 219
permafrost, 34, 37, 138
Peru, 188, 244, 298
  desert, 193–94, **193**, **195**
  earthquakes, 79
Peru-Chile Trench, 207
Peru (Humboldt) Current, 190, 193–94
pesticides, 223, 249
Peter and Paul Rocks, **160**
Peterson, Melvin N. A., 22
petroleum, 232, 252, 304
Peyto, Lake, **83**
Philippines, 120, 282
photochemical smog, 252–53
phytoplankton, 207, 223
Piccard, August, 210
Piccard, Jacques, 190, 210–14
Pickett, Jack, 179
piranha, 144

Pittsburgh, Pennsylvania, 249
planets, 10–13, **11**
plankton, 190, 207, 208, 212, 217
Plantation Key, Florida, **284**
plants
  beginnings of, 13, 16
  in deserts, 125–26, 193–94, **194–95**
  in fresh water, 144
  near geysers, 157
  microscopic marine, 190
  in oceans, 190, 192, 207, 208, 233
  oxygen from, 13
  pollen, 236, 245, **247**, 248
plaster of Paris, **300**, 301, 306
Plate, River, 120
Plateau d'Assay, France, 91
plate tectonics, 18, 19–20, 78–79, 99,
          102–3, 207–8
  see also continents—drifting of
platinum, 298, **299**
Plesiosaurus, **32**
polders, 224
pollen, 236, 245, **247**, 248
pollution
  air, 249–53, **249**, 273, **307**
  animal mutation, 253
  oceans, 220–23, **221–22**
Ponce de León, 188–89, 219
Portolá, Gaspar de, 38
Posso, Lake, 145
potash, 152, 306
Powell, John Wesley, 117
precipitation, *see* hail; rain; snow
primroses, **139**
Project Stormfury, 275
proteins, 14
Pteranodon, **33**
Pulido, Dionisio, 62
Pyramid of Khufu, 300

Pyrenees, 23
pyrite, 296, **296**

Q

quartz, 294, **295**, 296, **297**, 308
Quervain, Marcel de, 95
quicksand, 179–81, **180**

R

radioactive decay, 19
radioactivity, 223
radio waves, 240
ragweed pollen, **247**
rain, 14, **14–15**, 142, 157, 193, 270
  erosion, 108, 110, 308
  formation, 247, 267–69
  rainfall extremes, 272
  rainmaking, 273–76, **274**, **276**
rainbow, 258, **258**
rainbow, white, 256–57
rain forests, 120–25, **122–24**
Ramapo, U.S.S., 203
Rance River, 187
Rancho La Brea, 38–41, **38–40**
Ranieri, Mario, 154–56
Ray, John, 26
rayfish, 214
Rechnitzer, Andreas B., 210
Reckinger, Switzerland, 91
Red Sea, **21**
reptiles, 144
  extinct, **32–33**
Reykjanes Peninsula, 161
Rhine River, 226, **226**, 227
Rhone Glacier, **85**
rice, **308**, 309, 310
Richter, Charles, 74
Richter scale, 74
Rinehart, Joan, 159
Ring of Fire, 58, **60**, 75, 207
Riobamba, Ecuador, 75
rivers and streams, 111, 114, 142–44,
          157, 174
  origins, 82
  tidal bores, 187
  underground (subterranean), 129,
          153–56, **154–56**, 172
  waterfalls, **90**, 146–49, **146–49**
rocks, 161, 300–1, **300–1**, 308
  age of Earth from, 18, 114
  Ayers Rock, 164–68, **165–68**
  erosion, 108, 110, 308
  formation, 12, 13
  magnetism in, 20, 96
  spheroid, 168–71, **169–71**
Rocky Mountains, 82, **103**
Rognon, Pierre, 98
Ross, Sir John, 262
Rotorua, Lake, 163
rubies, 294, 295, **295**
Russia, 294, 298, 302, 304

S

sabertooths, 38–41, **38**
saguaro, 126
Sahara Desert, 95–99, **97–99**, 127, 246
  sand dunes, 106, 107, 127, **127**
St. Alfio, Sicily, **70**, 71, **71**
St. Helena Island, 160
St. John Island, 216, 230
St. Paul Rocks, 160

salt, 151–52, 306, **306–7**
  particles, 246–47
  salinity, 151, 153, 227, 309
Salt Lake City, Utah, 150
Saltstraum, 199–201, **200**
San Andreas Fault, 23, 76–81, **77–81**
sand, 308–9
  dunes, **105**, 106–7, **106–7**, 111, **127**,
                                    128, 178
  erosion, **110**, 111, **111**
  quicksand, 179–81
  sand spouts, 75
  singing, 177–79, **177, 178**
San Diego, California, 78, **80**
sandstone, 110, 113, 301
San Francisco, California, 188
  earthquakes, 75, 81
Sangay volcano, **59**
San Juan Mountains, 276
San Juan Parangaricutiro, Mexico, 63
San Juan y Martínez, Cuba, 282
Sanriku, Japan, 74
Santorini, Mount, 64–68, **64–67**
sapphires, 294, 295
Sargasso Sea, 190, 192, **192**, 208
satellites, artificial, 240–42, 244
  weather, 271, 277–80
Saudi Arabia, 304
Sautuola, Marcelino S. de, 51
Schaefer, Vincent J., 273–75
Scheldt River, 226, **226**
*Scientific Uses of Earth Satellites,
                                    The,* 241
Scotch Cap, 198
Scott, Comm. Robert Falcon, 264
scree, 108
Sea Around Us, The, 220
Sea Lab III, 232
seamounts, 206, 209
seaquakes, 75
seas, *see* oceans and seas
Seistan, 105–6
Senyürek, Muzaffer, 47
sequoias, California, 120
Serpens constellation, **12**
serpentine, 301
Sete Quedas Falls, 146
Severn River, 187, **187**
Shakespeare, William, 73
shale, 113, 296, **296**
Shanidar Cave, 47–50
sharks, 214, 233
  fossil, **25**
  refuges, **231**, 233
Sheenjek River, 136, **138, 140**

Ship Rock, **118–19**
shrimp, 150
  cleaning, 233
Shumaker, Lt. Lawrence, 210
Siberia, U.S.S.R., 34, 294
Sierra Nevadas, 23, 87, **88**, 89–90, 103
silver, 298, **299**
silver iodide, 273, **274**, 275
Simiutak Island, 186
Simpson, Joanne, 273, 275, 276
Sinai, Mount, 177–78
Sirbonis, Lake, 68
Siwalik Hills, 130
sleet, 269
sloths, 40, 125
Smith, Ernest Rice, 180
Smith, Phil, 48
Smith, Robert L., 171

Smith, William, 18, 27
smog, 252–53
snow, 142, 157, 269, 270
  avalanches, 91–95, **92–94**
  erosion, 108
  seeding, 275–76
Snowdon, Mount, 100
sodium salts, 306
Soendre Stroemfjord, 186
soil, 294, 308–10
Solar System, 240
  beginnings, 10–12, **11**
  gravitational attraction, 184, 239
  nebular (proto-planet) theory,
                                    12–13, 15
Solon, 67
South Africa, 174–76, 294, 298
South America, 79
  animals, 40, 41,
  climate extremes, 272
  continental drifting, 20, 22, 23,
                                    22–23, 96
  rain forests, 120, **123**, 144
South Pole, 95, 96, **96**, 99
  see also magnetic poles
space, 236, 240–44
  records, **237**
  weather photography, **277–79, 282**
specter of the Brocken, 256
speleology, 153
Spitsbergen Islands, 29–31
Sputnik I, **237**, 244
Sputnik II, 244
Stahl, Fred, 179
stalactites, 172, **173**
stalagmites, 172, **173**
starfish, 217
stars, 184
  shooting, 240
  see also meteors and meteorites
steel, 302, **302–3**
Steele, Mount, 82
Stegosaurus, 32
Stewart, T. Dale, 47, 49, **50**
stibnite, 298, **299**
Stirling, Matthew, 168–70
stone spheres, 168–71
Störmer, Carl, 244
storm surges, 284
Storstraum channel, 199–200
stratosphere, 238, 246
stratus clouds, 266, 270
streams, *see* rivers and streams
Strehlow, Ted, 164
sulfur, 152, 306, **307**
sumac, **25**
Sumatra Island, 120
Sumbawa Island, 273
Sun, 128, 142, 240
  coronas, **242**, 254–58
  cosmic rays, 240–44
  dust clouds, 271
  formation, 10–12, **11**
  halos, 254–58, **255, 256**, 266
  hurricanes, 280, 284
  light from, 238, 239, 248
  mock (sun dogs), **255, 256**, 257–58,
                                    257
  pollution, 252–53
  pseudhelion (false sun), 256
  radiation from, 13–15, 36, 85–86,
                                    259, 269, 273
  solar flares, 240, **241**, 259, 260
  sunspots, 259

tornadoes, 286
  ultraviolet light, 13, 238, 248, 288
Superior, Lake, 145, 302
Surtsey Island, **61**, 63, 162
swells, 203, 204
Switzerland, 91–94

T

Tabb, Durbin, 221
Tahiti Island, 184
Takakkaw Falls, **149**
talus, 108
tamarisk tree, 126
Tambora, Mount, 271
Tamerlane, 105
Tanganyika, Lake, 143, 145
Tanzania, 294
Tassili Plateau, 96, 98
Taupo, Lake, 163
Tektite I, 230–33, **230, 232, 233**
temperature extremes, 272
teratorn vultures, 38, **39**, 41
Terra Nova, 264
Tethys Sea, 131, 132, **132**
Thera Island, 64, **65, 66**
thermals, 269
thermocline, 212
Thomas, E. R., 179
thorium, 304
thunderstorms, **268**, 269, 285–86, 289
Thyanboche monastery, **134**
Tibetan Marginal Mountains, 130, 132
Tibetan Plateau, 130, 132
tidal waves, 196–99, 284
  tsunami, 75, 196–99, **196–97, 198**
  from volcanoes, 61, 64, 66, 68, 198
tidelands, 220, 223, 229
tides, 184–87, 198
  dams, **228**, 229, **229**
  as energy, 187
  hurricanes, 283
  tidal bores, **185**, 187, **187**, 197
Tigris River, 129, 309
Tillamook Lighthouse, 204
Tillandsia purpurea, **194**
Tiros, 277, 280
Titanic, 84
Titicaca Lake, **144**
Tokyo, Japan, **73**, 75, 76
Toledo, Ohio, 289
tools, 42, **46**
  of *Homo sapiens,* 50
  of Neanderthal man, 50

tornadoes, 269, 285–89, **286–88**
  man-made, 288
  waterspouts, 289, **291**
Toroweap Overlook, **112**
Torrey Canyon, 223
Torrey Pines State Park, 218–19
Toulon canyon, 214–16, **215**
Toyama Bay, 265
Trachodon, 33
trees, 120, 122, **123**, 253
Trent River, 187
Tribus, Milton, 276
Triceratops, 33
Trieste, 210–14, **211–13**
trilobite, 97, 117
Trinidad, Island of, **41**
Tristan da Cunha Island, 160, **160**
tropical cyclones, 280
troposphere, 236–38, **238–39**

Trunk Bay, 218
Tsientang River, 187
tsunamis, 75, 196–99, **196–97, 198,** 202
Tulsa, Oklahoma, 289
tundra, 98, 139–41, **137**
tungsten, 302
turbidity currents, 208, 209, 215, 216, 219
Turnagain Arm, 187
turtles, sea, 216, 218
Typhoon Bernice, **278**
typhoons, 278, **278, 279,** 280, 282–84, **283,** 285
Tyrannosaurus, 32, **33**

U

ultraviolet light, 13, 238, 248, 288
undertow, 205
Unimak Island, 198
United States, 304
U.S. Academy of Sciences, 275
U.S. Federal Volcanological Observatory, 60
U.S. Oceanic and Atmospheric Laboratory, 273
uranium, 176, 304
Urey, Harold, 14–15
Ussher, Archbishop, 17

V

Val d'Isère, France, 91
valleys, 111
  glaciers, 87–91, **88, 89**
  hanging, 88, 147
  rift, 161
  stream-cut, 87
  submarine, 199
Van Allen, James A., 240–42
Van Allen radiation belts, 240, 242–44, **243**
VanDerwalker, John, 230, 233
vegetables, **310, 311**
Venezuela, 302, 304
Venice Lagoon, **222**
Vesuvius, Mount, 58, 60

Victoria Falls, 146, 148, **149**
Vietnam, South, 120
Vine, F. J., 20
Virgin Islands, 230, **230**
Virgin Islands National Park, 216, 218
volcanoes, 58–71, 104, 174
  dust clouds, 271–73
  during Earth's formation, 13
  flooding caused by, 162
  formation, 20
  Iceland, 97, 161–62
  North American Southwest, 114, **114, 118–19**
  nucleic acids, 15
  ocean floor, 18, 63, 207, 209, 282–83
  Ring of Fire, 58, **60,** 75, 207
  spheroid stones, **169,** 171
  subglacial, 97–98, 162
  tidal waves, 61, 64, 66, 68, 198
  volcanic ash, 246, 271–73
  volcanic avalanches, 171
volcanology, 60–62
Vonnegut, Bernard, 275
vultures, teratorn, 38, **39,** 41

W

Wadi Taffassesset, valley of, 98
Walker, George, 174
Wallace, Alfred Russell, 122
Wallace, Robert, 77–78, 81
Waller, Richard A., 230, 233
Wandank, 210–14
warm front, 270
water, 142–44
  caverns, 172
  erosion, 108, 110, 111, 144
  groundwater, 129, 142, 157, 159, 308
waterfalls, **90,** 146–49, **146–49**
waterspouts, 75, 285, 289–91, **290, 291**
Wave Rock, **109**
waves, **182–83,** 202–5, **202–5,** 246
  hurricanes, 282–84
weather, *see* climate and weather
Wegener, Alfred, 20, 131
West Africa, 294
Weyer, James, 289

whales, gray, 219
wheat, 309, **309,** 310
wheat-stem rust, **246,** 248
whirlpools, 199–201, **199–201**
Wick, Scotland, 204
willy-willy, 280
wind
  airborne particles, 245, 246, 248
  clouds, 266, 269, 270
  desert, 105–7
  erosion, **108,** 110, 111
  in Himalayas, 130, **133**
  hurricanes, 280, 282–84
  jet streams, 192, 270, **270**
  navigation, 187
  ocean currents, 188–90
  pollution, **249**
  soil, 308, 309
  space photography, **278**
  tornadoes, 285, 286, 288
  waterspouts, 289, 290
  waves, 202–3
  whirlpools, 200
wolves, **35,** 38, 39, **39,** 41
Worcester, Massachusetts, 285, 287
World Conference on National Parks, first, 217
World Weather Watch, 278
Wrightwood, California, 78–80

Y

Yangtse valley, 309
Yellowstone National Park, **9,** 157–60
Yoho National Park, **149**
Yokohama, Japan, 75, 76
Yosemite valley, 87–91, **88–90**

Z

Zambezi River, 148
Zelfana, 129
Zinal, Switzerland, 91
zinc, 302
Zinjanthropus, 44–46, **44**
zooplankton, 207
Zuyder Zee, 224, **225, 227**

Illustration Credits

"PART" PHOTOGRAPHS: 8–9 (Yellowstone Geyser) Steven C. Wilson. 56–57: (Mount Etna eruption) Franz Lazi/Bruce Coleman, Inc. 118–119: (Ship Rock) Shelly Grossman/Woodfin Camp, Inc. 182–183: (Hawaiian surf) Werner Stoy/Camera Hawaii. 234–235: (African sky) Rhodes W. Fairbridge. 292–293: (Gems) Lee Boltin.

ALL OTHER ILLUSTRATIONS: 11: Helmut K. Wimmer. 12: Soren Noring. 13: Hale Observatories © California Institute of Technology and Carnegie Institution of Washington. 14–15: Helmut K. Wimmer. 16: (top) Walter Dawn; (bottom) Robert J. McCauley. 17: Soren Noring. 18: (top and bottom) NASA. 21: NASA. 22–23: George V. Kelvin after Robert Dietz and John C. Holden. 25: (top and center left) American Museum of Natural History; (center right and lower right) Bill Ratcliffe; (bottom left) SRD/Pierrhumbert, Museum d'Histoire Naturelle. 26: Bibliothèque Nationale. 27: Elso S. Barg-

hoorn. 28: Soren Noring. 29: Matthew Kalmenoff. 30: Natascha Heintz. 31: (lower left) George V. Kelvin; (lower right) Natascha Heintz. 32–33: Matthew Kalmenoff. 35: Matthew Kalmenoff. 36: Novosti Press Agency. 37: Soren Noring. 38–39: Matthew Kalmenoff. 40: Howard Koslow. 41: Rhodes W. Fairbridge. 43: Matthew Kalmenoff. 44–45: Adrian Boshier. 47: (top) Soren Noring; (bottom) Norman Myers/Bruce Coleman, Inc. 48: Howard Koslow. 49, 50, 51: Ralph S. Solecki. 52–53: Jean Vertut/Editions Mazenod—Prehistoire d'Art Occidental. 55: Howard Koslow. 59: Loren A. McIntyre. 60: George V. Kelvin. 61: Mats Wibe Lund, Jr. 62: Victor Englebert/DeWys. 63: (top) Takanori Ogawa; (center) World Photo Service/Madeline Grimoldi; (bottom) Norman Myers/Bruce Coleman, Inc. 64: George V. Kelvin. 65: James W. Mavor, Jr. 66: (top) Howard Koslow after James W. Mavor, Jr.; (bottom) H. F. Edgerton. 67: James W. Mavor, Jr. 68: George V. Kelvin. 69: J. G. Ramsey/Daily Telegraph Collection. 70: (top) Keystone/Madeline Grimoldi; (center) F. Scianna/Madeline Grimoldi; (bottom) Christian Bonington/Daily Telegraph Collection. 71: (top left and lower right) F. Scianna/Madeline Grimoldi; (top right) Christian Bonington/Daily Telegraph Collection. 72: Câmara Municipal de Lisboa. 73: American Museum of Natural History. 74, 76: Ward W. Wells. 77: George Hall/Woodfin Camp, Inc. 78–79: George V. Kelvin after Tanya Atwater. 80, 81: George Hall/Woodfin Camp, Inc. 83: World Photo Service/Madeline Grimoldi. 84: George V. Kelvin. 85: (top) Alan Ternes; (center) Allyn Baum/Rapho Guillumette Pictures; (bottom) R. Zanatta/Bruce Coleman, Inc. 86: Svend Galtt. 87: Lebe Photo Agency. 88: (top left) David Muench; (all others) Howard Koslow. 89, 90: Ansel Adams. 92, 93, 94: André Roch. 96: George V. Kelvin. 97, 98, 99: Rhodes W. Fairbridge. 101: Loren A. McIntyre/Alpha 102: Howard Koslow. 103: (top) Aerofilms Ltd.; (bottom) William Garnett. 104: (top) Frank Hurley/Rapho Guillumette Pictures; (bottom) Howard Koslow. 105, 106: Rhodes W. Fairbridge. 108: Esther Henderson/Rapho Guillumette Pictures. 109: (top left) David Moore/Black Star; (top right) Florita Botts/Madeline Grimoldi; (bottom) Australian News and Information Bureau. 110: (top) Luis F. Palafox; (bottom) Toni Schneider/Bruce Coleman, Inc. 111: (top) K. E. Kronefalk/Gotland Tourist Association; (bottom) John Rushmer/Carl E. Östman; (bottom right) Jack Cannon/Carl E. Östman. 112: David Muench. 113: Helga Teiwes. 114–115: Howard Koslow after Hal Shelton. 116, 117: Helga Teiwes. 121: Kingsley C. Fairbridge. 122: George V. Kelvin. 123: (left) Jacques Jangoux; (right) Werner Stoy/Camera Hawaii. 124: (top left) Norman Myers/Bruce Coleman, Inc.; (top right) Norman Tomalin/Bruce Coleman, Inc.; (bottom left) Kingsley C. Fairbridge. 125: Jacques Jangoux. 126: George V. Kelvin. 127: Victor Englebert/DeWys. 128: (top) Loren A. McIntyre; (bottom) Douglass Baglin. 129: R. Vroom/Bruce Coleman, Inc. 130: Ted Spiegel/Rapho Guillumette Pictures. 131: (top) Barry C. Bishop; (bottom) George V. Kelvin. 132: Howard Koslow after Tony Hagen. 133, 134: Barry C. Bishop. 135: Christian Bonington/Daily Telegraph Collection. 136: Barry C. Bishop © National Geographic Society. 137: Pete Martin. 138: (top) George V. Kelvin; (bottom) Robert Belous. 139, 140, 141: Robert Belous. 143: Freeman W. Patterson. 144: Jacques Jangoux. 145: Loren A. McIntyre. 146: Georg Gerster/Rapho Guillumette Pictures. 147: (top) Jen & Des Bartlett/Bruce Coleman, Inc.; (bottom left) South African Tourist Corp.; (bottom right) Bruno Pellegrini/Madeline Grimoldi. 148: (top) Martin Weaver/Woodfin Camp, Inc.; (bottom) Bill Belknap/Rapho Guillumette Pictures. 149: (top left) Freeman W. Patterson; (top right) Shelly Grossman/Woodfin Camp, Inc.; (bottom left) William Middleton/Editorial Photocolor Archives; (bottom right) World Photo Service/Madeline Grimoldi. 150, 151, 152: Bill Ratcliffe. 153: Louis Goldman/Rapho Guillumette Pictures. 154–155: (top) Howard Koslow; (bottom, both) Lamberto Ferri-Ricchi/Madeline Grimoldi. 156: Lamberto Ferri-Ricchi/Madeline Grimoldi. 157: (top) Paolo Koch/Rapho Guillumette Pictures; (bottom) Howard Koslow. 158: David Muench. 159: Soren Noring. 160: George V. Kelvin. 161: Kingsley C. Fairbridge. 162: Howard Koslow after J. G. Jones. 163: (left) V. M. D. Franszon/Carl E. Östman. 163: (right) Kingsley C. Fairbridge. 165: David Moore/Black Star. 166: R. D. Piesse. 167: John Brownlie/Bruce Coleman, Inc. 168: George Hall/Woodfin Camp, Inc. 169, 170: Victor Magallon. 171: David Moore/Black Star. 172: Adam Woolfitt/Woodfin Camp, Inc. 173: (top left and right, bottom left) Lamberto Ferri-Ricchi/Madeline Grimoldi; (bottom right) Adam Woolfitt/Woodfin Camp, Inc. 174, 175: Courtesy Anglo American Corp. of South Africa, Ltd. 176: (top) George V. Kelvin; (bottom) Courtesy Anglo American Corp. of South Africa, Ltd. 177: Art Clifton/Camera Hawaii. 178: Werner Stoy

/Camera Hawaii. 180, 181: Howard Koslow. 185: Loren A. McIntyre. 186: (top and center) Coleman C. Newman; (bottom) George V. Kelvin. 187: George Silk/Life Magazine © Time Inc. 188: NASA. 189: Penny Tweedie. 190: Courtesy Grumman Aerospace Corp. 191: (top) Courtesy Grumman Aerospace Corp.; (bottom) George V. Kelvin. 192: George V. Kelvin. 193, 194, 195: Loren A. McIntyre. 196–197: Howard Koslow. 198: Howard Koslow after Willard Bascom. 199: George Silk/Life Magazine © Time Inc. 200: Aage Larsen/Mittet Foto. 201: William Garnett. 202–203: Don James. 204, 205: Kingsley C. Fairbridge. 206–207: Antonio Petrucelli after Bruce C. Heezen. 208: Bruce C. Heezen. 209: (top) Bruce C. Heezen; (bottom) Conrad Limbaugh. 211: John Launois/Black Star. 212: George V. Kelvin. 213: George V. Kelvin. 215: Howard Koslow. 217, 218, 219, 220: Elgin Ciampi. 221: R. T. W./Carl E. Östman. 222: (top right) Bullaty-Lomeo Photography; (top left) Rick Strange/Carl E. Östman; (center right) S-E Hedin/Carl E. Östman; (bottom left) Menico Torchio; (bottom right) George Silk/Life Magazine © Time Inc. 224: Bart Hofmeester. 225: (left and bottom right) Bart Hofmeester; (top right) Ralph Turner/Carl E. Östman. 226: Bart Hofmeester. 227: (top) Adam Woolfitt/Woodfin Camp, Inc.; (bottom) George V. Kelvin. 228: Bart Hofmeester. 229: (top left and right) Bart Hofmeester; (bottom) Adam Woolfitt/Woodfin Camp, Inc. 230, 231: Flip Schulke/Black Star. 232: Howard Koslow. 233: Flip Schulke/Black Star. 237: Howard Koslow. 238–239: NASA. 241: NASA. 242: R. T. W./Carl E. Östman. 243: George V. Kelvin. 244: NASA. 245: Robin Smith. 246, 247: William R. Solomon. M.D. 249: George V. Kelvin. 250, 251: Glen E. McCaskey. 252: Howard Koslow. 253: Transworld Feature Syndicate. 255: Emil Schulthess/Black Star. 256: (top) Science and Technology Division, The New York Public Library Astor, Lenox and Tilden Foundations; (bottom) Charles Ott/Bruce Coleman, Inc. 257: (top) Freeman W. Patterson; (bottom) Paul Baich. 258: Sonja Bullaty. 259, 260: T. Lovgren/Carl E. Östman. 261: (top) James Quong; (bottom left) Peter G. Sanchez; (bottom right) T. Lovgren/Carl E. Östman. 263: Michael A. deCamp/Photo Trends. 264: George V. Kelvin. 265: (top) William Belknap, Jr./Rapho Guillumette Pictures; (bottom) Loren A. McIntyre. 267: Howard Koslow. 268: Lee Boltin. 269: (top) Jack Fields/Photo Researchers; (bottom) Lee Boltin. 270: NASA. 272: George V. Kelvin. 274: Howard Koslow. 276: Maurice Bennett/Photo Researchers. 277, 278, 279: NASA. 280: George V. Kelvin. 281: Daniell Farber/Photo Researchers. 282: NASA. 283: Fred Ayer/Photo Researchers. 284: Leslie F. Conover/Photo Researchers. 286, 287: Willis W. Wipf. 288: G. Wilson/Bureau of Meteorology, Australia. 290: Dukes County Historical Society. 291: Harry Pederson. 294, 295, 296, 297, 298, 299: Lee Boltin. 300: (left) Angelo Lomeo; (right) Lee Boltin. 301: (left top and bottom) Lee Boltin; (right) Elliott Erwitt/Magnum. 302: Loren A. McIntyre. 303: (top) William Garnett; (center left and bottom left) Loren A. McIntyre; (bottom right) Courtesy Bethlehem Steel Corp. 304: Dan Budnik/Woodfin Camp, Inc. 305: (top left and right) Thomas Höpker/Woodfin Camp, Inc.; (center) Dan Budnik/Woodfin Camp, Inc.; (bottom right) Billy Davis/Black Star. 306: Lee Boltin. 307: (top and left) Loren A. McIntyre; (bottom right) Black Star, courtesy Standard Oil of N.J. 308: (left) Fred Mayer/Woodfin Camp, Inc.; (right) Florita Botts/Madeline Grimoldi. 309: (top) Angelo Lomeo; (bottom) World Photo Service/Madeline Grimoldi. 310: (top right) Tony Howarth/Daily Telegraph Collection; (center) Patrick Thurston/Woodfin Camp, Inc.; (bottom left) Sonja Bullaty; (bottom right) Dan Budnik/Woodfin Camp, Inc. 311: (top) Florita Botts/Madeline Grimoldi; (bottom) Werner Stoy/Camera Hawaii.

PHOTO RESEARCH BY YVONNE R. FREUND.

Acknowledgments

THE EARTH'S BEGINNINGS by Gerald S. Hawkins, from *Splendor in the Sky,* revised edition, © 1961, 1969, by Gerald S. Hawkins. Reprinted by permission of Harper & Row, Publishers, Inc. THE FIRST STIRRINGS OF LIFE by George Gamow, from *A Planet Called Earth,* © 1963, by George Gamow. Reprinted by permission of The Viking Press, Inc. and Macmillan & Co. Ltd. UNRAVELING THE AGE OF EARTH by E. L. Simons, cond. from *Natural History,* © 1967, The American Museum of Natural History. REVELATIONS FROM FOSSIL FINDS by Ruth Moore. Reprinted from *Audubon,* the magazine of the National Audubon Society, © 1971. WHEN DINOSAURS ROAMED THE ARCTIC by Edwin H. Colbert, cond. from *Natural History,* © 1964, The American Museum of Natural History. THE RIDDLE OF THE QUICK-FROZEN MAMMOTHS by Ivan T. Sanderson, cond. from *The Saturday Evening Post,* © 1960, The Curtis Pub. Co. Reprinted by permission of Paul R. Reynolds, Inc. DEATH TRAP OF THE AGES by J. R. Macdonald, cond. from *National Wildlife,* © 1969, National Wildlife Federation. NEW FINDINGS ON THE ORIGIN OF MAN by Ronald Schiller, cond. from *Tuesday,* © 1973. NEANDERTHAL WASN'T ALWAYS A BRUTE by Ralph S. Solecki, cond. from *Smithsonian,* © 1971, Ralph S. Solecki. Reprinted by permission of Alfred A. Knopf, Inc. and Penguin Books Ltd. THOSE MYSTERIOUS CAVE PAINTINGS by Ronald Schiller, cond. from *Travel,* © 1971, Travel Magazine, Inc. THE SEETHING FURY OF VOLCANOES by Noel F. Busch, cond. from *Beacon Magazine of Hawaii,* © 1969, Kokohala Pub. Co. THE GREAT KILLER QUAKES by Ann and Myron Sutton, from the book *Nature on the Rampage,* © 1962, Ann and Myron Sutton. Reprinted as condensed and adapted by permission of the publishers, J. B. Lippincott Company. PERILS OF THE SAN ANDREAS FAULT by Sandra Blakeslee, cond. from *The New York Times,* © 1971, The New York Times Company. THE POWER AND THE GLORY OF GLACIERS by Paul Friggens, cond. from *Empire,* © 1969, The Denver Post, Inc. MOUNTAIN SCULPTING WITH A CHISEL OF ICE by John A. Shimer, adapted from *This Sculptured Earth,* New York; Columbia University Press, 1969, pp. 66–79, by permission of the publisher. ANTARCTIC GLACIERS IN THE SAHARA DESERT! by Rhodes W. Fairbridge, cond. from *Natural History,* © 1971, The American Museum of Natural History. WHY WE HAVE MOUNTAINS by Lowell Thomas, cond. from *Book of the High Mountains,* © 1964, Lowell Thomas. Reprinted by permission of Simon & Schuster, Inc. WASTELANDS OF WIND AND SAND by David I. Blumenstock, cond. from *The Ocean of Air,* © 1959, Rutgers University Press, New Brunswick, N.J. EARTH'S AWESOME BELT OF GREEN by Marston Bates, cond. from *The Forest and the Sea,* © 1960, Random House, Inc. Published in Great Britain by Museum Press Ltd. and reprinted by permission of A.M. Heath & Co. Ltd., © 1961, Marston Bates. HOW LIFE DEFIES THE DESERT by Edwin Muller, cond. from *Frontiers,* © 1957, The Academy of Natural Sciences of Philadelphia. THE ARCTIC—UNSPOILED, BUT NOT FOR LONG by John P. Milton, cond. from *Natural History,* © 1969, The American Museum of Natural History. THERE'S MAGIC IN A POND by Marston Bates, cond. from *The Forest and the Sea,* © 1960, Random House, Inc. Published in Great Britain by Museum Press Ltd. and reprinted by permission of A.M. Heath & Co. Ltd., © 1961, Marston Bates. WE EXPLORED AN UNDERGROUND RIVER by Lamberto Ferri-Ricchi, cond. from *Atlante,* © 1970. THE 10,000 GEYSERS OF YELLOWSTONE by Peter Farb, cond. from *The Face of North America,* © 1963, Peter Farb and reprinted by permission of Harper & Row, Publishers, Inc. and the Evelyn Singer Agency. Published in Great Britain by Constable & Co. Ltd. ICELAND'S HOT WATER WONDERLAND by Julian Kane, cond. from *Natural History,* © 1969, The American Museum of Natural History. AYERS ROCK—HOME OF THE PRIMEVAL SPIRITS by Victor Carell, cond. from *Walkabout,* © 1969. DOWN TO THE DEEPS FOR GOLD by J. D. Ratcliff, cond. from *Business and Finance,* © 1967, Business and Finance Ltd., Dublin. THE MYSTERY OF THE SINGING SANDS by Paul Brock, © 1970, The Hearst Corp. Reprinted with permission from *Science Digest.* QUICKSAND—NATURE'S TERRIFYING DEATH TRAP by Max Gunther, cond. from *True,* © 1964, Fawcett Publications. THE ETERNAL FORCE OF THE TIDES by Peter Freuchen, cond. from *Peter Freuchen's Book of the Seven Seas,* © 1957, Julian Messner, Inc. and reprinted by permission of Harold Matson Co., Inc. Published in Great Britain by Jonathan Cape Ltd., © 1958. FACE TO FACE WITH A TIDAL WAVE by Francis P. Shepard, cond. from *The Earth Beneath the Sea,* published by Johns Hopkins Press. THE WILD WHIRLPOOLS OF NORWAY by Olga Osing, © 1970, The International Oceanographic Foundation and reprinted by permission from *Sea Frontiers.* THE WAYS OF THE WAVES by James Nathan Miller, cond. from *Beacon Magazine of Hawaii,* © 1968, Kokohala Pub. Co., Inc. STRANGE LANDSCAPES BENEATH THE SEA by E. P. Lay, cond. from *Oceans,* © 1970, Oceans Magazine Co. WE MADE THE WORLD'S DEEPEST DIVE by Commander Don Walsh, USN., cond. from *Life,* © 1960, Time, Inc. Reprinted by permission of the author. TRAPPED IN AN UNDERSEA AVALANCHE by Jacques-Yves Cousteau, cond. from *The Living Sea,* © 1963, by Harper & Row, Publishers, Inc. Published in Great Britain by Hamish Hamilton Ltd. and © 1963. WE MUST STOP KILLING OUR OCEANS by Gaylord Nelson, cond. from *National Wildlife,* © 1970, National Wildlife Federation. SNUG HOUSE ON THE OCEAN FLOOR by Steven M. Spencer, cond. from *Popular Science Monthly,* © 1969, Popular Science Pub. Co., Inc. THE DEEP REALM OF THE ATMOSPHERE by Theo Loebsack, cond. from *Our Atmosphere,* translated from the German by E. L. and D. Rewald, © 1959, Pantheon Books and reprinted by permission of Random House, Inc. Published in Great Britain as *Earth's Envelope* and © 1959, William Collins Sons & Co. Ltd. STORMS THAT RAGE BEYOND THE SKY by Ralph E. Lapp. Reprinted by permission from *Think* Magazine, copyright 1961, International Business Machines Corp. WHAT YOU DON'T SEE ON A CLEAR DAY by Clyde Orr, Jr., cond. from *Between Earth and Space,* © 1959, The Macmillan Co. THE SKY'S RAREST SPECTACLES by Richard G. Beidleman, cond. from *Natural History,* © 1961, The American Museum of Natural History. THE WORLD THAT ISN'T THERE by Theo Loebsack, cond. from *Our Atmosphere,* translated from the German by E. L. and D. Rewald, © 1959, Pantheon Books and reprinted by permission of Random House, Inc. Published in Great Britain as *Earth's Envelope* and © 1959, William Collins Sons & Co. Ltd. THE PASSING PARADE ABOVE US by Richard M. Romin, cond. from *Natural History,* © 1968, The American Museum of Natural History. THE YEAR WITHOUT A SUMMER by George S. Fichter, © 1971, The Hearst Corp. Reprinted with permission from *Science Digest.* RAINMAKING COMES OF AGE by Ben Funk, cond. from *Frontiers: A Magazine of Natural History,* © 1971, The Academy of Natural Sciences of Philadelphia. THE WORLD'S MOST POWERFUL STORMS by David I. Blumenstock, cond. from *The Ocean of Air,* © 1959, Rutgers University Press, New Brunswick, N.J. INSIDE THE TORNADO by Bernard Vonnegut, cond. from *Natural History,* © 1968, The American Museum of Natural History. THE GREAT WATERSPOUT OF 1896 by Michael J. Mooney, cond. from *Oceans,* © 1970, Oceans Magazine Co.